D0699816

Fundamentals of Vehicle Dynamics

Thomas D. Gillespie

Published by:
Society of Automotive Engineers, Inc.
400 Commonwealth Drive
Warrendale, PA 15096-0001

Library of Congress Cataloging-in-Publication Data

Gillespie, T.D., (Thomas D.)
     Fundamentals of vehicle dynamics / T.D. Gillespie.
       p.   cm.
     Includes bibliographical references and index.
     ISBN 1-56091-199-9
     1. Motor vehicles--Dynamics. I. Title.
TL243.G548   1992
629.2--dc20   91-43595
                                CIP

ISBN 1-56091-199-9

SAE Order No. R-114

PREFACE

Throughout all of history it is doubtful that any invention has so effectively captured the interest and devotion of man as the automobile. The mobility enjoyed by humanity in the twentieth century has become an integral component of the modern lifestyle. In this first century of its history, more than a billion automobiles have been manufactured to satisfy the appetite for personal mobility. The marvels of mass production at times have reduced the cost of an automobile to only a few months of personal income. Most profoundly, however, for many people automobiles are a first love at some point in their lives, taking first priority with their interest and finances. In the words from a poem penned in earlier days:

> I drive my "Lizzie" every day,
> Up hill, down dale, and every way.
> A faithful auto it has been
> Even if it is of tin.
>
> I'll have to say — It's rattling good.
> — The engine, it's beneath the hood,
> — The wheels turn backward in reverse,
> — The paint it's looking worse and worse.
>
> When I have money in my jeans,
> I'll not ride in a can of beans,
> I'll buy what's called an "automobile."
> Won't I look fine behind the wheel.
>
> — T. N. Gillespie

Much of the infatuation with the automobile has centered around performance—acceleration, braking, cornering and ride. The art is practiced by the backyard mechanic, the racing enthusiast, and the automotive engineer. A library of books, magazine articles, and technical papers has been written to explain the engineering principles, the rules of thumb, and sometimes the "wrong way" to enhance the performance of an automobile. Most of the books written by practitioners from the racing circuits expound the wisdom of experience but without rigorous engineering explanation. A few textbooks have been written by those knowledgeable in automotive engineering, but the

i

books are often rather analytical and theoretical in nature. This book attempts to find the middle ground—to provide a foundation of engineering principles and analytical methods to explain the performance of an automotive vehicle, when those explanations are not too laborious, and to smooth the way between the doses of equations with practical explanations of the mechanics involved. The inclusion of engineering principles and equations biases the book to interest only the engineer, but it is hoped that the explanations are complete enough that those without a formal engineering degree can still comprehend and use most of the principles discussed.

Those responsible for the design and development in the manufacturing companies today are challenged by questions about the qualities desired in the product by the customer, and how these qualities are related to design and manufacturing processes. In recent years the complexity of the automotive design process has been increased by regulatory actions arising from the social and environmental consequences of the millions of motor vehicles operating on our highways. Added to this is the competitive pressure of the modern automotive manufacturing industry. In order to remain competitive in the future the manufacturers must seek ways to improve the efficiency of the design and development processes and shorten the time span from concept to production. Achievement of these goals requires a better understanding of the automobile as a system, so that qualities and performance of proposed designs can be predicted at an early stage in the design evolution, allowing refinements to be introduced while there is minimal impact to program costs.

Acceleration, braking, turning and ride are among the most fundamental properties of a motor vehicle and, therefore, should be well understood by every automotive engineer. Performance in one mode is closely linked to the others as a consequence of the dependence on a common set of vehicle mechanical properties. To understand the vehicle as a system it is necessary to acquire a knowledge of all the modes. Motion is the common denominator of all these modes; thus, the study of this field is denoted as vehicle dynamics.

The objectives in writing this book were:

1) *To introduce the basic mechanics governing vehicle dynamic performance* in the longitudinal (acceleration and braking modes), ride (vertical and pitch motions), and handling (lateral, yaw, and roll modes). Engineering analysis techniques will be applied to basic systems and subsystems to derive the controlling equations. The equations reveal which vehicle properties are influential to a given mode of performance and provide a tool for its prediction. By understanding the derivation of the equations, the practitioner

is made aware of the range of validity and limitations of the results.

2) *Familiarization with analytical methods available.* Over past decades analytical methods have been developed for predicting many aspects of automotive performance. Although the engineer has no need to master and utilize these techniques in daily activity, a knowledge of their existence greatly increases his/her value to the company. Awareness of these methods is the first step in knowing what is possible and where to find the necessary tools when the need arises.

3) *Familiarization with terminology.* Clarity in communication is vital to problem solving. Over the years, appropriate terminology for automotive engineering has been defined to facilitate communication. The study of vehicle dynamics provides the opportunity to become familiar with the terminology.

Thomas D. Gillespie

ACKNOWLEDGEMENTS

This book is dedicated to my wife, Susan, and our four wonderful children, Dave, Darren, Devin and Jessica. Throughout the long hours necessary to prepare these materials they have shown me patience and encouragement—two ingredients essential to any such endeavor.

The author is also indebted to many colleagues in the vehicle dynamics community who have provided comments and encouragement in preparation of this manuscript. Among those who have contributed their time and energy are Paul Fancher, Sam Clark, Charles MacAdam, Ray Murphy, James Bernard, Bill Fogarty, Manfred Rumple, Bill Stewart, Chuck Houser, Don Tandy, and the many dedicated staff members of the Society of Automotive Engineers.

LIST OF SYMBOLS

a	Tire cornering stiffness parameter
b	Tire cornering stiffness parameter
A	Frontal area of a vehicle
A_f	Lateral force compliance steer coefficient on the front axle
A_r	Lateral force compliance steer coefficient on the rear axle
a_x	Acceleration in the x-direction
a_y	Acceleration in the lateral direction
b	Longitudinal distance from front axle to center of gravity
c	Longitudinal distance from center of gravity to rear axle
C_α	Cornering stiffness of the tires on an axle
C_α'	Cornering stiffness of one tire
CC_α	Tire cornering coefficient
C_γ	Tire camber stiffness
C_D	Aerodynamic drag coefficient
C_h	Road surface rolling resistance coefficient
C_L	Aerodynamic lift coefficient
C_{PM}	Aerodynamic pitching moment coefficient
C_{RM}	Aerodynamic rolling moment coefficient
C_{YM}	Aerodynamic yawing moment coefficient
C_s	Suspension damping coefficient
C_s	Aerodynamic side force coefficient
CP	Center of pressure location of aerodynamic side force
d	Lateral distance between steering axis and center of tire contact at the ground
d_h	Distance from axle to the hitch point
d_{ns}	Distance from center of mass to the neutral steer point
D	Tire diameter
DI	Dynamic index
D_x	Linear deceleration
D_A	Aerodynamic drag force
e	Height of the pivot for an "equivalent torque arm"
	Drum brake geometry factor
$E[y^2]$	Mean square vibration response
f	Longitudinal length for an "equivalent torque arm"
f_a	Wheel hop resonant frequency (vertical)

f_n	Undamped natural frequency of a suspension system (Hz)
f_r	Rolling resistance coefficient
F_b	Braking force
	Vertical disturbance force on the sprung mass
F_i	Imbalance force in a tire
F_x	Force in the x-direction (tractive force)
F_{xm}	Maximum brake force on an axle
F_{xt}	Total force in the x-direction
F_y	Force in the y-direction (lateral force)
	Lateral force on an axle
F_y'	Lateral force on one tire
F_z	Force in the z-direction (vertical force)
F_{zi}	Vertical force on inside tire in a turn
F_{zo}	Vertical force on outside tire in a turn
F_w	Tire/wheel nonuniformity force on the unsprung mass
g	Acceleration of gravity (32.2 ft/sec^2, 9.81 m/sec^2)
G	Brake gain
G_o	Road roughness magnitude parameter
G_z	Power spectral density amplitude of road roughness
G_{zs}	Power spectral density amplitude of sprung mass acceleration
h	Center of gravity height
h_a	Height of the aerodynamic drag force
h_h	Hitch height
h_1	Height of the sprung mass center of gravity above the roll axis
h_r	Height of suspension roll center
h_t	Tire section height
HP	Engine or brake horsepower
HP_A	Aerodynamic horsepower
HP_R	Rolling resistance horsepower
HP_{RL}	Road load horsepower
H_v	Response gain function
I_d	Moment of inertia of the driveshaft
I_e	Moment of inertia of the engine
I_t	Moment of inertia of the transmission
I_w	Moment of inertia of the wheels
I_{xx}	Moment of inertia about the x-axis

I_{yy}	Moment of inertia about the y-axis
I_{zz}	Moment of inertia about the z-axis
k	Radius of gyration
K	Understeer gradient
K_{at}	Understeer gradient due to aligning torque
K_{llt}	Understeer gradient due to lateral load transfer on the axles
K_{lfcs}	Understeer gradient due to lateral force compliance steer
K_s	Vertical stiffness of a suspension
K_{ss}	Steering system stiffness
K_{strg}	Understeer gradient due to the steering system
K_t	Vertical stiffness of a tire
K_ϕ	Suspension roll stiffness
L	Wheelbase
L_A	Aerodynamic lift force
m	Drum brake geometry parameter
M	Mass of the vehicle
M_{AT}	Moment around the steer axis due to tire aligning torques
M_L	Moment around the steer axis due to tire lateral forces
M_r	Equivalent mass of the rotating components
M_{SA}	Moment around the steer axis due to front-wheel-drive forces and torques
M_T	Moment around the steer axis due to tire tractive forces
M_V	Moment around the steer axis due to tire vertical forces
M_ϕ	Rolling moment
n	Drum brake geometry parameter
N	Normal force
N_t	Numerical ratio of the transmission
N_f	Numerical ratio of the final drive
N_{tf}	Numerical ratio of the combined transmission and final drive
NSP	Neutral steer point
p	Pneumatic trail
P_a	Brake application pressure/effort
P_{atm}	Atmospheric pressure
P_f	Front brake application pressure
P_r	Rear brake application pressure
P_s	Static pressure
P_t	Total pressure

PM	Aerodynamic pitching moment
p	Roll velocity about the x-axis of the vehicle
q	Pitch velocity about the y-axis of the vehicle
q	Dynamic pressure
r	Yaw velocity about the z-axis of the vehicle
r	Rolling radius of the tires
r_k	Ratio of tire to suspension stiffness
R	Radius of turn
R_h	Hitch force
R_g	Grade force
R_x	Rolling resistance force
R_{RL}	Road load
RM	Aerodynamic rolling moment
RR	Ride rate of a tire/suspension system
R_ϕ	Roll rate of the sprung mass
s	Lateral separation between suspension springs
S_A	Aerodynamic side force
S_o	Spectral density of white-noise
SD	Stopping distance
t	Tread
t_s	Length of time of a brake application
T_a	Torque in the axle
T_b	Brake torque
T_c	Torque at the clutch
T_d	Torque in the driveshaft
T_e	Torque of the engine
T_{sf}	Roll torque in a front suspension
T_{sr}	Roll torque in a rear suspension
T_{amb}	Ambient temperature
T_x	Torque about the x-axis
V	Forward velocity
V_w	Ambient wind velocity
V_f	Final velocity resulting from a brake application
V_o	Initial velocity in a brake application
w	Tire section width
W	Weight of the vehicle

W_a	Axle weight
W_d	Dynamic load transfer
W_f	Dynamic weight on the front axle
W_r	Dynamic weight on the rear axle
W_{rr}	Dynamic weight on the right rear wheel
W_{fs}	Static weight on the front axle
W_{rs}	Static weight on the rear axle
W_y	Lateral weight transfer on an axle
x	Forward direction on the longitudinal axis of the vehicle
y	Lateral direction out the right side of the vehicle
YM	Aerodynamic yawing moment
z	Vertical direction with respect to the plane of the vehicle
X	Forward direction of travel
Y	Lateral direction of travel
Z	Vertical direction of travel
	Vertical displacement of the sprung mass
Z_r	Road profile elevation
Z_u	Vertical displacement of the unsprung mass
α	Tire slip angle
	Coefficient in the pitch plane equations
α_{cw}	Aerodynamic wind angle
α_d	Rotational acceleration of the driveshaft
α_e	Rotational acceleration of the engine
α_w	Rotational acceleration of the wheels
α_x	Rotational acceleration about the x-axis
β	Sideslip angle
	Rotation angle of a U-joint
	Coefficient in the pitch plane equations
γ	Camber angle
	Coefficient in the pitch plane equations
γ_g	Wheel camber with respect to the ground
γ_b	Wheel camber with respect to the vehicle body
δ	Steer angle
δ_c	Compliance steer
δ_i	Steer angle of the inside wheel in a turn

δ_o	Steer angle of the outside wheel in a turn
Δ	Off-tracking distance in a turn
ε	Roll steer coefficient
	Inclination of the roll axis
ζ	Moment arm related to tire force yaw damping
	Half-shaft angle on a front-wheel drive
ζ_s	Damping ratio of the suspension
η_b	Braking efficiency
η_t	Efficiency of the transmission
η_f	Efficiency of the final drive
η_{tf}	Combined efficiency of the transmission and final drive
θ	Pitch angle
	Angle of a U-joint
θ_p	Body pitch due to acceleration squat or brake dive
Θ	Grade angle
λ	Lateral inclination angle of the steer axis (kingpin inclination angle)
μ	Coefficient of friction
μ_p	Peak coefficient of friction
μ_s	Sliding coefficient of friction
ν	Wavenumber of road roughness spectrum
ξ	Fraction of the drive force developed on the front axle of a 4WD
	Fraction of the brake force developed on the front axle
	Rear steer proportioning factor on a 4WS vehicle
ρ	Density of air
υ	Caster angle of the steer axis
ϕ	Roll angle
φ	Road cross-slope angle
χ	Ratio of unsprung to sprung mass
ψ	Heading angle
	Yaw angle
ω	Rotational speed
ω_d	Damped natural frequency of a suspension system (radians/second)
	Rotational speed of the driveshaft
ω_e	Rotational speed of the engine
ω_i	Rotational speed at the input of a U-joint
ω_n	Undamped natural frequency of a suspension system (radians/second)

ω_O Rotational speed at the output of a U-joint

ω_u Natural frequency of the unsprung mass

ω_w Rotational speed of the wheels

TABLE OF CONTENTS

CHAPTER 1
INTRODUCTION

The next-generation Camaro. (Photo courtesy of Chevrolet Motor Division.)

DAWN OF THE MOTOR VEHICLE AGE

The dawn of the motor vehicle age occurred around 1769 when the French military engineer, Nicholas Joseph Cugnot (1725-1804), built a three-wheeled, steam-driven vehicle for the purpose of pulling artillery pieces [1]. Within a few years an improved model was built, only to cause the first automotive accident when it ran into a wall! This was followed by a steam-powered vehicle built in 1784 by the Scottish engineer, James Watt (1736-1819), which proved unworkable. By 1802, Richard Trevithick (1771-1833), an Englishman, developed a steam coach that traveled from Cornwall to London. The coach met its demise by burning one night after Trevithick forgot to extinguish the boiler fire. Nevertheless, the steam coach business thrived in England until about 1865 when competition from the railroads and strict antispeed laws brought it to an end [2].

Fig. 1.1 First motor vehicle, circa 1769, built by Cugnot. (Photo courtesy of Smithsonian Institution.)

The first practical automobiles powered by gasoline engines arrived in 1886 with the credit generally going to Karl Benz (1844-1929) and Gottlieb Daimler (1834-1900) working independently. Over the next decade, automotive vehicles were developed by many other pioneers with familiar names such as Rene Panhard, Emile Levassor, Armand Peugeot, Frank and Charles Duryea, Henry Ford, and Ransom Olds. By 1908 the automotive industry was well established in the United States with Henry Ford manufacturing the Model T and the General Motors Corporation being founded. In Europe the familiar companies like Daimler, Opel, Renault, Benz, and Peugeot were becoming recognized as automotive manufacturers. By 1909, over 600 makes of American cars had been identified [3].

In the early decades of the 1900s, most of the engineering energy of the automotive industry went into invention and design that would yield faster, more comfortable, and more reliable vehicles. The speed capability of motor vehicles climbed quickly in the embryonic industry as illustrated by the top speeds of some typical production cars, as shown in Figure 1.2.

In general, motor vehicles achieved high speed capability well before good paved roads existed on which to use it. With higher speeds the dynamics of the vehicles, particularly turning and braking, assumed greater importance as an engineering concern. The status of automotive engineering during this period was characterized in the reminiscences of Maurice Olley [4] as follows:

2

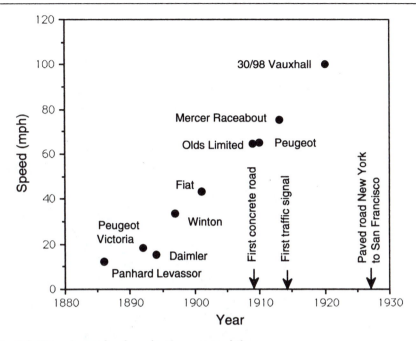

Fig. 1.2 Travel speeds of production automobiles.

"There had been sporadic attempts to make the vehicle ride decently, but little had been done. The rear passengers still functioned as ballast, stuck out behind the rear wheels. Steering was frequently unstable and the front axle with front brakes made shimmy almost inevitable. The engineers had made all the parts function excellently, but when put together the whole was seldom satisfactory."

One of the first engineers to write on automotive dynamics was Frederick William Lanchester (1868-1946). (In a 1908 paper [5] he observed that a car with tiller steering "oversteers" if the centrifugal force on the driver's hands pushes toward greater steer angle [6].) Steering shimmy problems were prevalent at that time as well [7, 8]. But, as described by Segel [6], the understanding of both turning behavior and the shimmy problems was hampered by a lack of knowledge about tire mechanics in these early years.

In 1931, a test device—a tire dynamometer—was built which could measure the necessary mechanical properties of the pneumatic tire for the understandings to be developed [9]. Only then could engineers like Lanchester [10], Olley [11], Rieckert and Schunk [12], Rocard [13], Segel [14] and others develop mechanistic explanations of the turning behavior of automobiles which lays the groundwork for much of our understanding today.

3

The industry has now completed its first century. Engineers have achieved dramatic advancements in the technologies employed in automobiles from the Model T to the Taurus (Figure 1.3). More than ever, dynamics plays an important role in vehicle design and development. A number of textbooks have been written to help the engineer in this discipline [15 - 24], but there remains a need for books that lay out the fundamental aspects of vehicle dynamics. This book attempts to fill that need.

Fig. 1.3 Eighty years of progress from the Model T to the Taurus. (Photos courtesy of Henry Ford Museum and Ford Motor Company.)

INTRODUCTION TO VEHICLE DYNAMICS

It has often been said that the primary forces by which a high-speed motor vehicle is controlled are developed in four patches—each the size of a man's hand—where the tires contact the road. This is indeed the case. A knowledge of the forces and moments generated by pneumatic (rubber) tires at the ground is essential to understanding highway vehicle dynamics. Vehicle dynamics in its broadest sense encompasses all forms of conveyance—ships, airplanes, railroad trains, track-laying vehicles, as well as rubber-tired vehicles. The principles involved in the dynamics of these many types of vehicles are diverse and extensive. Therefore, this book focuses only on rubber-tired vehicles. Most of the discussion and examples will concentrate on the automobile, although the principles are directly applicable to trucks and buses, large and small. Where warranted, trucks will be discussed separately when the functional design or performance qualities distinguish them from the automobile.

Inasmuch as the performance of a vehicle—the motions accomplished in accelerating, braking, cornering and ride—is a response to forces imposed, much of the study of vehicle dynamics must involve the study of how and why the forces are produced. The dominant forces acting on a vehicle to control performance are developed by the tire against the road. Thus it becomes necessary to develop an intimate understanding of the behavior of tires, characterized by the forces and moments generated over the broad range of conditions over which they operate. Studying tire performance without a thorough understanding of its significance to the vehicle is unsatisfying, as is the inverse. Therefore, the relevant properties of tires are introduced at appropriate points in the early chapters of the text, while the reader is referred to Chapter 10 for a more comprehensive discussion of tire properties.

At the outset it is worth noting that the term "handling" is often used interchangeably with cornering, turning, or directional response, but there are nuances of difference between these terms. Cornering, turning, and directional response refer to objective properties of the vehicle when changing direction and sustaining lateral acceleration in the process. For example, cornering ability may be quantified by the level of lateral acceleration that can be sustained in a stable condition, or directional response may be quantified by the time required for lateral acceleration to develop following a steering input. Handling, on the other hand, adds to this the vehicle qualities that feed back to the driver affecting the ease of the driving task or affecting the driver's ability to maintain control. Handling implies, then, not only the vehicle's explicit capabilities, but its contributions as well to the system performance of the driver/vehicle combination. Throughout the book the various terms will be used in a manner most appropriate to the discussion at hand.

Understanding vehicle dynamics can be accomplished at two levels—the empirical and the analytical. The empirical understanding derives from trial and error by which one learns which factors influence vehicle performance, in which way, and under what conditions. The empirical method, however, can often lead to failure. Without a mechanistic understanding of how changes in vehicle design or properties affect performance, extrapolating past experience to new conditions may involve unknown factors which may produce a new result, defying the prevailing rules of thumb. For this reason (and because they are methodical by nature), engineers favor the analytical approach. The analytical approach attempts to describe the mechanics of interest based on the known laws of physics so that an analytical model can be established. In the simpler cases these models can be represented by algebraic or differential equations that relate forces or motions of interest to control inputs and vehicle or tire properties. These equations then allow one to evaluate the role of each vehicle property in the phenomenon of interest. The existence of the model thereby provides a means to identify the important factors, the way in which they operate, and under what conditions. The model provides a predictive capability as well, so that changes necessary to reach a given performance goal can be identified.

It might be noted at this point that analytical methods also are not foolproof because they usually only approximate reality. As many have experienced, the assumptions that must be made to obtain manageable models may often prove fatal to an application of the analysis, and on occasion engineers have been found to be wrong. Therefore, it is very important for the engineer to understand the assumptions that have been made in modeling any aspect of dynamics to avoid these errors.

In the past, many of the shortcomings of analytical methods were a consequence of the mathematical limitations in solving problems. Before the advent of computers, analysis was only considered successful if the "problem" could be reduced to a closed form solution. That is, only if the mathematical expression could be manipulated to a form which allowed the analyst to extract relationships between the variables of interest. To a large extent this limited the functionality of the analytical approach to solution of problems in vehicle dynamics. The existence of large numbers of components, systems, sub-systems, and nonlinearities in vehicles made comprehensive modeling virtually impossible, and the only utility obtained came from rather simplistic models of certain mechanical systems. Though useful, the simplicity of the models often constituted deficiencies that handicapped the engineering approach in vehicle development.

Today with the computational power available in desktop and mainframe computers, a major shortcoming of the analytical method has been overcome. It is now possible to assemble models (equations) for the behavior of individual components of a vehicle that can be integrated into comprehensive models of the overall vehicle, allowing simulation and evaluation of its behavior before being rendered in hardware. Such models can calculate performance that could not be solved for in the past. In cases where the engineer is uncertain of the importance of specific properties, those properties can be included in the model and their importance assessed by evaluating their influence on simulated behavior. This provides the engineer with a powerful new tool as a means to test our understanding of a complex system and investigate means of improving performance. In the end we are forced to confront all the variables that may influence the performance of interest, and recognize everything that is important.

FUNDAMENTAL APPROACH TO MODELING

The subject of "vehicle dynamics" is concerned with the movements of vehicles—automobiles, trucks, buses, and special-purpose vehicles—on a road surface. The movements of interest are acceleration and braking, ride, and turning. Dynamic behavior is determined by the forces imposed on the vehicle from the tires, gravity, and aerodynamics. The vehicle and its components are studied to determine what forces will be produced by each of these sources at a particular maneuver and trim condition, and how the vehicle will respond to these forces. For that purpose it is essential to establish a rigorous approach to modeling the systems and the conventions that will be used to describe motions.

Lumped Mass

A motor vehicle is made up of many components distributed within its exterior envelope. Yet, for many of the more elementary analyses applied to it, all components move together. For example, under braking, the entire vehicle slows down as a unit; thus it can be represented as one lumped mass located at its center of gravity (CG) with appropriate mass and inertia properties. For acceleration, braking, and most turning analyses, one mass is sufficient. For ride analysis, it is often necessary to treat the wheels as separate lumped masses. In that case the lumped mass representing the body is the "sprung mass," and the wheels are denoted as "unsprung masses."

For single mass representation, the vehicle is treated as a mass concentrated at its center of gravity (CG) as shown in Figure 1.4. The point mass at the CG, with appropriate rotational moments of inertia, is dynamically equivalent to the vehicle itself for all motions in which it is reasonable to assume the vehicle to be rigid.

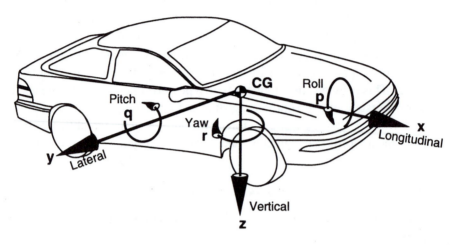

Fig. 1.4 SAE Vehicle Axis System.

Vehicle Fixed Coordinate System

On-board, the vehicle motions are defined with reference to a right-hand orthogonal coordinate system (the vehicle fixed coordinate system) which originates at the CG and travels with the vehicle. By SAE convention [25] the coordinates are:

x - Forward and on the longitudinal plane of symmetry

y - Lateral out the right side of the vehicle

z - Downward with respect to the vehicle

p - Roll velocity about the x axis

q - Pitch velocity about the y axis

r - Yaw velocity about the z axis

Motion Variables

Vehicle motion is usually described by the velocities (forward, lateral, vertical, roll, pitch and yaw) with respect to the vehicle fixed coordinate system, where the velocities are referenced to the earth fixed coordinate system.

Earth Fixed Coordinate System

Vehicle attitude and trajectory through the course of a maneuver are defined with respect to a right-hand orthogonal axis system fixed on the earth. It is normally selected to coincide with the vehicle fixed coordinate system at the point where the maneuver is started. The coordinates (see Figure 1.5) are:

X - Forward travel

Y - Travel to the right

Z - Vertical travel (positive downward)

ψ - Heading angle (angle between x and X in the ground plane)

v - Course angle (angle between the vehicle's velocity vector and X axis)

β - Sideslip angle (angle between x axis and the vehicle velocity vector)

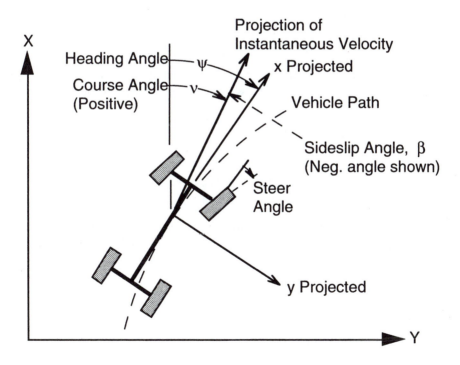

Fig. 1.5 Vehicle in an Earth Fixed Coordinate System.

Euler Angles

The relationship of the vehicle fixed coordinate system to the earth fixed coordinate system is defined by Euler angles. Euler angles are determined by a sequence of three angular rotations. Beginning at the earth fixed system, the axis system is first rotated in yaw (around the z axis), then in pitch (around the y axis), and then in roll (around the x axis) to line up with the vehicle fixed coordinate system. The three angles obtained are the Euler angles. It is necessary to adhere strictly to the defined sequence of rotations, because the resultant attitude will vary with the order of rotations.

Forces

Forces and moments are normally defined as they act on the vehicle. Thus a positive force in the longitudinal (x-axis) direction on the vehicle is forward. The force corresponding to the load on a tire acts in the upward direction and is therefore negative in magnitude (in the negative z-direction). Because of the inconvenience of this convention, the SAE J670e, "Vehicle Dynamics Terminology," gives the name Normal Force as that acting downward, and Vertical Force as the negative of the Normal Force. (See Appendix A.) Thus the Vertical Force is the equivalent of tire load with a positive convention in the upward direction. In other countries, different conventions may be used.

Given these definitions of coordinate systems and forces, it is now possible to begin formulating equations by which to analyze and describe the behavior of a vehicle.

Newton's Second Law

The fundamental law from which most vehicle dynamics analyses begin is the second law formulated by Sir Isaac Newton (1642-1727). The law applies to both translational and rotational systems [26].

Translational systems - The sum of the external forces acting on a body in a given direction is equal to the product of its mass and the acceleration in that direction (assuming the mass is fixed).

$$\Sigma F_x = M \cdot a_x \tag{1-1}$$

where:

F_x = Forces in the x-direction
M = Mass of the body
a_x = Acceleration in the x-direction

Rotational Systems - The sum of the torques acting on a body about a given axis is equal to the product of its rotational moment of inertia and the rotational acceleration about that axis.

$$\sum T_x = I_{xx} \cdot \alpha_x \qquad\qquad (1\text{-}2)$$

where:

T_x = Torques about the x-axis
I_{xx} = Moment of inertia about the x-axis
α_x = Acceleration about the x-axis

Newton's Second Law (NSL) is applied by visualizing a boundary around the body of interest. The appropriate forces and/or moments are substituted at each point of contact with the outside world, along with any gravitational forces. This forms a free-body diagram. A NSL equation may then be written for each of the three independent directions (normally the vehicle fixed axes).

DYNAMIC AXLE LOADS

Determining the axle loadings on a vehicle under arbitrary conditions is a first simple application of Newton's Second Law. It is an important first step in analysis of acceleration and braking performance because the axle loads determine the tractive effort obtainable at each axle, affecting the acceleration, gradeability, maximum speed, and drawbar effort.

Consider the vehicle shown in Figure 1.6, in which most of the significant forces on the vehicle are shown.

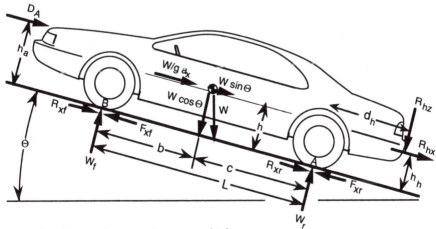

Fig. 1.6 Arbitrary forces acting on a vehicle.

- W is the weight of the vehicle acting at its CG with a magnitude equal to its mass times the acceleration of gravity. On a grade it may have two components, a cosine component which is perpendicular to the road surface and a sine component parallel to the road.

- If the vehicle is accelerating along the road, it is convenient to represent the effect by an equivalent inertial force known as a "d'Alembert force" (Jean le Rond d'Alembert, 1717-1783) denoted by $W/g \cdot a_x$ acting at the center of gravity opposite to the direction of the acceleration [26].

- The tires will experience a force normal to the road, denoted by W_f and W_r, representing the dynamic weights carried on the front and rear wheels.

- Tractive forces, F_{xf} and F_{xr}, or rolling resistance forces, R_{xf} and R_{xr}, may act in the ground plane in the tire contact patch.

- D_A is the aerodynamic force acting on the body of the vehicle. It may be represented as acting at a point above the ground indicated by the height, h_a, or by a longitudinal force of the same magnitude in the ground plane with an associated moment (the aerodynamic pitching moment) equivalent to D_A times h_a.

- R_{hz} and R_{hx} are vertical and longitudinal forces acting at the hitch point when the vehicle is towing a trailer.

The loads carried on each axle will consist of a static component, plus load transferred from front to rear (or vice versa) due to the other forces acting on the vehicle. The load on the front axle can be found by summing torques about the point "A" under the rear tires. Presuming that the vehicle is not accelerating in pitch, the sum of the torques at point A must be zero.

By the SAE convention, a clockwise torque about A is positive. Then:

$$W_f L + D_A h_a + \frac{W}{g} a_x h + R_{hx} h_h + R_{hz} d_h + W h \sin \Theta - W c \cos \Theta = 0$$

$$(1-3)$$

Note that an uphill attitude corresponds to a positive angle, Θ, such that the sine term is positive. A downhill attitude produces a negative value for this term.

From Eq. (1-3) we can solve for W_f and from a similar equation about point B we can solve for W_r. The axle load expressions then become:

12

$$W_f = (W\ c\ \cos\Theta - R_{hx}\,h_h - R_{hz}\,d_h - \frac{W}{g}\,a_x\,h - D_A\,h_a - W\,h\,\sin\Theta)/L$$

$$(1-4)$$

$$W_r = (W\ b\ \cos\Theta + R_{hx}\,h_h + R_{hz}\,(d_h + L) + \frac{W}{g}\,a_x\,h + D_a\,h_a + W\,h\,\sin\Theta)/L$$

$$(1-5)$$

Static Loads on Level Ground

When the vehicle sits statically on level ground, the load equations simplify considerably. The sine is zero and the cosine is one, and the variables R_{hx}, R_{hz}, a_x, and D_A are zero. Thus:

$$W_{fs} = W\,\frac{c}{L}$$

$$(1-6)$$

$$W_{rs} = W\,\frac{b}{L}$$

$$(1-7)$$

Low-Speed Acceleration

When the vehicle is accelerating on level ground at a low speed, such that D_A is zero (and presuming no trailer hitch forces), the loads on the axles are:

$$W_f = W\ (\frac{c}{L} - \frac{a_x}{g}\frac{h}{L}) = W_{fs} - W\,\frac{a_x}{g}\frac{h}{L}$$

$$(1-8)$$

$$W_r = W\ (\frac{b}{L} + \frac{a_x}{g}\frac{h}{L}) = W_{rs} + W\,\frac{a_x}{g}\frac{h}{L}$$

$$(1-9)$$

Thus when the vehicle accelerates, load is transferred from the front axle to the rear axle in proportion to the acceleration (normalized by the gravitational acceleration) and the ratio of the CG height to the wheelbase.

Loads on Grades

The influence of grade on axle loads is also worth considering. Grade is defined as the "rise" over the "run." That ratio is the tangent of the grade angle, Θ. The common grades on interstate highways are limited to 4 percent wherever possible. On primary and secondary roads they occasionally reach 10 to 12 percent. The cosines of angles this small are very close to one, and the sine is very close to the angle itself. That is:

13

$$\cos \Theta = 0.99^+ \cong 1$$

$$\sin \Theta \cong \Theta$$

Thus the axle loads as influenced by grades will be:

$$W_f = W(\frac{c}{L} - \frac{h}{L} \Theta) = W_{fs} - W\frac{h}{L}\Theta \qquad (1\text{-}10)$$

$$W_r = W(\frac{b}{L} + \frac{h}{L} \Theta) = W_{rs} + W\frac{h}{L}\Theta \qquad (1\text{-}11)$$

where a positive grade causes load to be transferred from the front to the rear axle.

EXAMPLE PROBLEMS

1) The curb weights of a Continental 4-door sedan without passengers or cargo are 2313 lb on the front axle and 1322 lb on the rear. The wheelbase, L, is 109 inches. Determine the fore/aft position of the center of gravity for the vehicle.

Solution:

The fore/aft position of the CG is defined by either parameter c or b in Eqs. (1-6) or (1-7), which apply to a vehicle sitting at rest on level ground. Using Eq. (1-7) we can solve for b:

$$b = L\frac{W_{rs}}{W} = 109" \ \frac{1322 \text{ lb}}{(2313 + 1322) \text{ lb}} = 39.64"$$

i.e., the CG of the vehicle is 39.64 inches aft of the front axle.

2) A Taurus GL sedan with 3.0L engine accelerates from a standing start up a 6 percent grade at an acceleration of 6 ft/sec^2. Find the load distribution on the axles at this condition.

Solution:

Assume that aerodynamic forces are negligible since the vehicle starts from zero speed, and there are no trailer hitch forces. Equations (1-4) and (1-5) are the fundamental equations from which to start, but to use them the values for the parameters b and c must be determined. To obtain these check the MVMA (Motor Vehicle Manufacturers Association) specification sheets for the Taurus GL sedan for relevant data. The curb weights are 1949 lb on the front axle and 1097 lb on the rear; the wheelbase is 106 inches; and front passenger's weight is distributed 49 percent on the front axle and 51 percent on the rear. Assuming a 200 lb driver this gives weights as follows:

$$W_{fs} = 2047 \text{ lb} \quad W_{rs} = 1199 \text{ lb} \quad W = 3246 \text{ lb}$$

Eqs. (1-6) and (1-7) are used to find that b = 39.15 inches and c = 66.85 inches. From a pocket calculator we determine that a 6 percent grade corresponds to a 3.433 degree angle (arctangent of 0.06). Lacking a source for information on CG height, h, assume it is 20 inches. Now all of the information is available to solve equation (1-4).

$$W_f = \frac{W(c \cos \Theta - h \, a_x/g - h \sin \Theta)}{L}$$

$$W_f = \frac{3246 \text{ lb} (66.85" \, 0.998 - 20" \, 6/32.2 - 20" \, 0.0599)}{106"} = 1892.2 \text{ lb}$$

Using the same approach the rear axle load is found to be 1347.3 lb. Curiously, these only add up to about 3239.5 lb, not the 3246 lb weight of the car. Why? The reason, of course, is that the vehicle is sitting on a slope. Only the cosine of the weight vector acts to produce load on the axles. Thus, the weight on the axles should only add up to 3246 lb times cos 3.433° = 3240 lb.

3) You are planning to buy a new mini-van to pull your boat trailer (see below) out to those long weekends at the lake. Although you like the new front-wheel-drive (FWD) vans available, you are not sure a FWD will be able to pull the boat up out of the water on some of the steep access ramps you must use.

a) Derive the expressions for the maximum grade it can climb without wheel slippage (traction-limited gradeability) for this vehicle combination in a front-wheel-drive (FWD), rear-wheel-drive (RWD), and four-wheel-drive (4WD) power train.

(In the analysis it is reasonable to assume the longitudinal acceleration is zero, neglect rolling resistance, assume the boat is clear of the water so that there are no bouyancy forces on it, ignore any change in hitch height as the forces are applied, and use the small angle approximations.)

b) Calculate the maximum gradeability for the three combinations on a ramp with a coefficient of friction of 0.3, given the following information on the vehicles.

Van properties:
Front axle weight = 1520 lb
Rear axle weight = 1150 lb
CG height = 24.5 inches
Hitch height = 14 inches
Hitch rear overhang = 23 inches
Wheelbase = 120 inches

Combined boat/trailer properties:
Axle weight = 1200 lb
Hitch load = 250 lb
Wheelbase = 110 inches
CG height = 35 inches

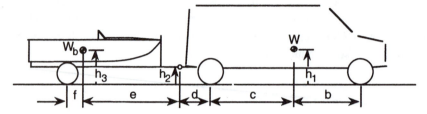

Solution:

In order to derive the equations for traction-limited gradeability, first perform a free-body analysis of the boat trailer to find the hitch forces as a function of grade.

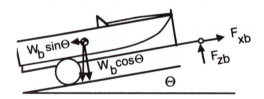

Taking moments about the point where the tire contacts the ground (counterclockwise moments are positive):

$$\Sigma T_y = 0 = W_b h_3 \sin \Theta + F_{zb}(e + f) - W_b f \cos \Theta - F_{xb} h_2 \qquad (1)$$

Also, the force balance along the longitudinal axis of the boat trailer gives:

$$\Sigma F_x = 0 = F_{xb} - W_b \sin \Theta \qquad (2)$$

Next we perform a similar analysis on the van.

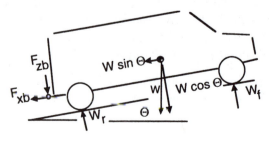

Taking moments about the rear tire contact point:

$$\Sigma T_y = 0 = W\,h_1\sin\Theta - W\,c\cos\Theta + F_{zb}\,d + F_{xb}\,h_2 + W_f(b + c)$$

$$(3)$$

And for moments about the front axle:

$$\Sigma T_y = 0 = W\,h_1\sin\Theta + W\,b\cos\Theta + F_{zb}\,(b + c + d) + F_{xb}\,h_2 - W_r(b + c)$$

$$(4)$$

There are four equations and four unknowns (F_{zb}, F_{xb}, W_f, W_r). We can therefore solve for any one of the unknowns desired. For the case of the FWD, the traction limit will be determined by the load on the front axle times the coefficient of friction, μ. The solution is obtained from Eq. (3), using Eqs. (1) and (2) to eliminate the hitch forces from the final equation. The tractive force will be equal to the combined weight of the van and boat times the grade angle. That is:

$$(W + W_b)\sin\Theta = F_{xf} = \mu\,W_f \tag{5}$$

$$= \mu\,[W\frac{c}{L}\cos\Theta - W\frac{h_1}{L}\sin\Theta - W_b\frac{h_2}{L}\sin\Theta$$

$$+ W_b\frac{d\,h_3}{L\,L_t}\sin\Theta - W_b\frac{d\,f}{L\,L_t}\cos\Theta - W_b\frac{d\,h_2}{L\,L_t}\sin\Theta]$$

The trigonometric functions in the equation make it complicated to obtain a simplified solution. Using the small angle approximations, $\sin\Theta$ can be replaced with Θ, and $\cos\Theta$ can be treated as 1. It is also convenient to define several alternate variables for use in the solution. These will be:

$L = b + c = $ Wheelbase of the van
$L_t = e + f = $ Wheelbase of the trailer (hitch to wheels)
$\zeta = W_b/W = $ Nondimensional weight of the trailer

Then solving the equation for Θ, we obtain the gradeability expression for the case of:

FWD

$$\Theta = \mu\,\frac{\dfrac{c}{L} - \zeta\dfrac{d}{L}\dfrac{f}{L_t}}{1 + \mu\dfrac{h_1}{L} + \zeta\left(1 + \mu\dfrac{h_2}{L} + \mu\dfrac{d}{L}\dfrac{h_2 - h_3}{L_t}\right)}$$

The numerator represents the static weight on the front axle, due to the weight of the van diminished by the vertical load of the trailer on the hitch (the hitch load decreases the front axle load, and hence the gradeability). The second term in the denominator reflects the effect of longitudinal transfer of load from the front axle on a grade due to the elevated position of the CG on the van. The terms in parentheses in the denominator represent the effects of the trailer. The first term in parentheses is the direct effect of the added weight of the trailer. The next term arises from the longitudinal transfer of load off of the front axle due to the towing force at the hitch. The last term is the effect of the change in vertical load on the hitch due to the tow force.

A similar analysis produces a different solution for the case of:

RWD

$$\Theta = \mu \cfrac{\dfrac{b}{L} + \zeta \dfrac{(L+d)}{L} \dfrac{f}{L_t}}{1 - \mu \dfrac{h}{L} + \zeta \left(1 - \mu \dfrac{h_2}{L} - \mu \dfrac{(L+d)}{L} \dfrac{h_2 - h_3}{L_t}\right)}$$

On the rear-wheel drive, static load of the trailer (second term in the numerator) increases gradeability because it increases load on the drive wheels. In the denominator, the terms representing longitudinal load transfer are negative (thereby decreasing the magnitude of the denominator and increasing the gradeability).

Finally, in the case of four-wheel drive, the performance that will be obtained depends on the type of drive system. The most effective utilizes a limited-slip differential on each axle and a limited-slip interaxle drive, so that the torque is distributed to all the wheels in proportion to their traction. Then the van can develop a traction force that is the coefficient of friction times its weight.

$$(W + W_b) \tan \Theta = \mu W$$

or:

$$\Theta = \mu \frac{W}{W + W_b} = \mu \frac{1}{1 + \zeta}$$

For four-wheel-drive systems that do not have full limited-slip features, the solution would require a more complex treatment based on an analysis of the drive forces that would be available from the individual axles.

Example calculations:

For the parameters given in the problem the solutions are:

FWD $\Theta = 0.1018 = 10.18$ % slope = 5.84 deg

RWD $\Theta = 0.1142 = 11.42$ % slope = 6.51 deg

4WD $\Theta = 0.1944 = 19.44$ % slope = 11.0 deg

Despite the fact that the assumed vehicle has a greater static load on the front axle (57% of the weight), the RWD configuration has better gradeability because of the longitudinal transfer of load on the grade.

REFERENCES

1. Roberts, P., Collector's History of the Automobile, Bonanza Books, New York, N.Y., 1978, 320 p.

2. Encyclopedia Americana, Vol. 2, 1966, 645 p.

3. American Cars Since 1775, Automobile Quarterly, Inc., New York, 1971, 504 p.

4. Olley, M., "Reminiscences - Feb 16/57," unpublished, 1957, 17 p.

5. Lanchester, F.W., "Some Reflections Peculiar to the Design of an Automobile," Proceedings of the Institution of Automobile Engineers, Vol. 2, 1908, pp.187-257.

6. Segel, L.,"Some Reflections on Early Efforts to Investigate the Directional Stability and Control of the Motor Car," unpublished, 1990, 7 p.

7. Broulhiet, G., "La Suspension de la Direction de la Voiture Automobile: Shimmy et Dandinement," *Societe des Ingenieurs Civils de France Bulletin*, Vol. 78, 1925.

8. Lanchester, F.W., "Automobile Steering Gear—Problems and Mechanism," Proceedings of the Institution of Automobile Engineers, Vol. 22, 1928, pp. 726-41.

9. Becker, G., *et al.*, "Schwingungen in Automobillernkung," Krayn Berlag, Berlin, 1931.

10. Lanchester, F.W., "Motor Car Suspension and Independent Springing," Proceedings of the Institution of Automobile Engineers, Vol. 30, 1936, pp. 668-762.

11. Olley, M., "Independent Wheel Suspensions—Its Whys and Wherefores," Society of Automotive Engineers Journal, Vol. 34, No. 3, 1934, pp. 73-81.

12. Rieckert, P., and Schunk, T.E., "Zur Fahrmechanik des Gummibereiften Kraftfahrzeugs," Ingenieur Archiv, Vol. 11, 1940.

13. Rocard, Y., "Les Mefaits du Roulement, Auto-Oscillations et Instabilite de Route," La Revue Scientifique, Vol. 84, No. 45, 1946.

14. Segel, L., "Research in the Fundamentals of Automobile Control and Stability," Transactions of the Society of Automotive Engineers, Vol. 65, 1956, pp. 527-40.

15. Ellis, J.R., Vehicle Dynamics, Business Books Limited, London, 1969, 243 p.

16. Ellis, J.R., Road Vehicle Dynamics, John R. Ellis, Inc., Akron, OH, 1988, 294 p.

17. Wong, J.Y., Theory of Ground Vehicles, John Wiley & Sons, New York, 1978, 330 p.

18. Fundamentals of Vehicle Dynamics, General Motors Institute, Flint, MI.

19. Cole, D., "Elementary Vehicle Dynamics," course notes in Mechanical Engineering, The University of Michigan, Ann Arbor, MI, 1972.

20. Fitch, J.W., Motor Truck Engineering Handbook (Third Edition), James W. Fitch, publisher, Anacortes, WA, 1984, 288 p.

21. Newton, K., Steeds, W., and Garrett, T.K., The Motor Vehicle (Tenth Edition), Butterworths, London, 1983, 742 p.

22. Automotive Handbook, 2nd Ed., Robert Bosch GmbH, Stuttgart, 1986, 707 p.

23. Bastow, D., Car Suspension and Handling, Second Edition, Pentech Press, London, 1990, 300 p.

24. Goodsell, D., Dictionary of Automotive Engineering, Butterworths, London, 1989, 182 p.

25. "Vehicle Dynamics Terminology," SAE J670e, Society of Automotive Engineers, Warrendale, PA (see Appendix A).

26. Den Hartog, J.P., Mechanics, McGraw-Hill Book Company, Inc., New York, NY, 1948, p. 174.

CHAPTER 2
ACCELERATION PERFORMANCE

3.4L Twin Dual Cam V6. (Photo courtesy of General Motors Corp.)

Maximum performance in longitudinal acceleration of a motor vehicle is determined by one of two limits—engine power or traction limits on the drive wheels. Which limit prevails may depend on vehicle speed. At low speeds tire traction may be the limiting factor, whereas at high speeds engine power may account for the limits.

POWER-LIMITED ACCELERATION

The analysis of power-limited acceleration involves examination of the engine characteristics and their interaction through the power train.

Engines

The source of propulsive power is the engine. Engines may be characterized by their torque and power curves as a function of speed. Figure 2.1 shows typical curves for gasoline and diesel engines. Gasoline engines typically have a torque curve that peaks in the mid-range of operating speeds controlled by the induction system characteristics. In comparison, diesel engines may have a torque curve that is flatter or even rises with decreasing speed. This character-istic, controlled by the programming of the injection system, has led to the high-torque-rise heavy-duty engines commonly used in commercial vehicles. (In some cases the torque rise may be so great as to provide nearly constant power over much of the engine operating speed range.)

The other major difference between the two types of engines is the specific fuel consumption that is obtained. At their most efficient, gasoline engines may achieve specific fuel consumption levels in the range near 0.4 lb/hp-hr, whereas diesels may be near 0.2 or lower.

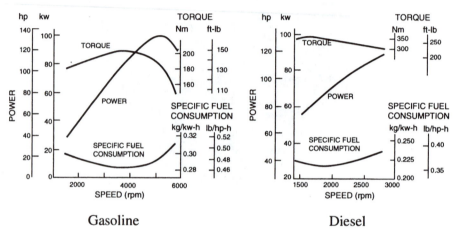

Gasoline   Diesel

Fig. 2.1 Performance characteristics of gasoline and diesel engines.

Power and torque are related by the speed. Specifically,

Power (ft-lb/sec) = Torque (ft-lb) x speed (radians/sec)

Horsepower = T (ft-lb) x ω_e (rad/sec) / 550 = T (ft-lb) x RPM / 5252

$$(2-1)$$

Also,

Power (kw) = 0.746 x HP   1 hp = 550 ft-lb/sec   (2-2)

The ratio of engine power to vehicle weight is the first-order determinant of acceleration performance. At low to moderate speeds an upper limit on acceleration can be obtained by neglecting all resistance forces acting on the vehicle. Then from Newton's Second Law:

22

$$M \, a_x = F_x \tag{2-3}$$

where:

M = Mass of the vehicle = W/g
a_x = Acceleration in the forward direction
F_x = Tractive force at the drive wheels

Since the drive power is the tractive force times the forward speed, Eq. (2-3) can be written:

$$a_x = \frac{1}{M} F_x = 550 \frac{g}{V} \frac{HP}{W} \quad (\text{ft/sec}^2) \tag{2-4}$$

where:

g = Gravitational constant (32.2 ft/sec^2)
V = Forward speed (ft/sec)
HP = Engine horsepower
W = Weight of the vehicle (lb)

Because of the velocity term in the denominator, acceleration capability must decrease with increasing speed. The general relationship of the above equation is shown in Figure 2.2 for cars and trucks. As might be expected, heavy trucks will have much lower performance levels than cars because of the less favorable power-to-weight ratio. Although this is a very simple representation of acceleration performance, it is useful to highway engineers responsible for establishing highway design policies with respect to the needs for climbing lanes on long upgrades, sight distances at intersections, and acceleration areas on entrance ramps [1].

Power Train

More exact estimation of acceleration performance requires modeling the mechanical systems by which engine power is transmitted to the ground. Figure 2.3 shows the key elements.

Starting with the engine, it must be remembered that engine torque is measured at steady speed on a dynamometer, thus the actual torque delivered to the drivetrain is reduced by the amount required to accelerate the inertia of the rotating components (as well as accessory loads not considered here). The torque delivered through the clutch as input to the transmission can be determined by application of NSL as:

23

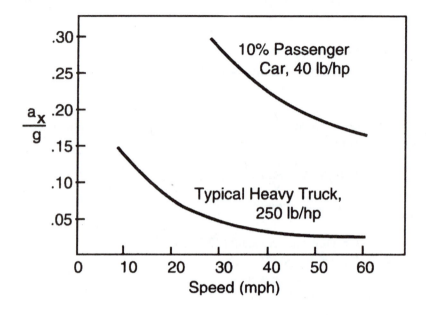

Fig. 2.2 Effect of velocity on acceleration capabilities of cars and trucks [2].

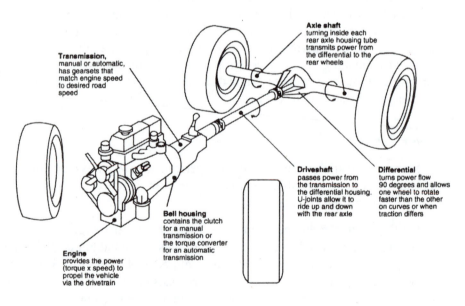

Fig. 2.3 Primary elements in the power train.

$$T_c = T_e - I_e \, \alpha_e \qquad\qquad (2\text{-}5)$$

where:

T_c = Torque at the clutch (input to the transmission)
T_e = Engine torque at a given speed (from dynamometer data)
I_e = Engine rotational inertia
α_e = Engine rotational acceleration

The torque delivered at the output of the transmission is amplified by the gear ratio of the transmission but is decreased by inertial losses in the gears and shafts. If the transmission inertia is characterized by its value on the input side, the output torque can be approximated by the expression:

$$T_d = (T_c - I_t \, \alpha_e) \, N_t \qquad\qquad (2\text{-}6)$$

where:

T_d = Torque output to the driveshaft
N_t = Numerical ratio of the transmission
I_t = Rotational inertia of the transmission (as seen from the engine side)

Similarly, the torque delivered to the axles to accelerate the rotating wheels and provide tractive force at the ground is amplified by the final drive ratio with some reduction from the inertia of the driveline components between the transmission and final drive. The expression for this is:

$$T_a = F_x \, r + I_w \, \alpha_w = (T_d - I_d \, \alpha_d) \, N_f \qquad\qquad (2\text{-}7)$$

where:

T_a = Torque on the axles
F_x = Tractive force at the ground
r = Radius of the wheels
I_w = Rotational inertial of the wheels and axles shafts
α_w = Rotational acceleration of the wheels
I_d = Rotational inertia of the driveshaft
α_d = Rotational acceleration of the driveshaft
N_f = Numerical ratio of the final drive

Now the rotational accelerations of the engine, transmission, and driveline are related to that of the wheels by the gear ratios.

$$\alpha_d = N_f \, \alpha_w \quad \text{and} \quad \alpha_e = N_t \, \alpha_d = N_t \, N_f \, \alpha_w \qquad\qquad (2\text{-}8)$$

25

The above equations (2-5) to (2-8) can be combined to solve for the tractive force available at the ground. Recognizing that the vehicle acceleration, a_x, is the wheel rotational acceleration, α_w, times the tire radius, yields:

$$F_x = \frac{T_e N_{tf}}{r} - \{(I_e + I_t) N_{tf}^2 + I_d N_f^2 + I_w\} \frac{a_x}{r^2} \qquad (2\text{-}9a)$$

where:

N_{tf} = Combined ratio of transmission and final drive

Thus far the inefficiencies due to mechanical and viscous losses in the driveline components (transmission, driveshaft, differential and axles) have not been taken into account. These act to reduce the engine torque in proportion to the product of the efficiencies of the individual components [3]. The efficiencies vary widely with the torque level in the driveline because viscous losses occur even when the torque is zero. As a rule of thumb, efficiencies in the neighborhood of 80% to 90% are typically used to characterize the driveline [5]. The effect of mechanical losses can be approximated by adding an efficiency value to the first term on the right-hand side of the previous equation, giving:

$$F_x = \frac{T_e N_{tf} \eta_{tf}}{r} - \{(I_e + I_t) N_{tf}^2 + I_d N_f^2 + I_w\} \frac{a_x}{r^2} \qquad (2\text{-}9b)$$

where:

η_{tf} = Combined efficiency of transmission and final drive

Thus Eq. (2-9b) provides an expression for the tractive force that can be obtained from the engine. It has two components:

1) The first term on the right-hand side is the engine torque multiplied by the overall gear ratio and the efficiency of the drive system, then divided by tire radius. This term represents the steady-state tractive force available at the ground to overcome the road load forces of aerodynamics and rolling resistance, to accelerate, or to climb a grade.

2) The second term on the right-hand side represents the "loss" of tractive force due to the inertia of the engine and drivetrain components. The term in brackets indicates that the equivalent inertia of each component is "amplified" by the square of the numerical gear ratio between the component and the wheels.

Knowing the tractive force, it is now possible to predict the acceleration performance of a vehicle. The expression for the acceleration must consider all the forces that were shown in Figure 1.6. The equation takes the form:

$$M a_x = \frac{W}{g} a_x = F_x - R_x - D_A - R_{hx} - W \sin \Theta \qquad (2\text{-}10)$$

where:

M = Mass of the vehicle = W/g
a_x = Longitudinal acceleration (ft/sec^2)
F_x = Tractive force at the ground (Eq. (2-9b))
R_x = Rolling resistance forces
D_A = Aerodynamic drag force
R_{hx} = Hitch (towing) forces

F_x includes the engine torque and rotational inertia terms. As a convenience, the rotational inertias from Eq. (2-9b) are often lumped in with the mass of the vehicle to obtain a simplified equation of the form:

$$(M + M_r)\, a_x = \frac{W + W_r}{g}\, a_x = \frac{T_e\, N_{tf}\, \eta_{tf}}{r} - R_x - D_A - R_{hx} - W \sin \Theta \qquad (2\text{-}11)$$

where:

M_r = Equivalent mass of the rotating components

The combination of the two masses is an "effective mass," and the ratio of $(M + M_r)/M$ is the "mass factor." The mass factor will depend on the operating gear, with typical values as below [9]:

		Mass Factor			
Vehicle	Gear:	High	Second	First	Low
Small Car		1.11	1.20	1.50	2.4
Large Car		1.09	1.14	1.30	—
Truck		1.09	1.20	1.60	2.5

A representative number is often taken as [4,9]:

$$\text{Mass Factor} = 1 + 0.04\, N_{tf} + 0.0025\, N_{tf}^{2} \qquad (2\text{-}12)$$

In the complete form of Eq. (2-11), there are no convenient explicit solutions for acceleration performance. Except for the grade term, all other forces vary with speed, and must be evaluated at each speed. An equation as shown above can be used to calculate acceleration performance by hand for a few speeds, but when repeated calculations are required (for example, to calculate acceleration from zero to a high speed), programming on a computer is most often the preferred method [6, 7, 8].

The tractive force generated by the engine/power train (the first term on the right side of Eq. (2-11)) is the effort available to overcome road load forces and accelerate the vehicle. This is shown for a four-speed manual transmission in Figure 2.4.

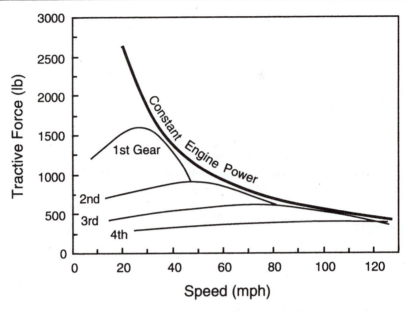

Fig. 2.4 Tractive effort-speed characteristics for a manual transmission.

The "Constant Engine Power" line is equal to the maximum power of the engine, which is the upper limit of tractive effort available, less any losses in the driveline. It is only approached when the engine reaches the speed at which it develops maximum power. The tractive force line for each gear is the image of the engine torque curve multiplied by the ratios for that gear. The curves illustrate visually the need to provide a number of gear ratios for operation of the vehicle (low gearing for start-up, and high gearing for high-speed driving).

For maximum acceleration performance the optimum shift point between gears is the point where the lines cross. The area between the lines for the different gears and the constant power curve is indicative of the deficiencies of the transmission in providing maximum acceleration performance.

Automatic Transmissions

Automatic transmissions provide somewhat different performance, more closely matching the ideal because of the torque converter on the input. Torque converters are fluid couplings that utilize hydrodynamic principles to amplify the torque input to the transmission at the expense of speed. Figure 2.5 shows the torque ratio and efficiency characteristics of a typical torque converter as a function of speed ratio (output/input speed). At zero output speed (speed ratio of zero) the output torque will be several times that of the input. Thus the torque

28

input to the transmission will be twice the torque coming out of the engine when the transmission is stalled, providing for good "off-the-line" acceleration performance. As speed builds up and the transmission input approaches engine speed, the torque ratio drops to unity.

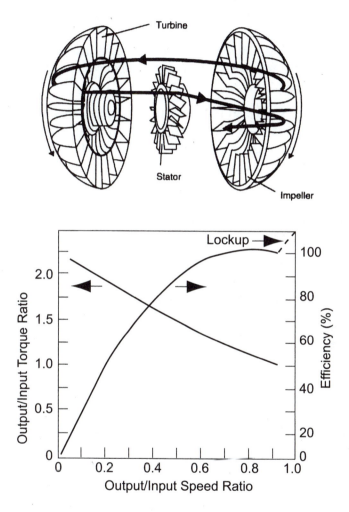

Fig. 2.5 Characteristics of a typical torque converter.

The torque amplification provides for more favorable tractive effort-speed performance as shown for a four-speed automatic transmission in Figure 2.6. Because of the slip possible with the fluid coupling, the torque curves in each gear can extend down to zero speed without stalling the engine. At low speed in first gear the effect of the torque converter is especially evident as the tractive effort rises down toward the zero speed condition.

Also shown on this figure are the road load forces arising from rolling resistance, aerodynamic drag, and road grade (0, 5, 10, 15 and 20%). At a given speed and gear the difference between the tractive effort curve and the appropriate road load curve is the tractive force available to accelerate the vehicle (and its rotating components). The intersection between the road load curves and any of the tractive effort curves is the maximum speed that can be sustained in that gear.

The actual ratios selected for a transmission may be tailored for performance in specific modes—an optimal first gear for starting, a second or third gear for passing, and a high gear for fuel economy at road speeds. The best gear ratios usually fall close to a geometric progression, in which the ratios change by a constant percentage from gear to gear. Figure 2.7 illustrates the relationship of engine speed to road speed obtained with geometric progression. Figure 2.8 shows the engine-road speed relationship for an actual production car. Note that although it is close to geometric progression, some variation occurs.

In these times the choices made in selection of transmission gear ratios must also reflect the realities of the pressures for fuel economy and emissions. The engine performance in both of these respects is quantified by mapping its characteristics. An example of a fuel consumption map for a V-8 engine is

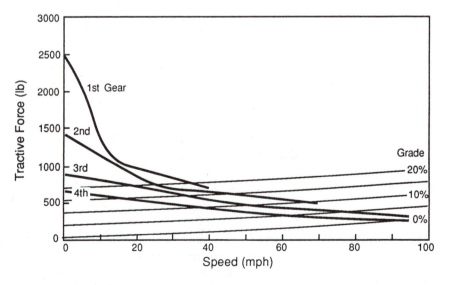

Fig. 2.6 Tractive effort-speed characteristics for an automatic transmission.

shown in Figure 2.9. The figure shows lines of constant fuel consumption (pounds per brake-horsepower-hour) as a function of brake-mean-effective-pressure (indicative of torque) and engine speed. Near the boundaries the specific fuel consumption is highest. In the middle is a small island of minimum fuel consumption at the rate of 0.46 lb/bhp-hr. To maximize

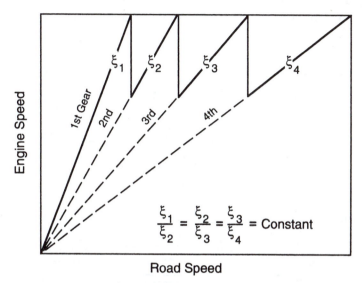

Fig. 2.7 Selection of gear ratios based on geometric progression.

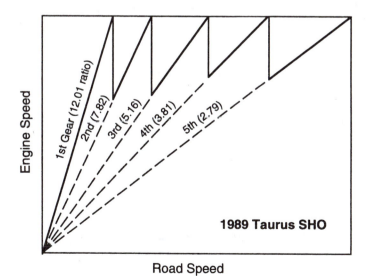

Fig. 2.8 Gear ratios on a typical passenger car.

highway fuel economy the vehicle and driveline should be designed to operate in this region. For best economy over the full driving range the transmission should be designed to operate along the bold line which stays within the valleys of minimum fuel consumption over the broadest range of engine speeds.

For emissions purposes, similar maps of engine performance can be developed to characterize the emissions properties, and a similar logic would be used to identify transmission characteristics that would minimize emissions.

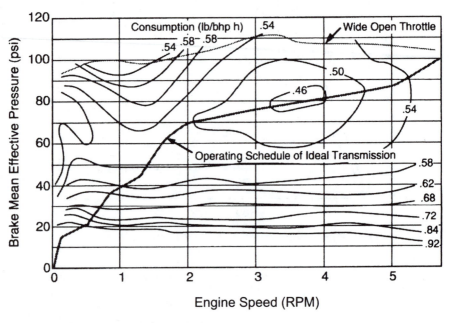

Fig 2.9 Specific fuel consumption map of a V-8 engine (300 cubic inch).

EXAMPLE PROBLEMS

1) We are given the following information about the engine and drivetrain components for a passenger car:

Engine inertia	0.8 in-lb-sec^2					
RPM/Torque (ft-lb)	800	120	2400	175	4000	200
	1200	132	2800	181	4400	201
	1600	145	3200	190	4800	198
	2000	160	3600	198	5200	180

Transmission Data - Gear	1	2	3	4	5
Inertias	1.3	0.9	0.7	0.5	0.3 in-lb-sec^2
Ratios	4.28	2.79	1.83	1.36	1.00
Efficiencies	0.966	0.967	0.972	0.973	0.970

Final drive -	Inertia	1.2 in-lb-sec^2
	Ratio	2.92
	Efficiency	0.99

Wheel inertias   Drive 11.0 in-lb-sec^2 Non-drive 11.0 in-lb-sec^2

Wheel size   801 rev/mile $\Rightarrow$ 6.59 ft circumference $\Rightarrow$ 12.59 inch radius

a) Calculate the effective inertia of the drivetrain components in first gear.

Solution:

The effective inertia is given by the second term on the right-hand side of Eq.(2-9b), which had the following form:

$$F_x = \frac{T_e N_{tf} \eta_{tf}}{r} - \{(I_e + I_t) N_{tf}^2 + I_d N_f^2 + I_w\} \frac{a_x}{r^2} \qquad (2\text{-}9b)$$

The term in the brackets is the effective inertia. It is calculated as follows:

$$I_{eff} = \{(I_e + I_t)(N_{tf})^2 + I_d N_f^2 + I_w\}$$
$$= (0.8 + 1.3) \text{ in-lb-sec}^2 (4.28 \times 2.92)^2 + 1.2 \times 2.92^2 + 2 \times 11.0 \text{ in-lb-sec}^2$$
$$= 328 + 10.2 + 22 = 360.2 \text{ in-lb-sec}^2$$

Notes:

1) The engine and first gear components are the largest inertia when operating in first gear. In fifth gear, the inertia of these components is about 9.7 in-lb-sec^2.

2) Only the inertia of the drive wheels was included in this solution because only they subtract from the tractive force available at the ground at the drive wheels. We must keep in mind that the non-driven wheels contribute an additional inertia when the vehicle is accelerated. The inertia of the non-driven wheels should be lumped in with the inertia (mass) of the total vehicle.

3) The rotational inertia, in units of in-lb-sec^2, is converted into translational inertia (mass) when divided by r^2 in Eq. (2-9). We can see its magnitude as follows:

$$M_{eff} = I_{eff}/r^2 = 360.2 \text{ in-lb-sec}^2 / 12.59^2 \text{ in}^2 = 2.27 \text{ lb-sec}^2/\text{in}$$

Perhaps the more familiar form is the effective weight:

$$W_{eff} = M_{eff}\, g = 2.27 \text{ lb-sec}^2/\text{in} \times 386 \text{ in/sec}^2 = 877 \text{ lb}$$

Comparing this figure to the weight of a typical passenger car (2500 lb), we see that it adds about 35% to the effective weight of the car during acceleration in first gear. The inertia of the non-driven wheels will add another 27 lb to the effective weight (1%).

2) Calculate the maximum tractive effort and corresponding road speed in first and fifth gears of the car described above when inertial losses are neglected.

Solution:

Maximum tractive effort will coincide with maximum torque, which occurs at 4400 rpm. So the problem reduces to finding the tractive effort from the first term in Eq. (2-9) for that value of torque.

$$F_x = T_e\, N_{tf}\, \eta_{tf}/r$$

$$= 201 \text{ ft-lb } (4.28 \times 2.92)\ (0.966 \times 0.99)/12.59 \text{ in} \times 12 \text{ in/ft}$$

$$= 2290 \text{ lb}$$

The road speed is determined by use of the relationships given in Eq. (2-8). Although the equation is written in terms of acceleration, the same relationships hold true for speed. That is:

$$\omega_d = N_f\, \omega_w \quad \text{and} \quad \omega_e = N_t\, \omega_d = N_t\, N_f\, \omega_w \qquad \text{(2-8a)}$$

The wheel rotational speed will be:

$$\omega_w = \omega_e/(N_t\, N_f) = 4400 \text{ rev/min} \cdot 2\pi \text{ rad/rev} \cdot 1 \text{ min/60 sec}/(4.28 \times 2.92)$$

$$= 36.87 \text{ rad/sec}$$

The corresponding ground speed will be found by converting the rotational speed to translational speed at the circumference of the tire.

$$V_x = \omega_w \cdot r = 36.87 \text{ rad/sec} \times 12.59 \text{ in} = 464.2 \text{ in/sec} = 38.7 \text{ ft/sec} = 26.4 \text{ mph}$$

The same method is used to calculate performance in high gear as well:

$$F_x = T_e \, N_{tf} \, \eta_{tf}/r$$

$$= 201 \text{ ft-lb } (1.0 \times 2.92) \, (0.99 \times 0.97)/12.59 \text{ in} \times 12 \text{ in/ft}$$

$$= 537 \text{ lb}$$

$$\omega_w = \omega_e /(N_t N_f) = 4400 \text{ rev/min} \cdot 2\pi \text{ rad/rev} \cdot 1 \text{ min/60 sec} /(1.0 \times 2.92)$$

$$= 157.8 \text{ rad/sec}$$

$$V_x = \omega_w \cdot r = 157.8 \text{ rad/sec} \times 12.59 \text{ in} = 1987 \text{ in/sec} = 165 \text{ ft/sec} = 113 \text{ mph}$$

TRACTION-LIMITED ACCELERATION

Presuming there is adequate power from the engine, the acceleration may be limited by the coefficient of friction between the tire and road. In that case F_x is limited by:

$$F_x = \mu \, W \qquad\qquad (2\text{-}13)$$

where:

μ = Peak coefficient of friction

W = Weight on drive wheels

The weight on a drive wheel then depends on the static plus the dynamic load due to acceleration, and on any transverse shift of load due to drive torque.

Transverse Weight Shift due to Drive Torque

Transverse weight shift occurs on all solid drive axles, whether on the front or rear of the vehicle. The basic reactions on a rear axle are shown in Figure 2.10. The driveshaft into the differential imposes a torque T_d on the axle. As will be seen, the chassis may roll compressing and extending springs on opposite sides of the vehicle such that a torque due to suspension roll stiffness, T_s, is produced. Any difference between these two must be absorbed as a difference in weight on the two wheels. If the axle is of the non-locking type, then the torque delivered to both wheels will be limited by the traction limit on the most lightly loaded wheel.

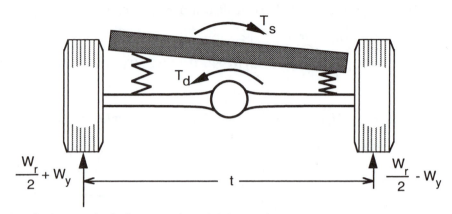

Fig. 2.10. Free-body diagram of a solid drive axle.

Writing NSL for rotation of the axle about its centerpoint allows the reactions to be related. When the axle is in equilibrium:

$$\Sigma T_o = (W_r/2 + W_y - W_r/2 + W_y)\, t/2 + T_s - T_d = 0 \qquad (2\text{-}14)$$

$$\text{or} \quad W_y = (T_d - T_s)/t$$

In the above equation, T_d can be related to the drive forces because:

$$T_d = F_x\, r/N_f \qquad (2\text{-}15)$$

where:

F_x = Total drive force from the two rear wheels

r = Tire radius

N_f = Final drive ratio

However, it is necessary to determine the roll torque produced by the suspension, which requires an analysis of the whole vehicle because the reaction of the drive torque on the chassis attempts to roll the chassis on <u>both</u> the front and rear suspensions. The entire system of interest is illustrated in Figure 2.11 for the case of a rear-wheel-drive car.

The drive torque reaction at the engine/transmission is transferred to the frame and distributed between the front and rear suspensions. It is generally assumed that the roll torque produced by a suspension is proportional to roll angle (Hooke's Law) of the chassis. Then:

36

$$T_{sf} = K_{\phi f}\, \phi \qquad\qquad (2\text{-}16a)$$

$$T_{sr} = K_{\phi r}\, \phi \qquad\qquad (2\text{-}16b)$$

$$K_{\phi} = K_{\phi f} + K_{\phi r} \qquad\qquad (2\text{-}16c)$$

where:

T_{sf} = Roll torque on the front suspension

T_{sr} = Roll torque on the rear suspension

$K_{\phi f}$ = Front suspension roll stiffness

$K_{\phi r}$ = Rear suspension roll stiffness

K_{ϕ} = Total roll stiffness

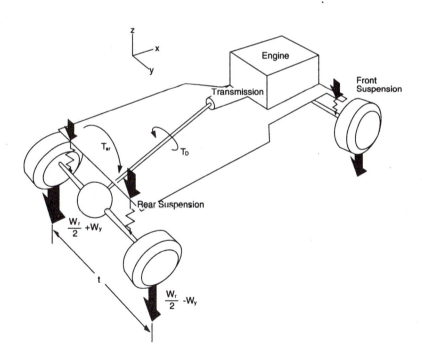

Fig. 2.11 Diagram of drive torque reactions on the chassis.

Now T_{sr} can be related to the roll angle, and the roll angle can be related to the drive torque as follows. The roll angle is simply the drive torque divided by the total roll stiffness:

$$\phi = T_d / K_\phi = T_d / (K_{\phi f} + K_{\phi r}) \tag{2-17}$$

Therefore, substituting in Eq. (2-16b):

$$T_{sr} = K_{\phi r} T_d / (K_{\phi f} + K_{\phi r})$$

This in turn can be substituted into Eq. (2-14), along with the expression for T_d obtained from Eq. (2-15):

$$W_y = \frac{F_x r}{N_f t} [1 - \frac{K_{\phi r}}{K_{\phi r} + K_{\phi f}}] \tag{2-18a}$$

The term in the brackets collapses to yield:

$$W_y = \frac{F_x r}{N_f t} \frac{K_{\phi f}}{K_\phi} \tag{2-18b}$$

This equation gives the magnitude of the lateral load transfer as a function of the tractive force and a number of vehicle parameters such as the final drive ratio, tread of the axle, tire radius, and suspension roll stiffnesses. The net load on the rear axle during acceleration will be its static plus its dynamic component (see Eq. (1-7)). For a rear axle:

$$W_r = W(\frac{b}{L} + \frac{a_x}{g} \frac{h}{L}) \tag{2-19}$$

Neglecting the rolling resistance and aerodynamic drag forces, the acceleration is simply the tractive force divided by the vehicle mass.

$$W_r = W(\frac{b}{L} + \frac{F_x}{Mg} \frac{h}{L}) \tag{2-20}$$

Then the weight on the right rear wheel, W_{rr}, will be $W_r/2 - W_y$, or:

$$W_{rr} = \frac{Wb}{2L} + \frac{F_x}{2} \frac{h}{L} - \frac{F_x r}{N_f t} \frac{K_{\phi f}}{K_\phi} \tag{2-21}$$

and

$$F_x = 2 \mu W_{rr} = 2 \mu (\frac{Wb}{2L} + \frac{F_x h}{2L} - \frac{F_x r}{N_f t} \frac{K_{\phi f}}{K_\phi}) \tag{2-22}$$

Traction Limits

Solving for F_x gives the final expression for the maximum tractive force that can be developed by a <u>solid rear axle with a non-locking differential</u>:

$$F_{xmax} = \frac{\mu \dfrac{Wb}{L}}{1 - \dfrac{h}{L}\mu + \dfrac{2\mu r}{N_f t}\dfrac{K_{\phi f}}{K_\phi}} \tag{2-23}$$

For a <u>solid rear axle with a locking differential</u>, additional tractive force can be obtained from the other wheel up to its traction limits such that the last term in the denominator of the above equation drops out. This would also be true in the case of an <u>independent rear suspension</u> because the driveline torque reaction is picked up by the chassis-mounted differential. In both of these cases the expression for the maximum tractive force is:

$$F_{xmax} = \frac{\mu \dfrac{Wb}{L}}{1 - \dfrac{h}{L}\mu} \tag{2-24}$$

Finally, in the case of a front axle, the fore/aft load transfer is opposite from the rear axle case. Since the load transfer is reflected in the second term of the denominator, the opposite direction yields a sign change. Also, the term "W b/L" arose in the earlier equations to represent the static load on the rear drive axle. For a front-wheel-drive vehicle the term becomes "W c/L." For the <u>solid front drive axle with non-locking differential</u>:

$$F_{xmax} = \frac{\mu \dfrac{Wc}{L}}{1 + \dfrac{h}{L}\mu + \dfrac{2\mu r}{N_f t}\dfrac{K_{\phi r}}{K_\phi}} \tag{2-25}$$

And for the <u>solid front drive axle with locking differential</u>, or the <u>independent front drive axle</u> as typical of most front-wheel-drive cars today:

$$F_{xmax} = \frac{\mu \dfrac{Wc}{L}}{1 + \dfrac{h}{L}\mu} \tag{2-26}$$

39

EXAMPLE PROBLEMS

1) Find the traction-limited acceleration for the rear-drive passenger car with and without a locking differential on a surface of moderate friction level. The information that will be needed is as follows:

Weights	Front - 2100 lb	Rear - 1850 lb	Total - 3950 lb
CG height	21.0 in	Wheelbase - 108 in	
Coefficient of friction	0.62	Tread - 59.0 in	
Final drive ratio	2.90	Tire size - 13.0 in	
Roll stiffnesses	Front - 1150 ft-lb/deg	Rear - 280 ft-lb/deg	

Solution:

The equation for the maximum tractive force of a solid axle rear-drive vehicle with a non-locking differential was given in Eq. (2-23):

$$F_{xmax} = \frac{\mu \, \dfrac{Wb}{L}}{1 - \dfrac{h}{L}\mu + \dfrac{2\,\mu\,r}{N_f t}\dfrac{K_{\phi f}}{K_\phi}} \qquad (2\text{-}23)$$

In this equation, W b/L is just the rear axle weight, which is known; therefore we do not have to find the value for the parameter "b." Likewise, all the other terms are known and can be substituted into the equation to obtain:

$$F_{xmax} = \frac{(0.62)\,1850\text{ lb}}{1 - \dfrac{21}{108}\,0.62 + \dfrac{2}{2.9}\dfrac{(0.62)}{}\dfrac{13\text{ in}}{59\text{ in}}\dfrac{1150}{1430}}$$

$$= \frac{1147\text{ lb}}{1 - 0.121 + 0.0758} = \frac{1147}{0.9548} = 1201\text{ lb}$$

$$a_x = \frac{F_{xmax}}{M\,g} = \frac{1201\text{ lb}}{3950\text{ lb}} = 0.3041\text{ g's} = 9.79\,\frac{\text{ft}}{\text{sec}^2}$$

With a locking differential the third term in the denominator disappears (Eq. (2-24)) so that we obtain:

$$F_{xmax} = \frac{(0.62)\ 1850\ \text{lb}}{1 - \frac{21}{108}\ 0.62} = \frac{1147\ \text{lb}}{1 - 0.121} = \frac{1147}{0.879} = 1305\ \text{lb}$$

$$a_x = \frac{F_{xmax}}{M\ g} = \frac{1305\ \text{lb}}{3950\ \text{lb}} = 0.330\ \text{g's} = 10.64\ \frac{\text{ft}}{\text{sec}^2}$$

Notes:

a) For both cases the numerator term is the weight on the drive axle times the coefficient of friction which is equivalent to 1147 lb of tractive force.

b) Similarly, the dynamic load transfer onto the rear (drive) axle from acceleration is accounted for by the second term in the denominator which diminishes the magnitude of the denominator by 12.1%, thereby increasing the tractive force by an equivalent percentage.

c) The lateral load transfer effect appears in the third term of the denominator, increasing its value by approximately 7.6%, which has the effect of decreasing the tractive force by about the same percentage. Comparing the two answers, the loss from lateral load transfer on the drive axle with a non-locking differential is 104 lb. On higher friction surfaces a higher loss would be seen.

2) Find the traction-limited performance of a front-wheel-drive vehicle under the same road conditions as the problem above. The essential data are:

Weights	Front - 1950	Rear - 1150	Total - 3100
CG Height	19.0 in	Wheelbase - 105 in	
Coefficient of friction	0.62	Tread - 60 inches	
Final drive ratio	3.70	Tire size - 12.59 inches	
Roll stiffnesses	Front - 950 ft-lb/deg	Rear - 620 ft-lb/deg	

Solution:

Most front-wheel-drive vehicles have an independent front suspension. Thus the equation for maximum tractive effort is given by Eq. (2-26), and we notice that all the data required to calculate lateral load transfer on the axle are not needed. The maximum tractive force is calculated by substituting in the equation as follows:

$$F_{xmax} = \frac{\mu \frac{W_c}{L}}{1 + \frac{h}{L}\mu} \tag{2-26}$$

$$F_{xmax} = \frac{(0.62)\ 1950\ lb}{1 + \frac{19}{105}\ 0.62} = \frac{1209\ lb}{1 + 0.1122} = 1087\ lb$$

$$a_x = \frac{F_{xmax}}{M\ g} = \frac{1087\ lb}{3100\ lb} = 0.3506\ g's = 11.29\ \frac{ft}{sec^2}$$

Note:

1) Even though the front-wheel-drive vehicle has a much higher percentage of its weight on the drive axle, its performance is not proportionately better. The reason is the loss of load on the front (drive) axle due to longitudinal weight transfer during acceleration.

REFERENCES

1. Gillespie, T.D., "Methods of Predicting Truck Speed Loss on Grades," The University of Michigan Transportation Research Institute, Report No. UM-85-39, November 1986, 169 p.

2. St. John, A.D., and Kobett, D.R., "Grade Effects on Traffic Flow Stability and Capacity," Interim Report, National Cooperative Highway Research Program, Project 3-19, December 1972, 173 p.

3. Marshall, H.P., "Maximum and Probable Fuel Economy of Automobiles," SAE Paper No. 800213, 1980, 8 p.

4. Cole, D., "Elementary Vehicle Dynamics," course notes in Mechanical Engineering, The University of Michigan, Ann Arbor, Michigan, 1972.

5. Smith, G.L., "Commercial Vehicle Performance and Fuel Economy," SAE Paper, SP-355, 1970, 23 p.

6. Buck, R.E., "A Computer Program (HEVSIM) for Heavy Duty Vehicle Fuel Economy and Performance Simulation," U.S. Department of Transportation, Research and Special Projects Administration, Transportation Systems Center, Report No. DOT-HS-805-912, September 1981, 26 p.

7. Zub, R.W., "A Computer Program (VEHSIM) for Vehicle Fuel Economy and Performance Simulation (Automobiles and Light Trucks)," U.S. Department of Transportation, Research and Special Projects Administration, Transportation Systems Center, Report No. DOT-HS-806-040, October 1981, 50 p.

8. Phillips, A.W., Assanis, D.N., and Badgley, P., "Development and Use of a Vehicle Powertrain Simulation for Fuel Economy and Performance Studies," SAE Paper No. 900619, 1990, 14 p.

9. Taborek, J.J., Mechanics of Vehicles, Towmotor Corporation, Cleveland, Ohio, 1957, 93 p.

CHAPTER 3
BRAKING PERFORMANCE

ABS test drive. (Photo courtesy of Robert Bosch GmbH.)

BASIC EQUATIONS

The general equation for braking performance may be obtained from Newton's Second Law written for the x-direction. The forces on the vehicle are generally of the type shown in Figure 1.6. Then, NSL is:

$$M\, a_x = -\frac{W}{g} D_x = -F_{xf} - F_{xr} - D_A - W \sin \Theta \qquad (3\text{-}1)$$

where:

W = Vehicle weight
g = Gravitational acceleration
$D_x = -a_x$ = Linear deceleration
F_{xf} = Front axle braking force
F_{xr} = Rear axle braking force
D_A = Aerodynamic drag
Θ = Uphill grade

45

The front and rear braking force terms arise from the torque of the brakes along with rolling resistance effects, bearing friction, and driveline drags. A comprehensive analysis of the deceleration requires detailed knowledge of all these forces acting on the vehicle.

Constant Deceleration

Simple and fundamental relationships can be derived for the case where it is reasonable to assume that the forces acting on the vehicle will be constant throughout a brake application. The simple equations that result provide an appreciation for the basic relationships that govern braking maneuvers. From Eq. (3-1):

$$D_x = \frac{F_{xt}}{M} = -\frac{dV}{dt} \tag{3-2}$$

where:

F_{xt} = The total of all longitudinal deceleration forces on the vehicle (+)

V = Forward velocity

This equation can be integrated (because F_{xt} is constant) for a deceleration (snub) from initial velocity, V_o, to final velocity, V_f:

$$\int_{V_o}^{V_f} dV = -\frac{F_{xt}}{M} \int_0^{t_s} dt \tag{3-3}$$

$$V_o - V_f = \frac{F_{xt}}{M} t_s \tag{3-4}$$

where:

t_s = Time for the velocity change

Because velocity and distance are related by $V = dx/dt$, we can substitute for "dt" in Eq. (3-2), integrate, and obtain the relationship between velocity and distance:

$$\frac{V_o^2 - V_f^2}{2} = \frac{F_{xt}}{M} X \tag{3-5}$$

where:

X = Distance traveled during the deceleration

In the case where the deceleration is a full stop, then V_f is zero, and X is the stopping distance, SD. Then:

$$SD = \frac{V_o^2}{2 \frac{F_{xt}}{M}} = \frac{V_o^2}{2 D_x} \tag{3-6}$$

and the time to stop is:

$$t_s = \frac{V_o}{\frac{F_{xt}}{M}} = \frac{V_o}{D_x} \tag{3-7}$$

Thus, all other things being equal, the time to stop is proportional to the velocity, whereas the distance is proportional to the velocity squared (i.e., doubling the velocity doubles the time to stop, but quadruples the distance required).

Deceleration with Wind Resistance

The aerodynamic drag on a vehicle is dependent on vehicle drag factors and the square of the speed. To determine stopping distance in such cases, a more complicated expression is necessary but can still be integrated. To analyze this case:

$$\Sigma F_x = F_b + C V^2 \tag{3-8}$$

where:

F_b = Total brake force of front and rear wheels

C = Aerodynamic drag factor

Therefore:

$$\int_0^{SD} dx = M \int_{V_0}^0 \frac{V \, dV}{F_b + C V^2} \tag{3-9}$$

47

This may be integrated to obtain the stopping distance:

$$SD = \frac{M}{2\,C} \ln \left| \frac{F_b + C\,V_o^2}{F_b} \right| \tag{3-10}$$

Energy/Power

The energy and/or power absorbed by a brake system can be substantial during a typical maximum-effort stop. The energy absorbed is the kinetic energy of motion for the vehicle, and is thus dependent on the mass.

$$\text{Energy} = \frac{M}{2} (V_o^2 - V_f^2) \tag{3-11}$$

The power absorption will vary with the speed, being equivalent to the braking force times the speed at any instant of time. Thus, the power dissipation is greatest at the beginning of the stop when the speed is highest. Over the entire stop, the average power absorption will be the energy divided by the time to stop. Thus:

$$\text{Power} = \frac{M}{2} \frac{V_o^2}{t_s} \tag{3-12}$$

Calculation of the power is informative from the standpoint of appreciating the performance required from a brake system. A 3000 lb car in a maximum-effort stop from 80 mph requires absorption of nearly 650,000 ft-lb of energy. If stopped in 8 seconds (10 mph/sec), the average power absorption of the brakes during this interval is 145 HP. An 80,000 lb truck stopped from 60 mph typically involves dissipation at an average rate of several thousands of horsepower!

BRAKING FORCES

The forces on a vehicle producing a given braking deceleration may arise from a number of sources. Though the brakes are the primary source, others will be discussed first.

Rolling Resistance

Rolling resistance always opposes vehicle motion; hence, it aids the brakes. The rolling resistance forces will be:

$$R_{xf} + R_{xr} = f_r (W_f + W_r) = f_r\,W \tag{3-13}$$

The parameter "f_r" is the rolling resistance coefficient, which will be discussed in the next chapter. Note that the total force is independent of the distribution of loads on the axles (static or dynamic). Rolling resistance forces are nominally equivalent to about 0.01 g deceleration (0.3 ft/sec^2).

Aerodynamic Drag

The drag from air resistance depends on the dynamic pressure, and is thus proportional to the square of the speed. At low speeds it is negligible. At normal highway speeds, it may contribute a force equivalent to about 0.03 g (1 ft/sec^2). More discussion of this topic is presented in the next chapter.

Driveline Drag

The engine, transmission, and final drive contribute both drag and inertia effects to the braking action. As discussed in the previous chapter on Acceleration Performance, the inertia of these components adds to the effective mass of the vehicle, and warrants consideration in brake sizing on the drive wheels. The drag arises from bearing and gear friction in the transmission and differential, and engine braking. Engine braking is equivalent to the "motoring" torque (observed on a dynamometer) arising from internal friction and air pumping losses. (It is worth noting that the pumping losses disappear if the engine is driven to a speed high enough to float the valves. Thus, engine braking disappears when an engine over-revs excessively. This can be a serious problem on low-speed truck engines where valve float may occur above 4000 rpm, and has been the cause of runaway accidents on long grades.) On a manual transmission with clutch engaged during braking, the engine braking is multiplied by the gear ratio selected. Torque-converter transmissions are designed for power transfer from the engine to the driveline, but are relatively ineffective in the reverse direction; hence, engine drag does not contribute substantially to braking on vehicles so equipped.

Whether or not driveline drag aids in braking depends on the rate of deceleration. If the vehicle is slowing down faster than the driveline components would slow down under their own friction, the drive wheel brakes must pick up the extra load of decelerating the driveline during the braking maneuver. On the other hand, during low-level decelerations the driveline drag may be sufficient to decelerate the rotating driveline components and contribute to the braking effort on the drive wheels as well.

Grade

Road grade will contribute directly to the braking effort, either in a positive sense (uphill) or negative (downhill). Grade is defined as the rise over the run (vertical over horizontal distance). The additional force on the vehicle arising from grade, R_g, is given by:

$$R_g = W \sin \Theta \qquad\qquad (3\text{-}14)$$

For small angles typical of most grades:

$$\Theta \text{ (radians)} \cong \text{Grade} = \text{Rise/run}$$

$$R_g = W \sin \Theta \cong W \Theta$$

Thus a grade of 4% (0.04) will be equivalent to a deceleration of $\pm$ 0.04 g (1.3 ft/sec^2).

BRAKES

Automotive brakes in common usage today are of two types—drum and disc [1, 2, 3] as shown in Figure 3.1.

Fig. 3.1   Drum brake and disc brake. (Photo courtesy of Chrysler Corp.)

Historically, drum brakes have seen common usage in the U.S. because of their high brake factor and the easy incorporation of parking brake features. On the negative side, drum brakes may not be as consistent in torque performance as disc brakes. The lower brake factors of disc brakes require higher actuation effort, and development of integral parking brake features has been required before disc brakes could be used at all wheel positions.

Brake Factor

Brake factor is a mechanical advantage that can be utilized in drum brakes to minimize the actuation effort required. The mechanism of a common drum brake is shown in simplified form in Figure 3.2. The brake consists of two shoes pivoted at the bottom. The application of an actuation force, P_a, pushes the lining against the drum generating a friction force whose magnitude is the normal load times the coefficient of friction (μ) of the lining material against the drum. Taking moments about the pivot point for shoe A:

$$\Sigma M_p = e\, P_a + n\, \mu\, N_A - m\, N_A = 0 \qquad (3\text{-}15)$$

where:

e = Perpendicular distance from actuation force to pivot
N_A = Normal force between lining A and drum
n = Perpendicular distance from lining friction force to pivot
m = Perpendicular distance from the normal force to the pivot

The friction force developed by each brake shoe is:

$$F_A = \mu\, N_A \quad \text{and} \quad F_B = \mu\, N_B$$

Then equation (3-15) can be manipulated to obtain:

$$\frac{F_A}{P_a} = \frac{\mu\, e}{(m - \mu\, n)} \quad \text{and} \quad \frac{F_B}{P_a} = \frac{\mu\, e}{(m + \mu\, n)} \qquad (3\text{-}16)$$

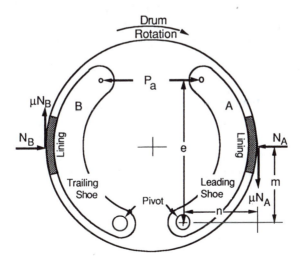

Fig. 3.2 Forces acting on the shoes of a simple drum brake.

The shoe on the right is a "leading" shoe. The moment produced by the friction force on the shoe acts to rotate it against the drum and increase the friction force developed. This "self-servo" action yields a mechanical advantage characterized as the "brake factor." The brake factor is not only proportional to μ in the numerator, but is increased by its influence in the denominator. (The expressions become more complicated with lining distributed over a larger arc, but show the same effect.) Clearly, if μ gets too large, the term "μn" may equal "m" and the brake factor goes to infinity, in which case the brake will lock on application.

Shoe B is a trailing shoe configuration on which the friction force acts to reduce the application force. The brake factor is much lower, and higher application forces are required to achieve the desired braking torque.

By using two leading shoes, two trailing shoes, or one of each, different brake factors can be obtained. The duo-servo brake has two leading shoes coupled together to obtain a very high brake factor. The consequences of using high brake factors is sensitivity to the lining coefficient of friction, and the possibility of more noise or squeal. Small changes in μ due to heating, wear, or other factors cause the brake to behave more erratically. Since disc brakes lack this self-actuation effect they generally have better torque consistency, although at the cost of requiring more actuation effort.

The difference between the two types of brakes can usually be seen in their torque properties during a stop. Brake torque performance can be measured in the laboratory using an inertial dynamometer, which is simply a large rotating mass attached to the drum with provisions to measure the torque obtained. The brake is applied with a constant actuation force to stop a rotating inertia nominally equivalent to the mass carried at the wheel on which it might be used. The torque measured during the stop typically looks like that shown in Figure 3.3.

On drum brakes, the torque will often exhibit a "sag" in the intermediate portion of the stop. It has been hypothesized that the effect is the combination of temperature fade and velocity effects (torque increases as velocity decreases). Disc brakes normally show less torque variation in the course of a stop. With an excess of these variations during a brake application, it can be difficult to maintain the proper balance between front and rear braking effort during a maximum-effort stop. Ultimately this can show up as less consistent deceleration performance in braking maneuvers resulting in longer stopping distances [6].

The torque from the brake can be modeled from the curves such as shown in Figure 3.3, but can be difficult to predict accurately over all conditions of

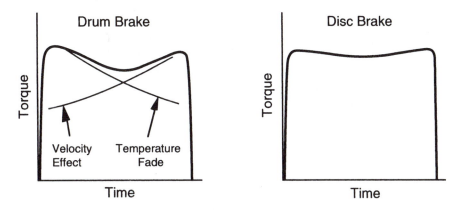

Fig. 3.3 Inertia dynamometer torque measurements.

operation. The torque normally increases almost linearly with the actuation effort, P_a, but to levels that vary with the speed and the energy absorbed (through the temperatures generated). Thus:

$$T_b = f (P_a, \text{Velocity, Temperature}) \qquad (3\text{-}17)$$

Efforts to model brakes by a general equation including each of the independent factors and the interrelated effects results in a torque equation which may require up to 27 coefficients. Because the equation depends on the brake temperature, which increases during a brake application, it is necessary to incorporate a thermal model of the brake in the calculation process [11]. Experience at The University of Michigan in trying to model brake torque performance in this fashion has been only partially successful. For moderate-level applications, good predictions can be obtained. However, a high-energy application (in which the temperature gets above 650°F) will permanently change the brake such that a new set of 27 coefficients must be determined.

The torque produced by the brake acts to generate a braking force at the ground and to decelerate the wheels and driveline components. Then:

$$F_b = \frac{(T_b - I_w \alpha_w)}{r} \qquad (3\text{-}18)$$

where:

      r = Rolling radius of the tires
      I_w = Rotational inertia of wheels (and drive components)
      α_w= Rotational deceleration of wheels

Except during a wheel lockup process, α_w is related to the deceleration of the vehicle through the radius of the wheel ($\alpha_w = a_x/r$), and I_w may be simply lumped in with the vehicle mass for convenience in calculation. In that case the torque and brake force are related by the relationship:

$$F_b = \frac{T_b}{r} \tag{3-19}$$

TIRE-ROAD FRICTION

As long as all wheels are rolling, the braking forces on a vehicle can be predicted using Eq. (3-19). However, the brake force can only increase to the limit of the frictional coupling between the tire and road.

There are two primary mechanisms responsible for friction coupling as illustrated in Figure 3.4. Surface adhesion arises from the intermolecular bonds between the rubber and the aggregate in the road surface. The adhesion component is the larger of the two mechanisms on dry roads, but is reduced substantially when the road surface is contaminated with water; hence, the loss of friction on wet roads.

The bulk hysteresis mechanism represents energy loss in the rubber as it deforms when sliding over the aggregate in the road. Bulk (or hysteretic) friction is not so affected by water on the road surface, thus better wet traction is achieved with tires that have high-hysteresis rubber in the tread.

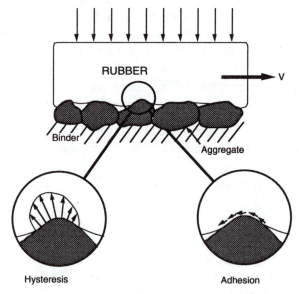

Fig. 3.4 Mechanisms of tire-road friction [4].

Both adhesive and hysteretic friction depend on some small amount of slip occurring at the tire-road interface. Additional slip is observed as a result of the deformation of the rubber elements of the tire tread as they deform to develop and sustain the braking force. This mechanism is illustrated in Figure 3.5. As the element enters the tire contact patch, it is undeformed. As it proceeds into the center of tire contact, deformation must occur for the tire to sustain a friction force. The deformation increases from the front to the back of the tire contact patch, and the force developed by each element increases proportionately from front to back. At high braking levels, the elements in the rear extreme of the contact patch begin to slide on the surface, and the braking force from the tire may begin to decrease.

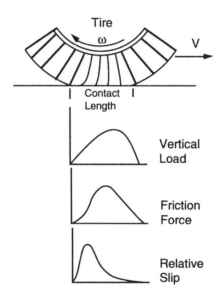

Fig 3.5 Braking deformations in the contact patch.

Because of these mechanisms, the brake force and slip are coexistent. Brake force (expressed as a coefficient F_x/F_z) is shown as a function of slip in Figure 3.6. Slip of the tire is defined by the ratio of slip velocity in the contact patch (forward velocity - tire circumferential speed) to forward velocity:

$$\text{Slip} = \frac{V - \omega r}{V} \tag{3-20}$$

where:

        V = Vehicle forward velocity

        ω = Tire rotational speed (radians/sec)

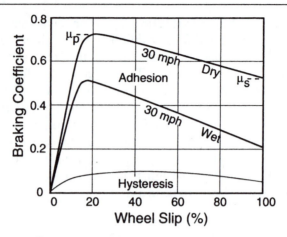

Fig. 3.6 Braking coefficient versus slip [4].

The brake coefficient deriving from adhesive and hysteretic friction increases with slip up to about 10 to 20% in magnitude depending on conditions. Under wet road conditions, the adhesive friction contribution is diminished such that the overall coefficient is lower. The peak coefficient is a key property, usually denoted by μ_p. It establishes the maximum braking force that can be obtained from the particular tire-road friction pair. At higher slip, the coefficient diminishes, reaching its lowest value at 100% slip, representing the full lock condition and denoted by μ_s. In a braking situation, μ_p corresponds to the highest brake force that can be generated, and is only theoretically possible to achieve because the system is unstable at this point. For a given brake torque output level, once the wheel is decelerated to achieve μ_p, any disturbance about this condition results in an excess of brake torque which causes further deceleration of the wheel. The increased slip reduces the brake force such that the wheel deceleration continues and the wheel goes to lock. Only a brake release (as in an anti-lock control) can return the wheel to operation at μ_p.

In addition to the tire and the road as key elements in determining the friction coupling available, other variables are important, as follows.

Velocity

On dry roads, both peak and slide friction decrease with velocity. Under wet conditions, even greater speed sensitivity prevails because of the difficulty of displacing water in the contact patch at high speeds. When the speed and water film thickness are sufficient, the tire tread will lift from the road creating a condition known as hydroplaning.

Inflation Pressure

On dry roads, peak and slide coefficients are only mildly affected by inflation pressure. On wet surfaces, inflation pressure increases are known to significantly improve both coefficients.

Vertical Load

Increasing vertical load is known to categorically reduce normalized traction levels (F_x/F_z) under both wet and dry conditions. That is, as load increases, the peak and slide friction forces do not increase proportionately. Typically, in the vicinity of a tire's rated load, both coefficients will decrease on the order of 0.01 for a 10% increase in load.

EXAMPLE PROBLEMS

1) Consider a light truck weighing 3635 lb, performing a full stop from 60 mph on a level surface with a brake application that develops a steady brake force of 2000 lb. Determine the deceleration, stopping distance, time to stop, energy dissipated and the brake horsepower at initial application and averaged over the stop. Neglect aerodynamic and rolling resistance forces.

Solution:

The deceleration may be calculated from NSL:

$$D_x = \frac{F_x}{M} = \frac{F_b}{M} = \frac{(2000 \text{ lb}) \ 32.2 \text{ ft/sec}^2}{3635 \text{ lb}} = 17.72 \ \frac{\text{ft}}{\text{sec}^2}$$

The deceleration can be computed directly in terms of g's by using the equational form:

$$D_x \ (g) = \frac{F_x}{W} = \frac{F_b}{W} = \frac{2000 \text{ lb}}{3635 \text{ lb}} = 0.55 \text{ g} = 12.08 \ \frac{\text{mph}}{\text{sec}}$$

Now that the deceleration is known, the stopping distance may be computed (Eq. (3-6)):

$$SD = \frac{V_o^2}{2 \dfrac{F_{xt}}{M}} = \frac{V_o^2}{2 D_x} \tag{3-6}$$

$$= \frac{(88 \text{ ft/sec})^2}{2 \ (17.72 \text{ ft/sec}^2)} = 218.51 \text{ ft}$$

The time to stop comes from Eq. (3-7):

$$t_s = \frac{V_o}{F_{xt}/M} = \frac{88 \text{ ft/sec}}{17.72 \text{ ft/sec}^2} = 4.966 \text{ sec}$$

The energy dissipated comes from Eq. (3-11):

$$\text{Energy} = \frac{M}{2}(V_o^2 - V_f^2) = \frac{3635 \text{ lb}}{2(32.2 \text{ ft/sec}^2)}(88 \text{ ft/sec})^2$$

$$= 437,103 \text{ ft-lb}$$

The power dissipation at the point of brake application is simply the brake force times the forward velocity, which is:

Power (initial) = (2000 lb) 88 ft/sec = 176,000 ft-lb/sec

$$\text{HP (initial)} = (176,000 \frac{\text{ft-lb}}{\text{sec}}) \frac{1 \text{ hp}}{550 \text{ ft-lb/sec}} = 320 \text{ hp}$$

On average over the stop, the power (from Eq. (3-12)) is:

$$\text{Power} = \frac{M}{2} \frac{V_o^2}{t_s} = \frac{3635 \text{ lb}}{2(32.2 \text{ ft/sec}^2)} \frac{(88 \text{ ft/sec})^2}{4.966 \text{ sec}}$$

$$= \frac{437,103 \text{ ft-lb}}{4.966 \text{ sec}} = 88,019 \frac{\text{ft-lb}}{\text{sec}} = 160 \text{ hp}$$

2) For the vehicle described in the previous problem, calculate the stopping distance taking aerodynamic drag into account. The aerodynamic drag force will be given by:

$$F_a = C V^2 = 0.00935 \, (\frac{\text{lb-sec}^2}{\text{ft}^2}) \, V^2 (\frac{\text{ft}^2}{\text{sec}^2})$$

The stopping distance may be computed from Eq. (3-10):

$$SD = \frac{M}{2C} \ln \left[\frac{(F_b + C V_o^2)}{F_b}\right] \qquad (3\text{-}10)$$

$$= \frac{3635 \text{ lb}}{2 (0.00935 \, \frac{\text{lb-sec}^2}{\text{ft}^2})(32.2 \, \frac{\text{ft}}{\text{sec}^2})} \ln \frac{2000 \text{ lb} + 0.00935 \, \frac{\text{lb-sec}^2}{\text{ft}^2}(88 \, \frac{\text{ft}}{\text{sec}})^2}{2000 \text{ lb}}$$

$$= 214.69 \text{ ft}$$

Thus, roughly 4 feet will be cut from the stopping distance when aerodynamic drag is included in the calculation. The drag itself is only 74.4 lb at the beginning of the stop and decreases with the square of the velocity, so its contribution becomes much less during the course of the stop.

FEDERAL REQUIREMENTS FOR BRAKING PERFORMANCE

Out of the public concern for automotive safety in the 1960s, the Highway Safety Act of 1965 was passed establishing the National Highway Traffic Safety Administration charged with promulgating performance standards for new vehicles which would increase safety on the highways. Among the many standards that have been imposed are Federal Motor Vehicle Safety Standard (FMVSS) 105 [5], establishing braking performance requirements for vehicles with hydraulic brake systems, and FMVSS 121 [6], establishing braking performance requirements for vehicles with air brake systems.

FMVSS 105 defines service brake and parking brake performance requirements over a broad range of conditions, such as:

- Lightly loaded to fully loaded at gross vehicle weight rating (GVWR)

- Preburnish to full burnish conditions

- Speeds from 30 to 100 mph

- Partially failed systems tests

- Failure indicator systems

- Water recovery

- Fade and recovery

- Brake control force limits

Although the standard is quite detailed and complex, the requirements for stopping distance performance can be summarized into five tests:

1) First effectiveness—A fully loaded passenger car with new, unburnished brakes must be able to stop from speeds of 30 and 60 mph in distances that correspond to average decelerations of 17 and 18 ft/sec^2, respectively.

59

2) Second effectiveness—A fully loaded passenger car with burnished[1] brakes must be able to stop from 30, 60 and 80 mph in distances that correspond to average decelerations of 17, 19 and 18 ft/sec^2, respectively.

3) Third effectiveness—A lightly loaded passenger car with burnished brakes must be able to stop from 60 mph in a distance that corresponds to an average deceleration of 20 ft/sec^2.

4) Fourth effectiveness—A fully loaded passenger car with burnished brakes must be able to stop from 30, 60, 80 and 100 mph in distances that correspond to average decelerations of 17, 18, 17 and 16 ft/sec^2, respectively.

5) Partial failure—A lightly loaded and fully loaded passenger car with a failure in the brake system must be able to stop from 60 mph in a distance that corresponds to an average deceleration of 8.5 ft/sec^2.

It is notable that the hydraulic brake standard (FMVSS 105) has stopping distance requirements only for dry surfaces of an 81 Skid Number. (Skid Number is the tire-road friction coefficient measured by American Society for Testing and Materials Method E-274-85 [8]. Although the Skid Number is measured with a special, standard tire, the Skid Number and coefficient of friction are generally assumed to be equivalent.) The air brake vehicle standard (FMVSS 121) has stopping distance performance requirements on both wet surfaces (30 Skid Number) and dry surfaces (81 Skid Number). Obviously, the prudent brake system designer considers a range of surface friction conditions, from at least 30 to 81 SN, despite any gaps in the Federal performance standards.

BRAKE PROPORTIONING

The braking decelerations achievable on a vehicle are simply the product of application level and the brake gains (torque/pressure) up to the point where lockup will occur on one of the axles. Lockup reduces the brake force on an axle, and results in some loss of ability to control the vehicle. It is well recognized that the preferred design is to bring both axles up to the lockup point simultaneously. Yet, this is not possible over the complete range of operating conditions to which a vehicle will be exposed. Balancing the brake outputs on both the front and rear axles is achieved by "proportioning" the pressure

---

[1] Burnish refers to a process in which new brakes are "worn in" by repeated brake applications according to a procedure defined in the standard.

appropriately for the foundation brakes installed on the vehicle. Proportioning then adjusts the brake torque output at front and rear wheels in accordance with the peak traction forces possible.

The first-order determinants of peak traction force on an axle are the instantaneous load and the peak coefficient of friction. During braking, a dynamic load transfer from the rear to the front axle occurs such that the load on an axle is the static plus the dynamic load transfer contributions. Thus for a deceleration, D_x:

$$W_f = \frac{c}{L} W + \frac{h}{L} \frac{W}{g} D_x = W_{fs} + W_d \qquad (3\text{-}21)$$

and

$$W_r = \frac{b}{L} W - \frac{h}{L} \frac{W}{g} D_x = W_{rs} - W_d \qquad (3\text{-}22)$$

where:

W_{fs} = Front axle static load

W_{rs} = Rear axle static load

$W_d = (h/L) (W/g) D_x$ = Dynamic load transfer

Then, on each axle the maximum brake force is given by:

$$F_{xmf} = \mu_p W_f = \mu_p (W_{fs} + \frac{h}{L} \frac{W}{g} D_x) \qquad (3\text{-}23)$$

and

$$F_{xmr} = \mu_p W_r = \mu_p (W_{rs} - \frac{h}{L} \frac{W}{g} D_x) \qquad (3\text{-}24)$$

where:

μ_p = Peak coefficient of friction

The maximum brake force is dependent on the deceleration, varying differently at each axle. Figure 3.7 shows graphically the maximum brake forces according to the above equations for a typical passenger car on both a high and low coefficient surface. The deceleration is shown in units of g's (equivalent to D/g). Attempts at braking on an axle above the boundary value results in lockup on the axle.

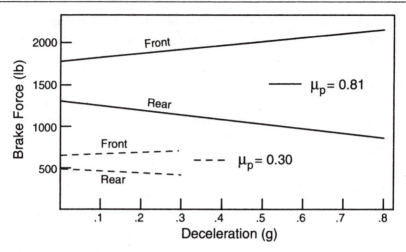

Fig. 3.7 Maximum brake forces as a function of deceleration.

Inasmuch as the equations above contain the deceleration as a variable, they do not provide an explicit solution for the maximum braking forces on an axle. These can be obtained by recognizing that the deceleration is a function of the total braking force imposed on the vehicle (neglecting for simplicity the other forces that may be present). To solve for F_{xmf}, we can use the relationship:

$$D_x = \frac{(F_{xmf} + F_{xr})}{M} \tag{3-25}$$

and for F_{xmr}:

$$D_x = \frac{(F_{xmr} + F_{xf})}{M} \tag{3-26}$$

Substituting into Eqs. (3-23) and (3-24) yields the following equations for the maximum braking force on each axle:

$$F_{xmf} = \frac{\mu_p (W_{fs} + \frac{h}{L} F_{xr})}{1 - \mu_p \frac{h}{L}} \tag{3-27}$$

$$F_{xmr} = \frac{\mu_p (W_{rs} - \frac{h}{L} F_{xf})}{1 + \mu_p \frac{h}{L}} \tag{3-28}$$

Thus the maximum braking force on the front axle is dependent on that present on the rear axle through the deceleration and associated forward load transfer resulting from the rear brake action. Conversely, the same effect is evident on the rear axle. These relationships can best be visualized by plotting the rear versus front brake forces as shown in Figure 3.8.

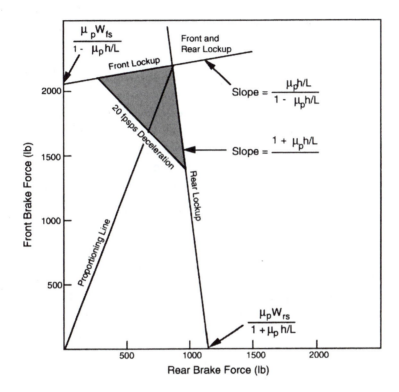

Fig. 3.8 Maximum braking forces on the front and rear axles.

The horizontal axis represents rear brake force, which is generally proportional to the rear brake pressure (related by the torque-to-pressure relationship for that foundation brake). The vertical axis is front brake force, again proportional to front brake pressure in accordance with the brake gain. The origin of each line is obtained from Eqs. (3-27) and (3-28) by setting the brake force of the opposite brake to zero.

Lines for the maximum front axle brake force slope upward and to the right (positive) at the slope of $\mu_p h/L /(1-\mu_p h/L)$. Lines for the rear axle maximum brake force slope downward to the right (negative) with a slope that is equal to $-\mu_p h/L /(1+\mu_p h/L)$. Increasing the surface coefficient or the CG height increases the slopes of the maximum brake force lines on the graph. Varying the load condition on the vehicle translates the origin of each of the lines on the graph. The intersection point for the front and rear brake boundaries can be determined by manipulating equations (3-27) and (3-28). Designating the points as F_{xfi} and F_{xri}, it can be shown that these coordinates are:

$$F_{xfi} = \mu \left(W_{fs} + \mu W \frac{h}{L} \right) \tag{3-29}$$

$$F_{xri} = \mu \left(W_{rs} - \mu W \frac{h}{L} \right) \tag{3-30}$$

An attempt to brake the vehicle to a level that goes above the front brake force boundary will cause front wheel lockup to occur, and steering control will be lost. Likewise, braking effort that falls to the right of the rear brake boundary causes rear wheel lockup, which places the vehicle in an unstable condition. The instability has safety implications and therefore warrants careful consideration in the design of the brake system. This issue is discussed in more detail in a later section.

In a graph of the form of Figure 3.8, the deceleration is proportional to the sum of the front and rear brake forces. Thus 2000 lb of front brake force with zero rear force, 1000 lb front with 1000 lb rear, and zero front with 2000 lb rear brake force, all correspond to the same deceleration level, and a line of constant deceleration can be plotted by connecting these points. If the same scale is used for the front and rear brake forces, lines of constant deceleration plot as 45-degree diagonals on the graph.

If a deceleration capability of 20 ft/sec^2 is required on the 0.81 μ surface, any combination of front to rear brake force would satisfy that requirement so long as it falls in the triangle bounded by the deceleration line and the maximum brake force lines for the 0.81 μ surface.

"Brake proportioning" describes the relationship between front and rear brake forces determined by the pressure applied to each brake and the gain of each. It is represented by a line on the graph starting at the origin and extending upward and to the right. A fixed, or constant, proportioning is a straight line.

The primary challenge in brake system design is the task of selecting a proportioning ratio (the slope of a line on the graph) that will satisfy all design goals despite the variabilities in surface friction, front/rear weight distribution, CG height, and brake condition. A number of these objectives are defined by the FMVSS 105 braking standard in the various effectiveness tests, although performance objectives for low coefficient surfaces should be included by the brake designer as well. To date, low coefficient performance has only been specified in FMVSS 121, which defines braking performance requirements for air-braked trucks.

The primary factor determining brake proportioning is the gain of the brakes used on the front and rear wheels. The brake force on individual wheels can be described by the equation:

$$F_b = \frac{T_b}{r} = G \frac{P_a}{r} \qquad (3\text{-}31)$$

where:

F_b = Brake force

T_b = Brake torque

r = Tire rolling radius

G = Brake gain (in-lb/psi)

P_a = Application pressure

Achieving good performance over the full range of conditions under which a vehicle operates can be difficult. As an example of the complexity that can be experienced in trying to identify appropriate brake proportioning, consider the case shown in Figure 3.9. The figure illustrates the range of variations that arise from vehicle loading (lightly loaded and GVWR) and surface friction (30 and 81 SN). On the graph the brake force boundaries and deceleration requirements have been plotted for the FMVSS 105 dry surface tests conditions. In addition, similar boundaries have been plotted for wet road conditions assuming a friction coefficient of 30 SN. Under the wet conditions, a deceleration performance goal of 8 ft/sec^2 (0.25 g) has been assumed.

To achieve all performance goals, a proportioning design must be selected that passes through all of the triangles shown. This cannot be achieved with a straight line providing a constant relationship between front and rear brake force. A solution to this problem is to incorporate a valve in the hydraulic system that changes the pressure going to the rear brakes over some portion of the operating pressure range. Such a valve is known as a pressure proportioning valve. Most pressure proportioning valves in common use today provide equal pressure to both front and rear brakes up to a certain pressure level, and then

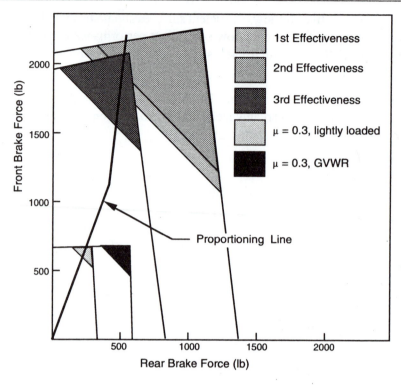

Fig. 3.9 Front/rear brake force graph for multiple braking conditions.

reduce the rate of pressure increase to one of the brakes thereafter. A proportioning valve identified as a "500/0.3" means that the pressure to the front and rear brakes is equal up to 500 psi. Above this level the pressure proportioned to the rear brakes increases at only 30% of the rate going to the front brakes. That is:

$$P_f = P_r = P_a = \text{Application pressure} \qquad \text{for } P_a < 500 \text{ psi} \qquad (3\text{-}32a)$$

$$P_f = P_a \quad \text{and} \quad P_r = 500 + 0.3 \, (P_a - 500) \quad \text{for } P_a > 500 \text{ psi} \qquad (3\text{-}32b)$$

With this proportioning it is seen that it is possible to achieve a front/rear brake balance satisfying all dry surface conditions as evidenced by the fact that the proportioning line passes through all of the performance triangles. The only exception is the fully loaded vehicle (GVWR) on the low coefficient surface, where the brake proportioning will not quite achieve 0.25 g. In every case, the plot indicates that front lockup will occur first.

Achieving good proportioning is especially difficult on trucks because of the disparity between loaded and empty conditions. Typically, the perfor-

66

mance triangles do not overlap in those cases, so no choice of proportioning will satisfy all goals. Several solutions are available. In Europe, load-sensing proportioning valves have been used on trucks for some years. These valves, installed on the axle(s), sense the load condition and adjust the brake proportioning appropriately. Less commonly used is the inertia-proportioning valve which senses the deceleration rate and can adjust proportioning in accordance with the deceleration level. Finally, anti-lock brake systems offer a versatile method of automatically proportioning brakes that is becoming well accepted in the automotive industry.

ANTI-LOCK BRAKE SYSTEMS

Rather than attempt to adjust the proportioning directly, anti-lock systems (ABS) sense when wheel lockup occurs, release the brakes momentarily on locked wheels, and reapply them when the wheel spins up again. Modern anti-lock brake systems are capable of releasing the brakes before the wheel goes to lockup, and modulating the level of pressure on reapplication to just hold the wheel near peak slip conditions.

The concept of ABS dates back to the 1930s, but has only become truly practical with electronics available on modern vehicles. An ABS consists of an electronic control unit (ECU), a solenoid for releasing and reapplying pressure to a brake, and a wheel speed sensor. The ECU normally monitors vehicle speed through the wheel speed sensors, and upon brake application begins to compute an estimate of the diminishing speed of the vehicle. Actual wheel speeds can be compared against the computed speed to determine whether a wheel is slipping excessively, or the deceleration rate of a wheel can be monitored to determine when the wheel is advancing toward lockup. Different ABS designs use different combinations of these variables to determine when lockup is imminent and brake release is warranted. At that point a command signal is sent to the solenoid to release the brake pressure, allowing the wheel to spin back up. Once the wheel regains speed, the pressure is increased again. Depending on the refinement of the control algorithms, the pressure rise rate and the final pressure may be controlled to minimize cycling of the brakes.

Figure 3.10 shows a typical plot of wheel speed cycling during the stop of a vehicle with ABS. When the brakes are first applied, wheel speeds diminish more or less in accordance with the vehicle speed in region 1 in the plot. If the brakes are applied to a high level, or the road is slippery, the speed of one or

67

more wheels begins to drop rapidly (point 2), indicating that the tire has gone through the peak of the μ-slip curve and is heading toward lockup. At this point the ABS intervenes and releases the brakes on those wheels before lockup occurs (point 3). Once the wheel speed picks up again the brakes are reapplied. The objective of the ABS is to keep each tire on the vehicle operating near the peak of the μ-slip curve for that tire. This is illustrated in Figure 3.11.

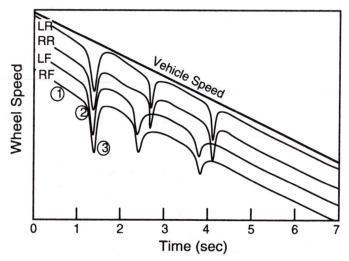

Fig. 3.10 Wheel speed cycling during ABS operation.

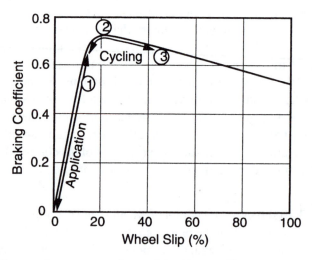

Fig 3.11 ABS operation to stay at the peak braking coefficient.

BRAKING EFFICIENCY

Recognizing that braking performance of any vehicle will vary according to the friction of the road surface on which it is attempted, the concept of braking efficiency has been developed as a measure of performance. Braking efficiency, η_b, may be defined as the ratio of actual deceleration achieved to the "best" performance possible on the given road surface. It can be shown with the use of the equations presented earlier that the best performance any vehicle can achieve is a braking deceleration (in g's) equivalent to coefficient of friction between the tires and the road surface. That is:

$$\eta_b = \frac{D_{act}}{\mu_p} \qquad (3\text{-}33)$$

The braking efficiency concept is useful as a design tool for the designer to assess success in optimizing the vehicle braking system [7]. Yet implementation of braking standards using the braking efficiency approach (to avoid the problems of designating surfaces with standard friction levels via the ASTM Skid Number [8]) has been unsuccessful. The main problem has been the difficulty of defining an effective friction level for a tire-road surface pair because of the variations in friction with velocity, wheel load, tire type and other factors.

Braking efficiency is determined by calculating the brake forces, deceleration, axle loads, and braking coefficient on each axle as a function of application pressure. The braking coefficient is defined as the ratio of brake force to load on a wheel or axle. The braking efficiency at any level of application pressure is the deceleration divided by the highest braking coefficient of any axle. Since the axle with the highest braking coefficient defines the required level of road friction, the braking efficiency is also equal to the ratio of deceleration to the required road surface coefficient.

Braking efficiency is a useful method for evaluating the performance of brake systems, especially on heavy trucks where multiple axles are involved. Figure 3.12 shows the braking efficiency calculated for a five-axle tractor-semitrailer.

Contributions to braking from individual axles are better assessed by examining the braking coefficient developed on each. A plot of these curves (sometimes known as a friction utilization plot) is shown in Figure 3.13. Five curves representing the five axles of the combination are shown. Brake coefficient is defined for an axle as the ratio of brake force to load. Ideally, all

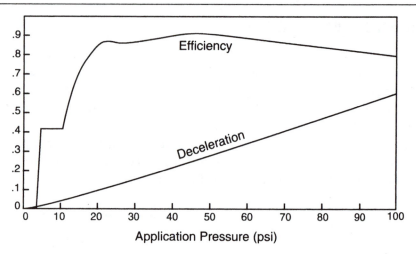

Fig 3.12 Efficiency plot for a tractor-semitrailer.

axles would have the same braking coefficient at a given application pressure, indicating that they all brake in proportion to their load. However, the diverse load conditions, longitudinal load transfer during braking, and shift of load between tandem axles due to brake reactions (inter-axle load transfer) preclude perfect harmony of the system. This is the reason the braking efficiency falls below the maximum theoretical value of 1. In the case of the tractor-semitrailer shown here, the braking efficiency rises quickly to a value of 0.9 at low brake application pressures, but drops off again at higher pressure due to the spread in braking coefficient among the axles at high deceleration conditions.

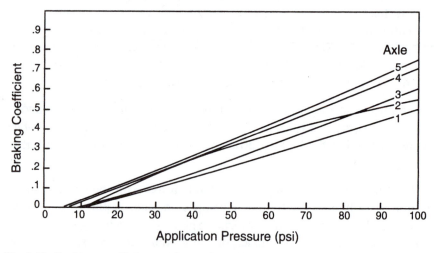

Fig 3.13 Braking coefficient on five axles of a tractor-semitrailer.

REAR WHEEL LOCKUP

In the discussion so far, wheel lockup has been considered only as a boundary on braking performance. However, it has great impact on the handling behavior of the vehicle as well, and must be considered by the brake designer. Once a wheel locks up it loses its ability to generate the cornering forces needed to keep the vehicle oriented on the road.

Lockup of front wheels causes loss of the ability to steer the vehicle, and it will generally continue straight ahead despite any steering inputs, drifting to the side only in response to cross-slope or side winds.

It is well recognized that rear wheel lockup places a motor vehicle in an unstable condition. Once the wheels lock up, any yaw disturbances (which are always present) will initiate a rotation of the vehicle. The front wheels, which yaw with the vehicle, develop a cornering force favoring the rotation, and the yaw angle continues to grow. Only when the vehicle has completely "switched ends" is it again stable. On long vehicles (some trucks and buses) the rotational accelerations are usually slow enough that the driver can apply corrective steer and prevent the full rotation. However, on smaller passenger cars, it is generally accepted that the average driver cannot readily control the vehicle in such a driving situation. Thus there is a philosophy among automotive designers that a front brake bias constitutes the preferred design.

The preference for a front brake lockup first cannot easily be achieved in a brake system design under all circumstances because of in-use variations in brake gain, CG height (particularly on light trucks), pavement friction, and parking brake requirements. The potential consequences in the hands of the motorist have been estimated using the braking efficiency as the measure of performance [9]. The basis for it arises from studies of driver behavior that show that brake applications occur on the average about 1.5 times per mile. Though most of the brake applications are executed at a moderate level, high decelerations are required in a certain percentage of the brake applications. Braking level demands of motorists are shown in Figure 3.14, which plots the percent of decelerations exceeding given deceleration levels. Twenty percent of all brake applications exceed 0.2 g, only 1% exceed 0.35 g, and less than 0.1% go up to 0.5 g.

The comparison of deceleration demands in normal driving to the available friction level of roads is shown in Figure 3.15. The distribution of road friction coefficients is estimated from numerous surveys of "skid resistance" routinely made by many highway departments. By and large, most roads have friction levels sufficient to accommodate the deceleration demands of the motorists if

71

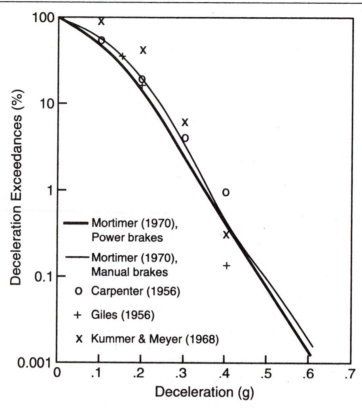

Fig. 3.14 Distribution of braking decelerations with passenger cars.

the friction is efficiently utilized. That is, if the brake systems on all vehicles were 100% efficient under all conditions, little overlap would occur in the braking "demand" and friction "available" curves, and there would be few braking instances in which wheel lockup occurs.

However, when braking efficiency is less than 100%, higher friction is required to achieve a desired deceleration level. With lower efficiency the "friction demand" curve shifts to the right. Thus the overlap and frequency at which braking demand will exceed the friction available will increase. Using the average figure of 1.5 brake applications per mile and 10,000 miles per year for a typical passenger car, the frequency of wheel lockup in braking can be estimated for different braking system efficiencies as shown in Figure 3.16. Clearly, it illustrates the acute sensitivity of lockup frequency to braking efficiency. If the inefficiency is due to a rear bias in the brake force distribution, the lockups will occur on the rear axle, and directional instability will result. Most occurrences will be on roads with lower friction levels, which are

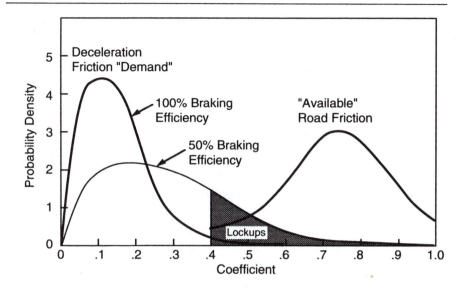

Fig. 3.15 Comparison of friction demand and availability.

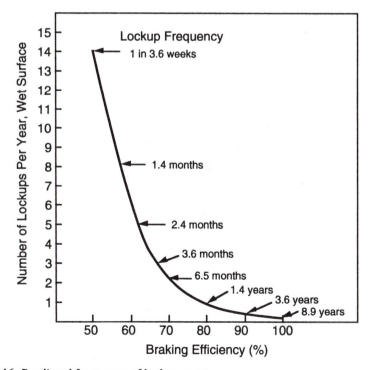

Fig. 3.16 Predicted frequency of lockup events.

normally wet road conditions. Since the majority of these instances will occur on roads with friction coefficients in the range of 0.4 to 0.6, particular emphasis should be placed on obtaining good braking system efficiency in this road friction range.

PEDAL FORCE GAIN

Ergonomics in the design of a brake system can play an important role in the ease with which the driving public can optimally use the braking capabilities built into a vehicle. Aside from positioning of the brake pedal, the effort and displacement properties of the pedal during braking are recognized as influential design variables. In the 1950s when power brake systems first came into general use, there was little uniformity among manufacturers in the level of effort and pedal displacement properties of the systems. In 1970 the National Highway Traffic Safety Administration sponsored research to determine ergonomic properties for the brake pedal that would give drivers the most effective control [10]. The research identified an optimum range for pedal force gain—the relationship between pedal force and deceleration. Figure 3.17 shows the results from the NHTSA study indicating the optimal gain values by the shaded area.

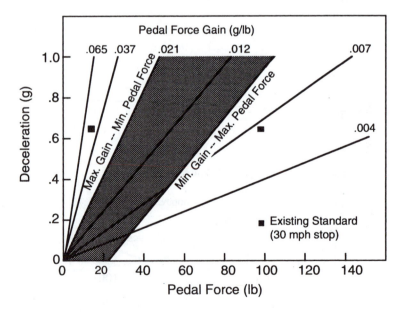

Fig. 3.17 Optimal pedal force gain properties.

EXAMPLE PROBLEM

Calculate the braking coefficients and braking efficiency for a passenger car in 100 psi increments of application pressure up to 700 psi, given the following information:

Wheelbase = 108.5 in	CGH = 20.5 in	Tire radius = 12.11 in
Weights: W_f = 2210 lb	W_r = 1864 lb	Total = 4074 lb
Front brake gain = 20 in-lb/psi	Rear brake gain = 14 in-lb/psi	

Proportioning valve design = 290/0.3

Solution:

The easiest way to visualize the answer is to tabulate data in columns as shown below. The calculation steps are the following:

1) The front application pressure is the reference, so we list values from 100 and up.

2) The rear application pressure is calculated from the front using the relationship similar to that given in Eq. (3-32). Namely,

$$P_r = P_a \qquad \text{for } P_a < 290 \text{ psi} \qquad (3\text{-}32a)$$

$$P_r = 290 + 0.3 (P_a - 290) \qquad \text{for } P_a > 290 \text{ psi} \qquad (3\text{-}32b)$$

3) The front and rear brake forces are the product of the application pressure on that brake times the torque gain times two brakes per axle divided by tire radius.

$$F_{xf} = 2 G_f \frac{P_f}{r} \qquad \text{and} \qquad F_{xr} = 2 G_r \frac{P_r}{r}$$

4) The deceleration is the sum of the brake forces divided by total vehicle weight (this results in deceleration in units of g).

$$D_x = \frac{F_{xf} + F_{xr}}{W}$$

5) The front and rear axle loads are calculated from Eqs. (3-21) and (3-22).

$$W_f = W_{fs} + (h/L) (W/g) D_x \qquad (3\text{-}21)$$

and

$$W_r = W_{rs} - (h/L) (W/g) D_x \qquad (3\text{-}22)$$

where "D_x" is in units of ft/sec^2.

75

6) The braking coefficients (μ_f and μ_r) are the ratio of axle brake force to axle load.

$$\mu_f = \frac{F_{xf}}{W_f} \qquad \text{and} \qquad \mu_r = \frac{F_{xr}}{W_r}$$

7) The braking efficiency, η_b, is the deceleration divided by the highest of the two braking coefficients from the axles.

P_f	P_r	F_f	F_r	D_x	W_f	W_r	μ_f	μ_r	η_b
100 psi	100 psi	330 lb	231 lb	.138 g	2316 lb	1758 lb	.142	.131	97%
200	200	661	462	.276	2422	1652	.273	.280	99
300	293	991	677	.409	2525	1549	.393	.437	94
400	323	1321	747	.508	2601	1473	.508	.507	100
500	353	1651	816	.606	2676	1398	.617	.583	98
600	383	1982	886	.704	2752	1322	.720	.670	98
700	413	2312	955	.802	2827	1247	.818	.766	98

Notes:

a) The braking efficiency starts high (97 - 99%) by the match of the brake gains and axle loads, but begins to diminish with deceleration because of the decreasing load on the rear axle.

b) When the application pressure reaches 290 psi, the proportioning valve "kicks in" reducing the pressure rise rate on the rear axle. This brings things back into balance providing 100% efficiency at 400 psi.

REFERENCES

1. Newcomb, T.P., and Spurr, R.T., Braking of Road Vehicles, Chapman and Hall, Ltd., London, England, 1967, 292 p.

2. Limpert, R., "Analysis and Design of Motor Vehicle Brake Systems," The University of Michigan, May 1971, 466 p.

3. Engineering Design Handbook, Analysis and Design of Automotive Brake Systems, DARCOM-P 706-358, US Army Material Develop-

ment and Readiness Command, Alexandria, VA, December 1976, 252 p.

4. Meyer, W.E., and Kummer, H.W., "Mechanism of Force Transmission between Tire and Road," Society of Automotive Engineers, Paper No. 620407 (490A), 1962, 18 p.

5. "Standard No. 105; Hydraulic Brake Systems," Code of Federal Regulations, Title 49, Part 571.105, October 1, 1990, pp. 199-215.

6. "Standard No. 121; Air Brake Systems," Code of Federal Regulations, Title 49, Part 571.121, October 1, 1990, pp. 366-382.

7. Gillespie, T.D., and Balderas, L., "An Analytical Comparison of a European Heavy Vehicle and a Generic U.S. Heavy Vehicle," The University of Michigan Transportation Research Institute, Report No. UMTRI-87-17, August 1987, 374 p.

8. "Test Method for Skid Resistance of Paved Surfaces Using a Full-Scale Tire," Method E274-85, 1986 Annual Book of ASTM Standards, American Society for Testing and Materials, Philadelphia, PA.

9. Ervin, R.D., and Winkler, C.B., "Estimation of the Probability of Wheel Lockup," IAVD Congress on Vehicle Design and Components, Geneva, March 3-5, 1986, pp. D145-D165.

10. Mortimer, R.E., Segel, L., Dugoff, H., Campbell, J.O., Jorgeson, C.M., and Murphy, R.W., "Brake Force Requirement Study: Driver-Vehicle Braking Performance as a Function of Brake System Design Variables," The University of Michigan Highway Safety Research Institute, Report No. HuF-6, April 1979, 22 p.

11. Johnson, L., Fancher, P.S., and Gillespie, T.D., "An Empirical Model for the Prediction of the Torque Output of Commercial Vehicle Air Brakes," Highway Safety Research Institute, University of Michigan, Report No. UM-HSRI-78-53, December 1978, 83 p.

CHAPTER 4
ROAD LOADS

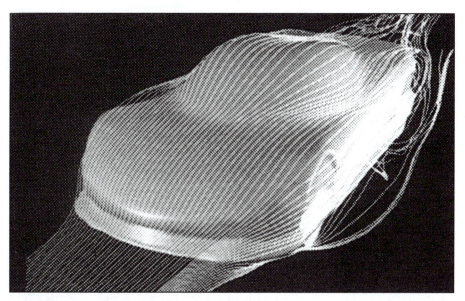

Flow field around the HSR II. (SAE Paper No. 910597.)

AERODYNAMICS

Aerodynamics makes its major impact on modern cars and trucks through its contribution to "road load." Aerodynamic forces interact with the vehicle causing drag, lift (or down load), lateral forces, moments in roll, pitch and yaw, and noise. These impact fuel economy, handling and NVH.

The aerodynamic forces produced on a vehicle arise from two sources—form (or pressure) drag and viscous friction. First, the mechanics of air flow will be examined to explain the nature of the flow around the body of the vehicle. Then, vehicle design features will be examined to show the qualitative influence on aerodynamic performance.

Mechanics of Air Flow Around a Vehicle

The gross flow over the body of a car is governed by the relationship between velocity and pressure expressed in Bernoulli's Equation [1,2].

(Bernoulli's Equation assumes incompressible flow, which is reasonable for automotive aerodynamics, whereas the equivalent relationship for compressible flow is the Euler Equation.) The equation is:

$$P_{static} + P_{dynamic} = P_{total} \qquad (4\text{-}1)$$

$$P_s + 1/2 \, \rho \, V^2 = P_t$$

where:

ρ = Density of air

V = Velocity of air (relative to the car)

This relationship is derived by applying Newton's Second Law to an incremental body of fluid flowing in a well-behaved fashion. For purposes of explanation, "well-behaved" simply means that the flow is moving smoothly and is experiencing negligible friction—conditions that apply reasonably to the air stream approaching a motor vehicle. In deriving the equation, the sum of the forces brings in the pressure effect acting on the incremental area of the body of fluid. Equating this to the time rate of change of momentum brings in the velocity term.

Bernoulli's equation states that the static plus the dynamic pressure of the air will be constant (P_t) as it approaches the vehicle. Visualizing the vehicle as stationary and the air moving (as in a wind tunnel), the air streams along lines, appropriately called "streamlines." A bundle of streamlines forms a streamtube. The smoke streams used in a wind tunnel allow streamtubes to be visualized as illustrated in Figure 4.1.

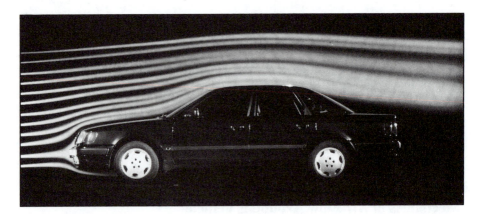

Fig. 4.1 Streamtubes flowing over an aerodynamic body. (Photo courtesy of Audi Division.)

At a distance from the vehicle the static pressure is simply the ambient, or barometric, pressure (P_{atm}). The dynamic pressure is produced by the relative velocity, which is constant for all streamlines approaching the vehicle. Thus the total pressure, P_t, is the same for all streamlines and is equal to $P_s + 1/2 \, \rho \, V^2$.

As the flow approaches the vehicle, the streamtubes split, some going above the vehicle, and others below. By inference, one streamline must go straight to the body and stagnate (the one shown impinging on the bumper of the car). At that point the relative velocity has gone to zero. With the velocity term zero, the static pressure observed at that point on the vehicle will be P_t. That is, if a pressure tap is placed on the vehicle at this point, it will record the total pressure.

Consider what must happen to the streamlines flowing above the hood. As they first turn in the upward direction, the curvature is concave upward. At a distance well above the vehicle where the streamlines are still straight, the static pressure must be the same as the ambient. In order for the air stream to be curved upward, the static pressure in that region must be higher than ambient to provide the force necessary to turn the air flow. If the static pressure is higher, then the velocity must decrease in this region in order to obey Bernoulli's Equation.

Conversely, as the flow turns to follow the hood (downward curvature at the lip of the hood) the pressure must go below ambient in order to bend the flow, and the velocity must increase. These points are illustrated in Figure 4.2, showing flow over a cylinder.

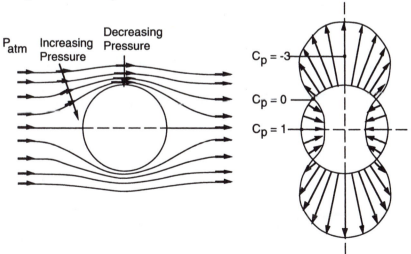

Fig 4.2 Pressure and velocity gradients in the air flow over a body.

Thus Bernoulli's Equation explains how the pressure and velocity must vary in the gross air flow over a car body. In the absence of friction the air would simply flow up over the roof and down the back side of the vehicle, exchanging pressure for velocity as it did at the front. In that case, the pressure forces on the back side of the vehicle would exactly balance those on the front, and there would be no drag produced.

From experience, however, we know that drag is produced. The drag is due in part to friction of the air on the surface of the vehicle, and in part to the way the friction alters the main flow down the back side of the vehicle. Its explanation comes about from understanding the action of boundary layers in the flow over an object. Consider a uniform flow approaching a sharp-edged body as shown in Figure 4.3.

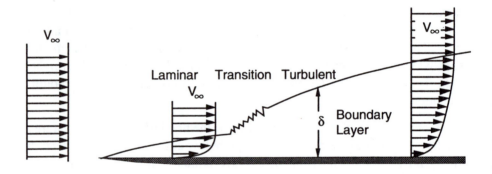

Fig 4.3 Development of a boundary layer.

Approaching the body, all air is traveling at a uniform velocity (and is assumed to be well-behaved, laminar flow). As it flows past the body, the air contacting the surface must drop to zero velocity due to friction on the surface. Thus a velocity profile develops near the surface, and for some distance, δ, the velocity is less than that of the main flow. This region of reduced velocity is known as the "boundary layer." The boundary layer begins with zero thickness and grows with distance along the body. Initially, it too is laminar flow, but will eventually break into turbulent flow.

On the front face of a vehicle body, the boundary layer begins at the point where the stagnation streamline hits the surface. In the boundary layer the velocity is reduced because of friction. The pressure at the stagnation point is the total pressure (static plus dynamic) and decreases back along the surface.

The pressure gradient along the surface thus acts to push the air along the boundary layer, and the growth of the layer is impeded. Pressure decreasing in the direction of flow is thus known as a "favorable pressure gradient," because it inhibits the boundary layer growth.

Unfortunately, as the flow turns again to follow the body, the pressure again increases. The increasing pressure acts to decelerate the flow in the boundary layer, which causes it to grow in thickness. Thus it produces what is known as an "adverse pressure gradient." At some point the flow near the surface may actually be reversed by the action of the pressure as illustrated in Figure 4.4. The point where the flow stops is known as the "separation point." Note that at this point, the main stream is no longer "attached" to the body but is able to break free and continue in a more or less straight line. Because it tries to entrain air from the region behind the body, the pressure in this region drops below the ambient. Vortices form and the flow is very irregular in this region. Under the right conditions, a von-Karman Vortex Street may be formed, which is a periodic shedding of vortices. Their periodic nature can be perceived as aerodynamic buffeting. The vortex action in flow over a cylinder is shown in Figure 4.5.

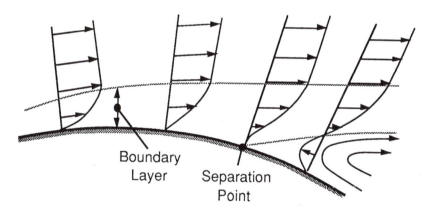

Fig. 4.4 Flow separation in an adverse pressure gradient.

The phenomenon of separation prevents the flow from simply proceeding down the back side of a car. The pressure in the separation region is below that imposed on the front of the vehicle, and the difference in these overall pressure forces is responsible for "form drag." The drag forces arising from the action of viscous friction in the boundary layer on the surface of the car is the "friction drag."

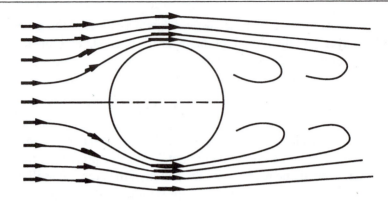

Fig. 4.5 Vortex shedding in flow over a cylindrical body.

Pressure Distribution on a Vehicle

These basic mechanisms account for the static pressure distribution along the body of a car. Figure 4.6 shows experimentally measured pressures [3] plotted perpendicular to the surface. The pressures are indicated as being negative or positive with respect to the ambient pressure measured some distance from the vehicle.

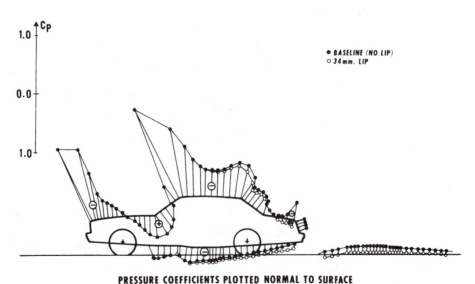

Fig. 4.6 Pressure distribution along the centerline of a car.

Note that a negative pressure is developed at the front edge of the hood as the flow rising over the front of the vehicle attempts to turn and follow horizontally along the hood. The adverse pressure gradient in this region has the potential to stall the boundary layer flow creating drag in this area. In recent years, styling detail in the front hood line has been given high priority to avoid separation on the hood and the drag penalty that results.

Near the base of the windshield and cowl, the flow must be turned upward, thus high pressure is experienced. The high-pressure region is an ideal location for inducting air for climate control systems, or engine intake, and has been used for this purpose in countless vehicles in the past. The high pressures are accompanied by lower velocities in this region, which is an aid to keeping the windshield wipers from being disturbed by aerodynamic forces.

Over the roof line the pressure again goes negative as the air flow tries to follow the roof contour. Evidence of the low pressure in this region is seen in the billowing action of the fabric roof on convertibles. The pressure remains low down over the backlite and on to the trunk because of the continuing curvature. It is in this area that flow separation is most likely. Design of the angles and details of the body contour in this region require critical concern for aerodynamics. Because of the low pressure, the flow along the sides of the car will also attempt to feed air into this region [4] and may add to the potential for separation. The general air flow patterns over the top and sides of a car are shown in Figure 4.7. The flow along the sides is drawn up into the low-pressure region in the rear area, combining with flow over the roof to form vortices trailing off the back of the vehicle.

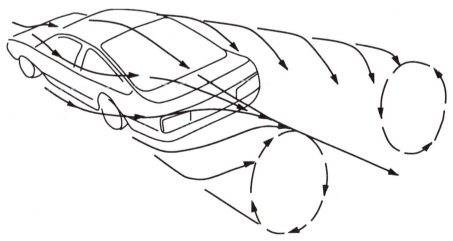

Fig. 4.7 Vortex systems in the wake of a car.

85

The choice of the backlite angles and deck lid lengths on the back of a car has a direct impact on aerodynamic forces through control of the separation point. Separation must occur at some point, and the smaller the area, generally the lower the drag. Theoretically, the ideal from an aerodynamic viewpoint is a teardrop rear shape, i.e., a conical shape that tapers off to a point with shallow angles of 15 degrees or less. It was recognized as early as the 1930s that because the area toward the point of the cone is quite small, the end of the ideal vehicle can be cut off without much penalty of a large separation area [5, 6, 7]. The blunt rear end shape allows greater head room in the back seat without substantially increasing drag. This characteristic shape has acquired the name "Kamm-back."

While the size of the separation area affects the aerodynamic drag directly, the extent to which the flow is forced to turn down behind the vehicle affects the aerodynamic lift at the rear. Figure 4.8 illustrates the effect on lift and drag for four styles of vehicle [4]. Flow control that minimizes the separation area generally results in more aerodynamic lift at the rear because of the pressure reduction as the flow is pulled downward.

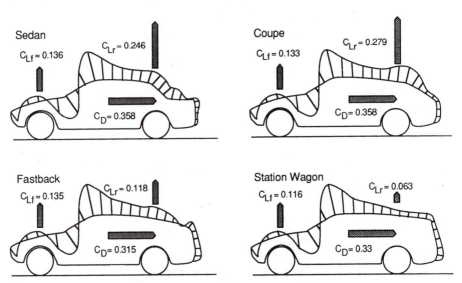

Fig 4.8 Aerodynamic lift and drag forces with different vehicle styles.

Another consideration in aerodynamic design at the rear is the potential for dirt deposition on the backlite and tail lights. The high degree of turbulence in the separation zone entrains moisture and dirt kicked up from the roadway by the tires. If the separation zone includes these items, dirt will be deposited on these areas and vision will be obstructed. Figure 4.9 illustrates this phenomenon.

Whether separation will occur at the rear edge of the roof line is strongly dependent on the shape at that location and the backlite angle. For the vehicle on the left, the sharp edge at the roof line promotes separation at this point. While a well-defined separation boundary helps minimize aerodynamic buffeting, the inclusion of the backlite in the separation area promotes dirt deposition on the window.

Although the vehicle on the right has a comparable backlite angle, the smooth transition at the rear of the roof and the addition of a modest trunk extension encourages the air stream to follow the vehicle contours down the rear deck. The separation region is well defined by the sharp contours at the end of the deck, helping to stabilize the separation zone and minimize buffeting. Only the tail light region is exposed to road dirt with this design.

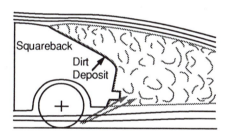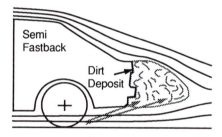

Fig 4.9 Effect of separation point on dirt deposition at the rear.

Aerodynamic Forces

As a result of the air stream interacting with the vehicle, forces and moments are imposed. These may be defined systematically as the three forces and three moments shown in Figure 4.10, acting about the principal axes of the car [8]. The reactions are as follows:

Direction	Force	Moment
Longitudinal (x-axis, positive rearward)	Drag	Rolling moment
Lateral (y-axis, positive to the right)	Sideforce	Pitching moment
Vertical (z-axis, positive upward)	Lift	Yawing moment

The origin for the axis system is defined in SAE J1594 [9]. Inasmuch as the aerodynamic reactions on a vehicle are unrelated to its center of gravity location (and the CG location may not be known in wind tunnel tests), the origin for force measurement is in the ground plane at the mid-wheelbase and mid-track position.

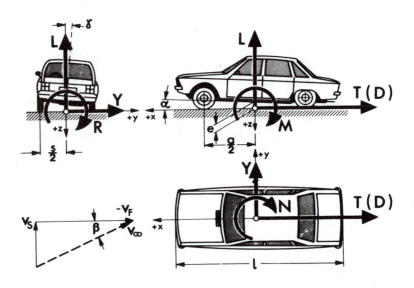

Fig 4.10 Aerodynamic forces and moments acting on a car [14].

Drag Components

Drag is the largest and most important aerodynamic force encountered by passenger cars at normal highway speeds. The overall drag on a vehicle derives from contributions of many sources. Various aids may be used to reduce the effects of specific factors. Figure 4.11 lists the main sources of drag and the potential for drag reductions in these areas estimated for cars in the 1970s.

For the vehicle represented in the figure, approximately 65% (.275/.42) of the drag arises from the body (forebody, afterbody, underbody and skin friction). The major contributor is the afterbody because of the drag produced by the separation zone at the rear. It is in this area that the maximum potential for drag reduction is possible. Figure 4.12 shows the influence of rear end inclination angle on the drag for various lengths of rear extension (beyond the rear edge of the roof line) [10]. Slope angles up to 15 degrees consistently reduce drag. As the angles increase, the drag again increases because of flow separation. (In practice, higher drop angles have been achieved without separation.)

DRAG COEFFICIENT COMPONENT	TYPICAL VALUE
Forebody	0.05
Afterbody	0.14
Underbody	0.06
Skin Friction	0.025
Total Body Drag	0.275
Wheels and wheel wells	0.09
Drip rails	0.01
Window recesses	0.01
External mirrors	0.01
Total Protuberance Drag	0.12
Cooling system	0.025
Total Internal Drag	0.025
Overall Total Drag	0.42[1]
VEHICLE OF THE 1980s	
Cars	0.30 - 0.35
Vans	0.33 - 0.35
Pickup trucks	0.42 - 0.46

[1] Based on cars of 1970s vintage.

Fig. 4.11 Main sources of drag on a passenger car.

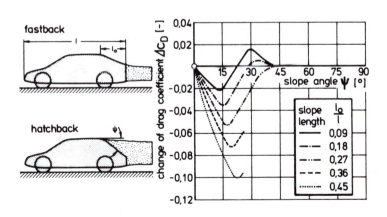

Fig 4.12 Influence of rear end inclination on drag.

Forebody drag is influenced by design of the front end and windshield angle. Generally the "roundness" of the front end establishes the area over which the dynamic pressure can act to induce drag. Figure 4.13 shows the influence of the height of the front edge of the vehicle [10]. The location of this point determines the location of the streamline flowing to the stagnation point. This streamline is important as it establishes the separation of flow above and below the body. Minimum drag is obtained when the stagnation point is kept low on the frontal profile of the vehicle. A well-rounded shape, in contrast to the crisp lines traditionally given to the frontal/grill treatment of passenger cars, is equally important to aerodynamics. A rounded low hood line can yield reductions of 5 to 15% in the overall drag coefficient [11].

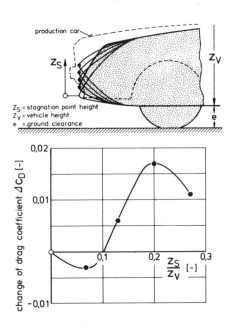

Fig 4.13 Influence of front end design on drag.

The windshield establishes the flow direction as it approaches the horizontal roof. Thus its angle has a direct influence on drag, particularly on trucks. Shallow angles reduce drag, but complicate vehicle design by allowing increased solar heating loads and placing more critical demands on the manufacturer of the windshield to minimize distortion at shallow angles. Figure 4.14 shows the change in drag as the windshield angle is increased from the nominal angle of 28 degrees [10]. With a steep angle, the air velocity

90

approaching the windshield is reduced by the high pressure in that region. With a shallow angle, the wind speed will be higher, adding to the aerodynamic loads on the windshield wipers.

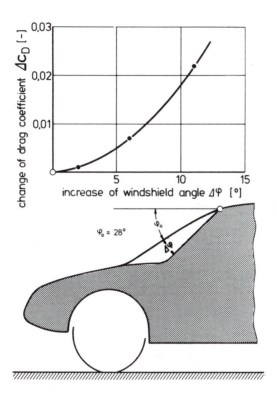

Fig. 4.14 Influence of windshield angle on drag.

The underbody is a critical area generating body drag. Suspensions, exhaust systems and other protruding components on the underbody are responsible for the drag. The air flow in this area is a shear plane controlled by zero air speed on the road surface, and induced flow by drag of the underbody components. The recognized fix for minimizing underbody drag is the use of a smooth underbody panel.

Protuberances from the body represent a second area where careful design can reduce drag. The wheels and wheel wells are a major contributor in this class. Significant drag develops at the wheels because of the turbulent, recirculating flow in the cavities. Figure 4.15 illustrates the complex flow patterns that occur around a wheel [13]. The sharp edges of the wheel cutout provide opportunities to induce flow in the horizontal plane, while the rotating wheel tends to induce circulation in the vertical plane. These effects allow the wheel to influence more flow than simply that which is seen because of its frontal area presented to the flow. The obvious improvement is aerodynamic shielding of the wheels and wheel well areas. While this is possible to some extent on rear wheels, steer rotation on the front wheels complicates the use of such treatment at the front. Experimental research has shown that decreasing the clearance between the underside and the ground and minimizing the wheel cavity decreases the total aerodynamic drag contribution from the wheel [12].

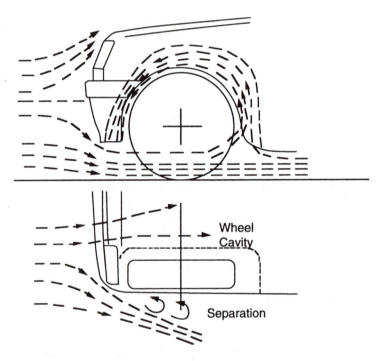

Fig 4.15 Air flow recirculation in a wheel well.

The cooling system is the last major contributor to drag. Air flow passing through the radiator impacts on the engine and the firewall, exerting its dynamic pressure as drag on the vehicle. The air flow pattern inside a typical engine compartment may be very chaotic due to the lack of aerodynamic treatment in this area. Figure 4.16 illustrates this situation [12]. With no attention to the need for air flow management, the air entering through the radiator dissipates much of its forward momentum against the vehicle compo-

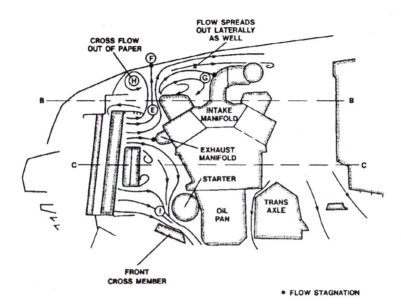

Fig 4.16 Air flow pattern inside a typical engine compartment. (Source: Williams, J., Ohler, W., Hackett, J., and Hammar, L., "Water Flow Simulation of Automotive Underhood Air Flow Phenomena," SAE Paper No. 910307, SP-855, 1991, 31 p.)

nents in the engine compartment before spilling out through the underside openings. The momentum exchange translates directly into increased drag.

Flow management in the cooling system can affect the drag coefficient by as much as 0.025 [10]. The drag contribution from this source is normally taken to be the difference in drag measured with the cooling system inlets open and covered. As seen in Figure 4.17, careful design to direct the flow (allowing it to maintain its velocity so that the static pressure remains low) can reduce the drag produced. Although these various arrangements may not be feasible within the styling theme of a given car, the potential for aerodynamic improvements is evident in the drag reductions shown. In order to reduce drag on modern cars, cooling inlet size is held to the practical minimum.

Aerodynamic Aids
Bumper Spoilers

Front bumper spoilers are aerodynamic surfaces extending downward from the bumper to block and redirect the shear flow that impacts on the underbody components. While the spoiler contributes pressure drag, at least

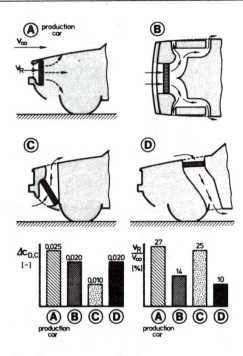

Fig. 4.17 Influence of cooling system on drag.

with a shallow depth the reduction in underbody drag is more significant. As spoiler depth is increased, eventually the increasing pressure drag outweighs further reduction in underbody drag and the overall drag increases. The low pressure produced also has the effect of reducing front-end lift.

Air Dams

Air dams are flow-blocking surfaces installed at the perimeter of the radiator to improve flow through the radiator at lower vehicle speeds. The improvement derives from the decreased pressure behind the radiator/fan, and may reduce drag by reduction of pressure on the firewall.

Deck Lid Spoilers

Spoilers and air foils on the rear deck may serve several purposes. By deflecting the air upward, as shown in Figure 4.18, the pressure is increased on the rear deck creating a down force at the most advantageous point on the vehicle to reduce rear lift. The spoilers may also serve to stabilize the vortices in the separation flow, thus reducing aerodynamic buffeting. In general, they tend to increase drag.

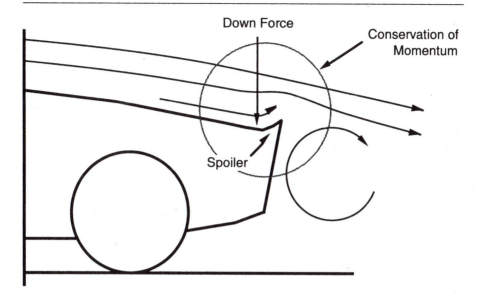

Fig. 4.18 Influence of a spoiler on flow over the rear.

Window and Pillar Treatments

Drip rails and offsets between windows and pillars on a car body are always sources of drag. Disturbance to the air flow in these regions may cause small separation zones. The disturbance to the air in the high-velocity air stream causes momentum loss which creates drag. Smooth contours are important not only for drag reduction, but also for reduction of aerodynamic noise.

Optimization

The development of automotive aerodynamics has been described as occurring in three stages [14]:

1) Adaptation of streamlined shapes from other disciplines (e.g., ship-building) at the turn of the century.

2) Application of the knowledge of fluid mechanics from aircraft aero-dynamics around the 1930s.

3) Current efforts to optimize the numerous details of the design to obtain good air flow characteristics.

The optimization is founded on the premise that the styling concept of the car is established and aerodynamic improvement can only be attempted in the form of changes to detail in the styling. An example of the optimization is shown in Figure 4.19. The sketches show minor modifications in detail such as a change in the air dam (A), hood line (B), A-pillar shape (C) and D-pillar shape (D and E). The graph illustrates the magnitude of drag reduction obtained from various combinations of these changes. The power of drag reduction by attention to details is illustrated by the fact that an overall drag reduction of 21% is achieved.

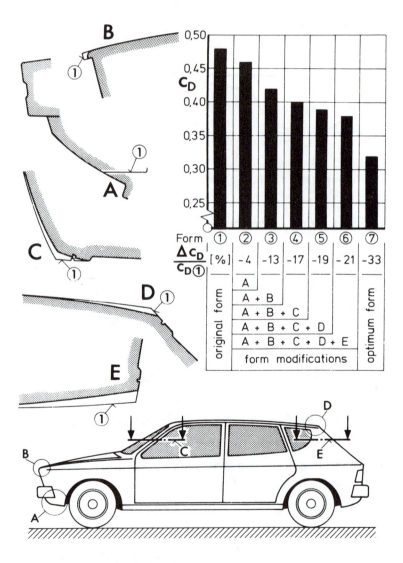

Fig. 4.19 Optimization of body detail.

Drag

Because air flow over a vehicle (or any other body for that matter) is so complex, it is necessary to develop semi-empirical models to represent the effect. Therefore, aerodynamic drag is characterized by the equation:

$$D_A = 1/2 \, \rho \, V^2 \, C_D \, A \qquad (4\text{-}2)$$

where:

C_D = Aerodynamic drag coefficient
A = Frontal area of the vehicle
ρ = Air density

(Note: The SAE symbol for drag, "D," is subscripted with an "A" in the text to denote it as aerodynamic drag to distinguish it from symbols used elsewhere in the text. The same convention will be used with aerodynamic lift and side force.)

The term $1/2 \, \rho \, V^2$ in the above equation is the dynamic pressure of the air, and is often referred to as the "q," typically expressed in units of pounds per square foot. The drag coefficient, C_D, is determined empirically for the car. The frontal area, A, is the scale factor taking into account the size of the car. (A half-scale model of a car, which has one-fourth of the area, will have one-fourth of the drag.) Because the size of a vehicle has a direct influence on drag, the drag properties of a car are sometimes characterized by the value of " $C_D \, A$."

Air Density

The air density is variable depending on temperature, pressure, and humidity conditions. At standard conditions (59°F and 29.92 inches of Hg) the density is 0.076 lb/ft^3. As used in this equation, the air density must be expressed as mass density, obtained by dividing by the acceleration of gravity; thus the value for standard atmospheric conditions is $\rho = 0.076/32.2 = 0.00236$ lb-sec^2/ft^4. Density at other conditions can be estimated for the prevailing pressure, P_r, and temperature, T_r, conditions by the equation:

$$\rho = 0.00236 \left(\frac{P_r}{29.92}\right) \left(\frac{519}{460 + T_r}\right) \qquad (4\text{-}3a)$$

where:

P_r = Atmospheric pressure in inches of mercury
T_r = Air temperature in degrees Fahrenheit

In the metric system the equivalent equation for air density in kg/m^3 is:

$$\rho = 1.225 \left(\frac{P_r}{101.325}\right)\left(\frac{288.16}{273.16 + T_r}\right) \qquad (4\text{-}3b)$$

where:

P_r = Atmospheric pressure in kiloPascals

T_r = Air temperature in degrees Celsius

Drag Coefficient

The drag coefficient is determined experimentally from wind tunnel tests or coast down tests. The definition of C_D comes from Eq. (4-2):

$$C_D = \frac{D_A}{\frac{1}{2}\rho V^2 A} = \frac{\text{Drag force}}{(\text{Dynamic pressure}) (\text{area})} \qquad (4\text{-}4)$$

The drag coefficient varies over a broad range with different shapes. Figure 4.20 shows the coefficients for a number of shapes. In each case it is presumed that the air approaching the body has no lateral component (i.e., it is straight along the longitudinal axis of the vehicle). Note that the simple flat plate has a drag coefficient of 1.95. This coefficient means that the drag force

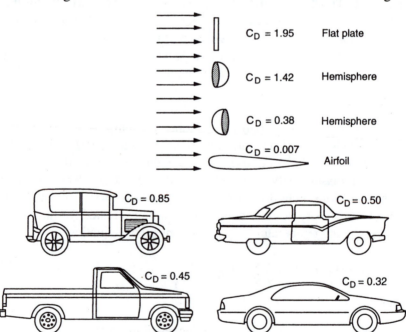

Fig. 4.20 Drag coefficients of various bodies.

is 1.95 times as large as the dynamic pressure acting over the area of the plate. The extreme drag produced by the plate results from the fact that the air spilling around the plate creates a separation area much larger than the plate itself.

In practice, a vehicle driving along a road experiences atmospheric winds in addition to the wind component arising from its speed. Atmospheric winds vary in intensity throughout the United States, with typical mean values of 10-20 mph, and gusty winds to 50 and 60 mph. The atmospheric wind will be random in direction with respect to the vehicle's direction of travel. Thus the relative wind seen by the vehicle will consist of the large component due to its speed, plus a smaller atmospheric wind component in any direction. Figure 4.21 illustrates how the relative wind will vary randomly.

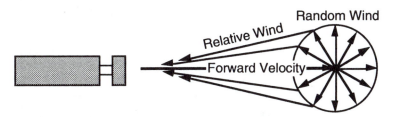

Fig. 4.21 Relative wind seen by a motor vehicle on the road.

When the atmospheric wind blows toward the vehicle a "headwind" is present, and the total velocity used in Eq. (4-2) is:

$$V = V_v + V_w \tag{4-5}$$

where:

V_v = Vehicle speed

V_w = Wind speed

Blowing in the direction of travel is a "tailwind," and the velocities are subtracted. Because the velocity is squared in Eq. (4-2), the increase in drag from a headwind is much greater than the decrease in drag from a tailwind of the same velocity.

In an average sense, the relative wind can be represented as a vector emanating from any point on the perimeter of the circle, and the average drag on the road will not be equivalent to simply the mean speed of the vehicle. Particularly important in this regard is the way in which the drag coefficient varies with a side wind component. On tractor-trailers side winds are particularly important because they disturb the aerodynamic flow field. Figure 4.22 shows the air flow around a tractor-trailer when the relative wind is at a 30-degree angle. Note that the flow is well attached on the right side of the vehicle, but a huge separation region occurs on the downwind side. In addition to the drag created by the wind impinging on the front of the truck, the large momentum change of the wind hitting the trailer adds another large drag component. Thus with trucks and cars, the change in drag coefficient with yaw angle of the wind is very important.

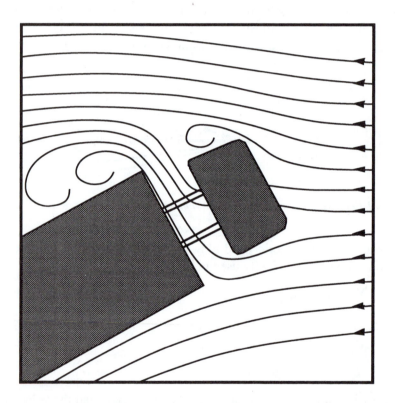

Fig. 4.22 Air flow around a tractor-semitrailer with 30-degree wind angle.

In contrast, with the much better aerodynamic design of cars, their drag coefficient is not as sensitive to yaw angle because the flow will not separate so readily. Normally, the drag coefficient increases by 5 to 10% with yaw angles in the range typical of on-road driving for passenger cars. Figure 4.23 shows the typical influence of yaw angle on the drag coefficients of several different types of vehicles.

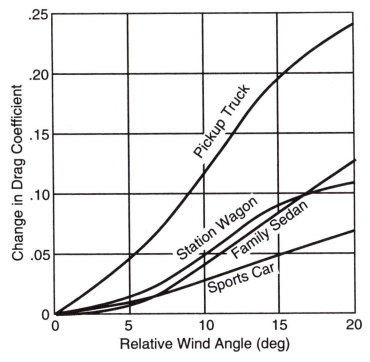

Fig. 4.23 Influence of yaw angle on drag coefficients of typical vehicle types.

Side Force

The lateral wind components will also impose a side force on the vehicle attempting to change its direction of travel. The exact effect depends both on the vehicle and the nature of the wind. In strong crosswinds, the side force is typically greater than the drag force, such that the angle of the overall wind force is much greater than the relative wind angle [15].

When the vehicle first encounters a crosswind condition on the road (a transient crosswind), the lateral force is first imposed on the front of the vehicle and may divert it in the downwind direction. The aerodynamic shape of the vehicle and even the steering system characteristics affect performance in this sense. Crosswind behavior is an important enough aspect of aerodynamics that it is discussed separately in a later section.

Under steady-state wind conditions, the side force imposed on a vehicle in a crosswind is given by:

$$S_A = 1/2 \, \rho \, V^2 \, C_S \, A \qquad (4\text{-}6)$$

where:

S_A = Side force
V = Total wind velocity
C_S = Side force coefficient (function of the relative wind angle)
A = Frontal area

Note that the frontal area, rather than the side area, is used in the equation. Figure 4.24 shows typical characteristics of C_S as a function of wind angle. The side force coefficient is zero at zero relative wind angle, and grows nearly linearly with the angle for the first 20 to 40 degrees. The slope of the gradient varies somewhat with vehicle type, but will typically be in the range of 0.035/ deg to 0.06/deg [16].

The side force acts on the body at the center of pressure, which is normally located ahead of the center of gravity such that the vehicle is turned away from the wind. In wind tunnel measurements the side force is measured in the ground plane at the mid-wheelbase position. The difference between this location and the center of pressure results in an overturning moment and a yaw moment whenever a side force is present.

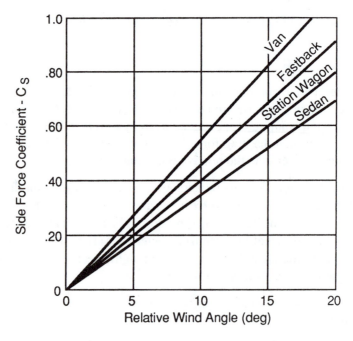

Fig. 4.24 Side force coefficient as a function of yaw angle for typical vehicles.

102

Lift Force

The pressure differential from the top to the bottom of the vehicle causes a lift force. These forces are significant concerns in aerodynamic optimization of a vehicle because of their influence on driving stability. The lift force is measured at the centerline of the vehicle at the center of the wheelbase. The force, L_A, is quantified by the equation:

$$L_A = 1/2 \, \rho \, V^2 \, C_L \, A \qquad\qquad (4-7)$$

where:

L_A = Lift force
C_L = Lift coefficient
A = Frontal area

As was seen in Figure 4.8, the lift force is dependent on the overall shape of the vehicle. At zero wind angle, lift coefficients normally fall in the range of 0.3 to 0.5 for modern passenger cars [17], but under crosswind conditions the coefficient may increase dramatically reaching values of 1 or more [18].

In aerodynamic studies the combined effect of lift and the pitching moment may be taken into account simultaneously by determining a lift coefficient for both the front and rear wheels [15]. In that case an equation similar to Eq. (4-7) is used to describe the lift effect at each axle.

Lift can have a negative impact on handling through the reduced control forces available at the tires. Front lift, which reduces steering controllability, is reduced by front bumper spoilers and by rearward inclination of front surfaces. Lift at the rear of the vehicle, which also reduces stability, is the most variable with vehicle design. In general, designs that cause the flow to depart with a downward angle at the rear of the vehicle create rear lift. Lift can be decreased by use of underbody pans, spoilers, and a change in the angle of attack of the body (a 3-degree cant on the body can decrease lift force by 40 percent).

Pitching Moment

While the lift force acts to decrease (or increase) the weight on the axles, the pitching moment acts to transfer weight between the front and rear axles. Pitching moment arises from the fact that the drag does not act at the ground plane (thus it accounts for the elevation of the drag force) and the lifting force may not act exactly at the center of the wheelbase. Pitching moment is described by the equation:

$$PM = 1/2\ \rho\ V^2\ C_{PM}\ A\ L \qquad (4\text{-}8)$$

where:

PM = Pitching moment
C_{PM}= Pitching moment coefficient
A   = Frontal area
L   = Wheelbase

Because it is a moment equation, a characteristic length is needed to achieve dimensional consistency in the equation. The vehicle wheelbase is used for this purpose. A moment can be translated about without changing its effect, so there is no need for a "point of action." Most modern cars have a pitching moment in the range of 0.05 to 0.2, and it is quite sensitive to the angle of attack on the vehicle. Figure 4.25 shows how the pitching moment coefficient varies with body pitch angle on several vehicles.

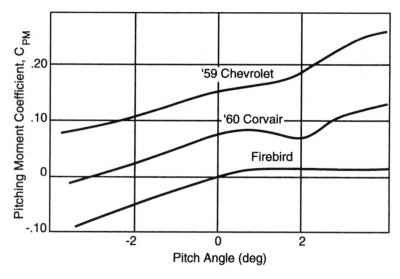

Fig. 4.25 Variation of Pitching Moment Coefficient with body pitch angle [16].

Yawing Moment

The lateral force caused by a side wind does not normally act at the mid-wheelbase position. Thus a yawing moment, YM, is produced. The yawing moment is quantified by the equation:

$$YM = 1/2 \, \rho \, V^2 \, C_{YM} \, A \, L \qquad\qquad (4-9)$$

where:

YM   = Yawing moment
C_{YM} = Yawing moment coefficient
A   = Frontal area
L   = Wheelbase

The yawing moment coefficient varies with wind direction, starting at zero with zero relative wind angle and growing almost linearly up to 20-degree angle. Figure 4.26 shows the coefficient for some typical vehicles. The slope of the coefficient at small angles ranges from 0.007/deg to 0.017/deg [18].

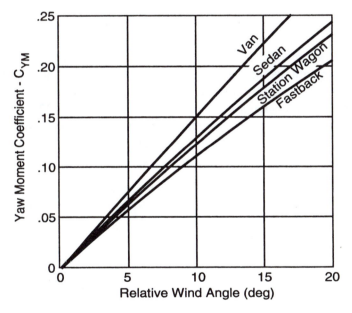

Fig 4.26 Yawing Moment Coefficient for typical vehicles.

Rolling Moment

The lateral force caused by a side wind acts at an elevated point on the vehicle. Thus a rolling moment, RM, is produced. The moment has only a minor influence on vehicle stability, depending largely on the roll steer properties of the suspensions. The rolling moment is quantified by the equation:

$$RM = 1/2\,\rho\,V^2\,C_{RM}\,A\,L \qquad (4\text{-}10)$$

where:

RM = Rolling moment
C_{RM} = Rolling moment coefficient
A = Frontal area
L = Wheelbase

The rolling moment coefficient is sensitive to wind direction much like the yawing moment coefficient, being quite linear over the first 20 degrees of relative wind angle. The slope of the rolling moment coefficient ranges from 0.018/deg to 0.04/deg [18].

Crosswind Sensitivity

The growing sophistication of aerodynamic design in motor vehicles in combination with the increased sensitivities to crosswinds often accompanying drag reductions has stimulated interest in understanding and controlling the factors that affect behavior in a crosswind [19, 20, 21, 22]. "Crosswind sensitivity" generally refers to the lateral and yawing response of a vehicle in the presence of transverse wind disturbances which affect the driver's ability to hold the vehicle in position and on course.

Crosswind sensitivity is dependent on more than just the aerodynamic properties of the vehicle. In the literature [19] the key elements that have been identified are:

• Aerodynamic properties

• Vehicle dynamic properties (weight distribution, tire properties, and suspensions)

• Steering system characteristics (compliances, friction and torque assist)

• Driver closed-loop steering behavior and preferences

Crosswind behavior is studied using instrumented vehicles in natural random (ambient) wind conditions, under exposure to crosswind generators (fans that produce a crosswind in an experimental test area), and in driving simulators. The primary variables of interest are the yaw response, lateral acceleration response, steering corrections when holding a specified course, and the subjective judgments of test drivers.

Good crosswind behavior is most strongly correlated with yaw rate response. Figure 4.27 illustrates the correlation that has been obtained between subjective ratings in a "gauntlet" crosswind (fans alternately blowing in opposite directions) and yaw rate response [19]. The high degree of correlation in these particular tests suggests that yaw rate response in a crosswind nearly explains all variation in subjective ratings from vehicle to vehicle. Other measures of response that correlate well with subjective ratings in order of importance are lateral acceleration at the driver's seat headrest, steering wheel displacement, and lateral acceleration.

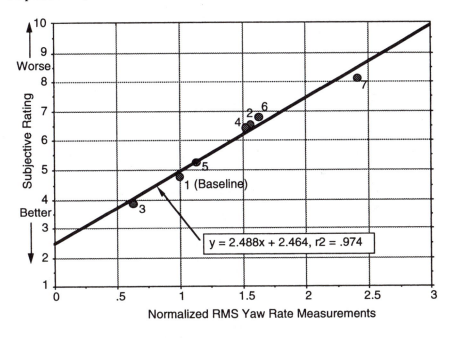

Fig. 4.27 Correlation of subjective ratings with normalized RMS yaw rate response [19].

107

The aerodynamic property of primary importance to crosswind sensitivity is the center of pressure (CP) location and its relative distance ahead of the vehicle's neutral steer point. The neutral steer point (NSP) is the point on the vehicle at which a lateral force produces equal sideslip angles at both front and rear axles.

The CP is the resultant action point of the combined lateral force and yaw moment reactions on the vehicle. In general, more rearward center of pressure locations, which are closer to the NSP, minimize lane deviations in a crosswind and are subjectively more acceptable. The effect of fore/aft CP location is seen in the lateral acceleration responses of three vehicles given in Figure 4.28. A forward CP location induces a large lateral acceleration response because the effective action point is near the front of the vehicle and the vehicle is turned strongly away from the wind. With a rearward CP position, the vehicle yaws less and resists the tendency to be displaced sideways.

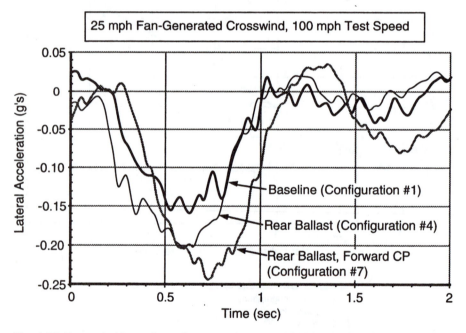

Fig. 4.28 Crosswind lateral acceleration response with variation of CP location [19].

Other vehicle dynamic properties come into play in determining how the vehicle responds to a given crosswind force. For example, the weight distribution on the front and rear axles determines the center of gravity location and the NSP location. Tire properties (such as cornering stiffness) also affect the location of the NSP and, hence, the degree to which the vehicle resists the yawing moment disturbance from the aerodynamic side force.

A first estimate of crosswind sensitivity can be obtained from a calculation of static yaw rate response to a steady crosswind with no steering input [19]. Under static conditions, a vehicle's passive crosswind response is given by:

$$\frac{r}{\alpha_{cw}} = \frac{q\,C_y\,A}{M\,V}\,[\frac{d_{ns} + (b - L/2 + L\,C_{YM})}{d_{ns} + \zeta_d}] \tag{4-11}$$

where:

r = Yaw rate
α_{cw} = Aerodynamic wind angle
q = Dynamic pressure = $\rho V^2/2$
C_y = Side force coefficient
C_{YM} = Yawing moment coefficient
A = Frontal area
M = Mass of the vehicle
V = Forward velocity
d_{ns} = Distance from center of mass to neutral steer point
b = Distance from front axle to center of mass
L = Wheelbase
ζ_d = Moment arm proportional to the tire force yaw damping moment about the neutral steer point

$$= \frac{L^2}{M\,V^2}\frac{C_f C_r}{(C_f + C_r)}$$

C_f = Effective total tire cornering stiffness of the front axle
C_r = Effective total tire cornering stiffness of the rear axle

As noted above, the neutral steer point identifies that fore/aft point on the vehicle where an external side force will not cause the vehicle to yaw. This point is affected by tire force properties, steering system compliance, suspension kinematics and weight distribution. The numerator in the second term on the right-hand side of Eq. (4-11) is the distance from the neutral steer point to the aerodynamic center of pressure. Thus a large distance between these points contributes to crosswind sensitivity.

The denominator in this same term contains the moment arm of the tire force yaw damping moment. This term can be increased to reduce crosswind sensitivity by increasing wheelbase or the effective tire cornering stiffness.

Effective tire cornering stiffness is favorably influenced directly by using tires with high cornering stiffness and by eliminating compliances in the steering or suspension which allow the vehicle to yield to the crosswind. However, the moment arm is strongly diminished by forward speed, thereby tending to increase crosswind sensitivity as speed goes up.

The static analysis discussed above may overlook certain other vehicle dynamic properties that can influence crosswind sensitivity. Roll compliance, particularly when it induces suspension roll steer effects, may play a significant role not included in the simplified analysis. Thus a more comprehensive analysis using computer models of the complete dynamic vehicle and its aerodynamic properties may be necessary for more accurate prediction of a vehicle's crosswind sensitivity.

ROLLING RESISTANCE

The other major vehicle resistance force on level ground is the rolling resistance of the tires. At low speeds on hard pavement, the rolling resistance is the primary motion resistance force. In fact, aerodynamic resistance becomes equal to the rolling resistance only at speeds of 50-60 mph. For off-highway, level ground operation, the rolling resistance is the only significant retardation force.

While other resistances act only under certain conditions of motion, rolling resistance is present from the instant the wheels begin to turn. Rolling resistance, in addition, has another undesirable property—a large part of the power expended in a rolling wheel is converted into heat within the tire. The consequent temperature rise reduces both the abrasion resistance and the flexure fatigue strength of the tire material, and may become the limiting factor in tire performance.

There are at least seven mechanisms responsible for rolling resistance:

1) Energy loss due to deflection of the tire sidewall near the contact area
2) Energy loss due to deflection of the tread elements
3) Scrubbing in the contact patch
4) Tire slip in the longitudinal and lateral directions
5) Deflection of the road surface
6) Air drag on the inside and outside of the tire
7) Energy loss on bumps

Considering the vehicle as a whole, the total rolling resistance is the sum of the resistances from all the wheels:

$$R_x = R_{xf} + R_{xr} = f_r W \qquad (4\text{-}12)$$

where:

R_{xf} = Rolling resistance of the front wheels

R_{xr} = Rolling resistance of the rear wheels

f_r = Rolling resistance coefficient

W = Weight of the vehicle

For theoretically correct calculations, the dynamic weight of the vehicle, including the effects of acceleration, trailer towing forces and the vertical component of air resistance, is used. However, for vehicle performance estimation, the changing magnitude of the dynamic weight complicates the calculations without offering significant improvements in accuracy. Furthermore, the dynamic weight transfer between axles has minimal influence on the total rolling resistance (aerodynamic lift neglected). For these reasons, static vehicle weight is sufficiently accurate for computation of rolling resistance in most cases.

All of these considerations apply, in a strict sense, only for straight-line motion. For vehicles which are subjected to lateral forces (cornering or aerodynamic loading), the direction of rolling resistance deviates from the direction of actual travel, and the tractive force must overcome the vectorial resultant of the side force and rolling resistance.

Factors Affecting Rolling Resistance

The coefficient of rolling resistance, f_r, is a dimensionless factor that expresses the effects of the complicated and interdependent physical properties of tire and ground. Establishment of standardized conditions for measurement of the effects of variables like the structure of the ground material, composition of the rubber, design elements of the tire, temperature, etc., proves difficult if not impossible. Some of the more important factors are discussed below.

Tire Temperature

Because much of the rolling resistance on paved surfaces arises from deflection and energy loss in the tire material, the temperature of the tire can have a significant effect on the resistance experienced. In the typical situation where a tire begins rolling from a cold condition, the temperature will rise and the rolling resistance will diminish over a first period of travel. Figure 4.29 shows the relative changes in temperature and rolling resistance that will occur [18]. In the figure the tire must roll a distance of at least 20 miles before the system approaches stable operation. In typical tire tests it is therefore common to warm up the tire for 20 minutes or more before taking measurements that may be affected by the warm-up condition. For the short trips representative of much automotive travel, the tires never warm up to benefit from the lowest possible levels of rolling resistance.

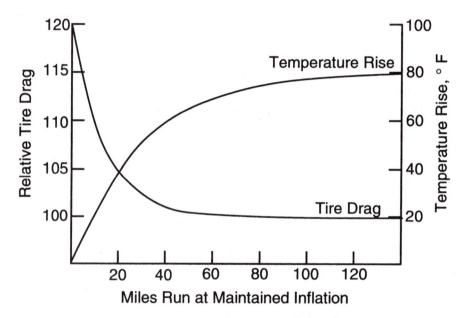

Fig. 4.29 *Relative tire temperature and rolling resistance during warm-up.*

Tire Inflation Pressure/Load

To a large extent, the tire inflation pressure determines the tire elasticity and, in combination with the load, determines the deflection in the sidewalls and contact region. The overall effect on rolling resistance also depends on the elasticity of the ground. Figure 4.30 shows how the coefficient changes with inflation pressure on different types of surfaces.

On soft surfaces like sand, high inflation pressures result in increased ground penetration work and therefore higher coefficients. Conversely, lower

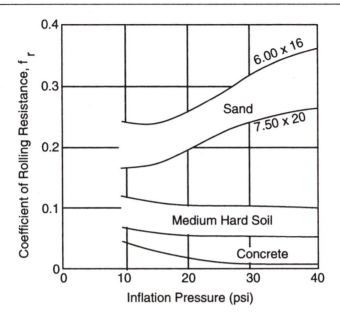

Fig. 4.30 Coefficient of rolling resistance versus inflation pressure [16].

inflation pressure, while decreasing ground penetration, increases tire-flexure work. Thus the optimum pressure depends on the surface deformation characteristics. In general, the "increased traction" obtained by lowering tire pressure on a sand surface is actually achieved through a reduction in rolling resistance.

On medium plastic surfaces such as dirt, the effects of inflation pressure on tire and ground approximately balance, and the coefficient remains nearly independent of inflation pressure. On hard (paved) surfaces, the coefficient decreases with higher inflation pressure since flexure work of the tire body will be greatly reduced.

Velocity

The coefficient is directly proportional to speed because of increased flexing work and vibration in the tire body, although the effect is small at moderate and low speeds and is often assumed to be constant for calculation. The influence of speed becomes more pronounced when speed is combined with lower inflation pressure. Figure 4.31 shows the rolling resistance versus speed for a radial, bias-belted, and bias-ply tires [23]. The sharp upturn in coefficient at high speeds is caused by a high-energy standing wave developed in the tire carcass just behind the tire contact patch. If allowed to persist for even moderate periods of time, catastrophic failure can result. Thus formation of a standing wave is one of the primary effects limiting a tire's rated speed. Modern tires rated for high speed normally include stabilizers in the shoulder area to control the development of standing waves.

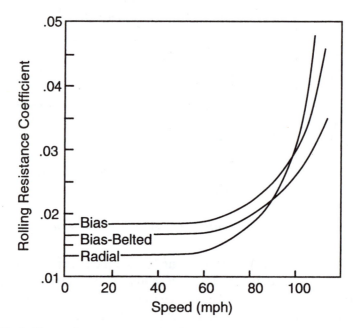

Fig. 4.31 Rolling resistance versus speed.

Tire Material and Design

The materials and thickness of both the tire sidewalls (usually expressed in plies) and the tread determine the stiffness and energy loss in the rolling tire. Figure 4.32 shows the rolling resistance of experimental tires constructed of different types of rubber in the sidewalls and tread areas [23]. The plot vividly illustrates the losses deriving from hysteresis in the tread material. Although hysteresis in the tread rubber is important for good wet traction, it degrades rolling resistance performance.

Worn-out, smooth-tread tires show coefficient values up to 20 percent lower than new tires. Fine laminations, on the other hand, increase the coefficient as much as 25 percent. The cord material in the sidewall has only a small effect, but the cord angle and tire belt properties (belted versus radial-ply tires) have a significant influence.

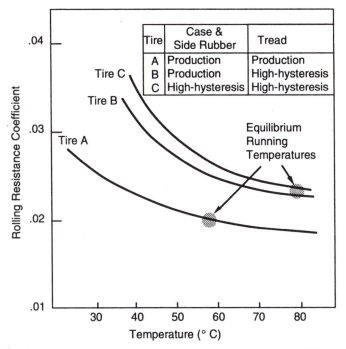

Tire	Case & Side Rubber	Tread
A	Production	Production
B	Production	High-hysteresis
C	High-hysteresis	High-hysteresis

Fig. 4.32   Rolling resistance versus temperature for tires with different polymers.

Tire Slip

Wheels transferring tractive or braking forces show higher rolling resistance due to wheel slip and the resulting frictional scuffing. Cornering forces produce the same effects. Figure 4.33 shows the rolling resistance effects as a function of slip angle [23]. At a few degrees of slip, equivalent to moderate-high cornering accelerations, the rolling resistance coefficient may nearly double in magnitude. The effect is readily observed in normal driving when one will "scrub" off speed in a corner.

Typical Coefficients

The multiple and interrelated factors affecting rolling resistance make it virtually impossible to devise a formula that takes all variables into account. Before a value for rolling resistance coefficient can be chosen for a particular application, the overall degree of accuracy required for the calculations should be established.

Several equations for estimating rolling resistance have been developed over the years. Studies on the rolling loss characteristics of solid rubber tires

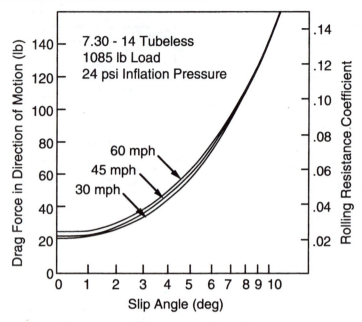

Fig. 4.33 Rolling resistance coefficient versus slip angle.

[23] led to an equation of the form:

$$f_r = \frac{R_x}{W} = C\frac{W}{D}\sqrt{\frac{h_t}{w}}$$ (4-13)

where:

R$_x$ = Rolling resistance force
W = Weight on the wheel
C = Constant reflecting loss and elastic characteristics of the tire material
D = Outside diameter
h$_t$ = Tire section height
w = Tire section width

From this formulation, rolling resistance is seen to be load sensitive, increasing linearly with load. Larger tires reduce rolling resistance, as do low aspect ratios (h/w). Some confirmation of the general trends from this equation appear in the literature from studies of the rolling resistance of conventional passenger car tires of different sizes under the same load conditions [23].

Other equations for the rolling resistance coefficient for passenger car tires rolling on concrete surfaces have been developed. The variables in these equations are usually inflation pressure, speed and load. The accuracy of a calculation is naturally limited by the influence of factors that are neglected.

At the most elementary level, the rolling resistance coefficient may be estimated as a constant. The table below lists some typical values that might be used in that case.

Vehicle Type	Concrete	Surface Medium Hard	Sand
Passenger cars	0.015	0.08	0.30
Heavy trucks	0.012	0.06	0.25
Tractors	0.02	0.04	0.20

At lower speeds the coefficient rises approximately linearly with speed. Thus equations have been developed which include a linear speed dependence, such as below:

$$f_r = 0.01 \ (1 + V/100) \tag{4-14}$$

where:

V = Speed in mph

Over broader speed ranges, the coefficient rises in a manner that is closer to a speed-squared relationship. The Institute of Technology in Stuttgart has developed the following equation for rolling on a concrete surface [16]:

$$f_r = f_0 + 3.24 \ f_s \ (V/100)^{2.5} \tag{4-15}$$

where:

V = Speed in mph

f_0 = Basic coefficient

f_s = Speed effect coefficient

The two coefficients, f_0 and f_s, depend on inflation pressure and are determined from the graph shown in Figure 4.34.

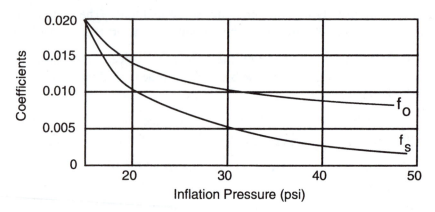

Fig. 4.34 Coefficients for Eq. (4-15).

At the University of Michigan Transportation Research Institute, similar equations for estimating rolling resistance of heavy truck tires of both the radial and bias-ply types were developed [24]. These are:

$$f_r = (0.0041 + 0.000041 \ V) \ C_h \quad \text{Radial tires} \qquad (4\text{-}16a)$$

$$f_r = (0.0066 + 0.000046 \ V) \ C_h \quad \text{Bias-ply tires} \qquad (4\text{-}16b)$$

where:

$\quad$ V $\ =$ Speed in mph

$\quad$ $C_h =$ Road surface coefficient

$\quad\quad = 1.0$ for smooth concrete

$\quad\quad = 1.2$ for worn concrete, brick, cold blacktop

$\quad\quad = 1.5$ for hot blacktop

Rolling resistance is clearly a minimum on hard, smooth, dry surfaces. A worn-out road almost doubles rolling resistance. On wet surfaces, higher rolling resistance is observed probably due to the cooler operating temperature of the tire which reduces its flexibility.

TOTAL ROAD LOADS

The summation of the rolling resistance and aerodynamic forces (and grade forces, if present) constitutes the propulsion load for the vehicle, and is normally referred to as "road load." The road load force is thus:

$$R_{RL} = f_r\, W + 1/2\, \rho\, V^2\, C_D\, A + W \sin\theta \qquad (4\text{-}17)$$

The sum of these forces is plotted for a typical large vehicle in Figure 4.35. The rolling resistance has been assumed constant with a coefficient of 0.02 and a vehicle weight of 3650 lb. The aerodynamic drag assumes a vehicle of 23.3 ft^2 frontal area and a drag coefficient of 0.34. The total road load curve rises with the square of the speed due to the aerodynamic component. Rolling resistance and grade simply slide the whole curve upward in proportion to their size.

The road load horsepower is computed by multiplying Eq. (4-16) by the vehicle velocity and applying the appropriate conversion factor to obtain horsepower. In that case:

$$HP_{RL} = R_{RL}\, V/550 = (f_r\, W + 1/2\, \rho\, V^2\, C_D\, A + W \sin\theta)\, V/550 \quad (4\text{-}18)$$

The road load power corresponding to the road load forces in Figure 4.35 is shown in Figure 4.36 for a level road condition. Note that the power increases much more rapidly with velocity due to the cubic relationship in Eq. (4-18). Thus at high speeds a small increase in speed results in a large increase in vehicle power required, with an associated penalty to fuel economy.

Fuel Economy Effects

Today, aerodynamic and rolling resistance forces are of particular interest for their effect on fuel consumption. Aerodynamic drag, of course, is the most

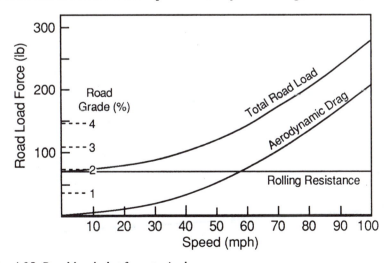

Fig. 4.35 Road load plot for a typical passenger car.

important of the aerodynamic properties. In the decade of the 1970s, drag coefficients of 0.4 to 0.5 were common on relatively large cars. In the 1980s, drag coefficients are commonly less than 0.4 with some cars less than 0.3. In addition, the smaller cross-sectional areas contribute to lower overall drag.

The exact improvements in fuel economy that may be expected from improvements in road loads are difficult to predict because of the uncertainty about the ways cars are used and driven. Figure 4.37 shows an estimate of where the energy is used in EPA driving cycles and in steady highway driving [25].

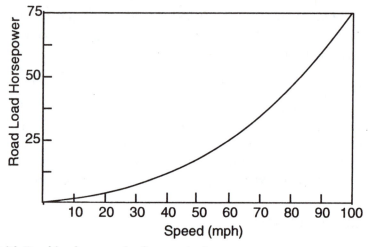

Fig. 4.36 Road load power plot for a typical passenger car.

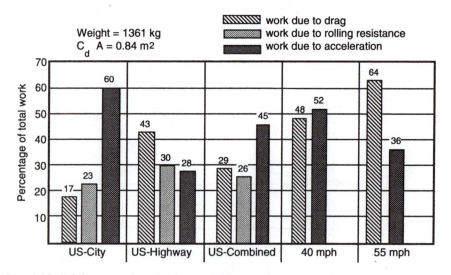

Fig. 4.37 Subdivision of work of a compact car in driving cycles.

EXAMPLE PROBLEMS

1) A heavy truck weighing 72,500 lb rolls along I70 in Denver at a speed of 67 mph. The air temperature is 55°F and the barometric pressure is 26.01 in Hg. The truck is 8' wide by 13.5' high, and has an aerodynamic drag coefficient of 0.65. The truck has radial-ply tires. Calculate the aerodynamic drag, the rolling resistance (according to the SAE equations) and the road load horsepower at these conditions.

Solution:

The aerodynamic drag may be calculated using Eq. (4-2). The temperature and barometric pressure conditions are not standard, so we must first calculate the local air density using Eq. (4-3a).

$$\rho = 0.00236 \frac{P_r}{29.92} \frac{519}{460 + T_r}$$

$$\rho = 0.00236 \frac{26.01}{29.92} \frac{519}{(460 + 55)} = 0.00207 \frac{lb\text{-}sec^2}{ft^4} = 0.0667 \frac{lb}{ft^3}$$

Now the aerodynamic drag can be calculated:

$$D_A = 0.5 \, (0.00207 \frac{lb\text{-}sec^2}{ft^4}) \, (\frac{67 \text{ mph}}{0.682 \text{ mph/ft/sec}})^2 0.65 \, (8 \text{ ft}) \, (13.5') = 702 \text{ lb}$$

The rolling resistance comes from a combination of Eqs. (4-12) and (4-16a). First from the SAE equation, we calculate the rolling resistance coefficient, assuming a surface coefficient of unity:

$$f_r = [0.0041 + 0.000041 \, (67 \text{ mph})] = 0.00685$$

Then the rolling resistance is:

$$R_x = 0.00685 \, (72,500 \text{ lb}) = 497 \text{ lb}$$

At the speed of 67 mph (98.3 ft/sec) the horsepower required to overcome aerodynamic drag is:

$$HP_A = 702 \text{ lb} \, (98.3 \text{ ft/sec}) \, 1 \text{ hp}/(550 \text{ ft-lb/sec}) = 125 \text{ hp}$$

And the horsepower to overcome rolling resistance is:

$$HP_R = 497 \text{ lb} \, (98.3 \text{ ft/sec}) \, 1 \text{ hp}/(550 \text{ ft-lb/sec}) = 88.8 \text{ hp}$$

Notes:

a) A total of nearly 215 hp is required to keep the truck rolling at this speed.

b) Highway trucks typically have diesel engines rated at 350 to 600 horse-power. These engines are designed to run continuously at maximum power, so it is not unreasonable for them to run at this output level for mile after mile.

c) At a typical brake specific fuel consumption of 0.35 lb per brake-horsepower-hour, the engine will burn 82.6 lb of diesel fuel per hour (13 gallons/hour) getting about 5.25 miles per gallon. It is not unusual to have a fuel tank capacity of 300 gallons on-board a highway tractor, so they can run for nearly 24 hours or 1500 miles without having to stop for fuel.

2) A passenger car has a frontal area of 21 square feet and a drag coefficient of 0.42. It is traveling along at 55 mph. Calculate the aerodynamic drag and the associated horsepower requirements if it is driving into a 25 mph headwind, and with a 25 mph tailwind.

Solution:

The drag can be calculated from Eq. (4-2), although the relative velocity must take into account the headwinds and tailwinds as given in Eq. (4-4). We will assume that the air temperature and pressure conditions are effectively near standard conditions so that the standard value for air density can be used.

Headwind condition:

$$D_A = 0.5 \; (0.00236 \frac{\text{lb-sec}^2}{\text{ft}^4}) \; (\frac{(55 + 25) \text{ mph}}{0.682 \text{ mph/ft/sec}})^2 0.42 \; (21 \text{ ft}^2) = 143 \text{ lb}$$

Tailwind condition:

$$D_A = 0.5 \; (0.00236 \frac{\text{lb-sec}^2}{\text{ft}^4}) \; (\frac{(55 - 25) \text{ mph}}{0.682 \text{ mph/ft/sec}})^2 0.42 \; (21 \text{ ft}^2) = 20 \text{ lb}$$

Notes:

a) The normal aerodynamic drag on this vehicle in the absence of any headwind or tailwind would be 68 lb.

b) The headwind more than doubles the drag because the drag increases with the square of the relative headwind velocity, which goes from 55 to 80 mph.

c) The tailwind reduces the drag considerably due to the speed square effect.

REFERENCES

1. Li, W.H., and Lam, S.H., Principles of Fluid Mechanics, Addison-Wesley Publishing Company, Inc., Reading, Massachusetts, 1964, 374 p.

2. Shepherd, D.G., Elements of Fluid Mechanics, Harcourt, Brace and World, Inc., New York, 1965, 498 p.

3. Schenkel, F.K., "The Origins of Drag and Lift Reductions on Automobiles with Front and Rear Spoilers," SAE Paper No. 770389, 1977, 11 p.

4. Kramer, C., "Introduction to Aerodynamics," Lecture notes for Short Course 1984-01, von Karman Institute for Fluid Dynamics, Jan. 1984, 60 p.

5. Lay, W.E., "Is 50 Miles per Gallon Possible with Correct Streamlining?" *SAE Journal,* Vol. 32, 1933, pp 144-156, pp 177-186.

6. Hoerner, S., Fluid-Dynamic Drag, Published by the author, Midland Park, NJ, 1965.

7. Kamm, W., "Einfluss der Reichsautobahn auf die Gestaltung von Kraftfahrzeugen," *ATZ,* Vol. 37, 1943, pp 341-354.

8. "SAE Vehicle Dynamics Terminology," SAE J670e, Society of Automotive Engineers, Warrendale, PA (see Appendix A).

9. "Vehicle Aerodynamics Terminology," SAE J1594, Society of Automotive Engineers, Warrendale, PA, June 1987, 5 p.

10. Buchheim, R., Deutenback, K.-R., and Luckoff, H.-J., "Necessity and Premises for Reducing the Aerodynamic Drag of Future Passenger Cars," SAE Paper No. 810185, 1981, 14 p.

11. Hucho, W.-H., and Janssen, L.J., "Beitrage der Aerodynamik im Rahmen einer Scirocco," *ATZ,* Vol. 77, 1975, pp 1-5.

12. Scibor-Rylski, A.J., Road Vehicle Aerodynamics, Second Edition, Pentech Press, London, 1984, 244 p.

13. Sardou - M.S.W.T , M., and Sardou, S.A., "Why to Use High Speed Moving Belt Wind Tunnel for Moving Ground Surface Vehicles Development," Lecture notes for Short Course 1984-01, von Karman Institute for Fluid Dynamics, Jan. 1984, 59 p.

14. Hucho, W.H., Janssen, L.J., and Emmelmann, H.J., "The Optimization of Body Details—A Method for Reducing Aerodynamic Drag of Road Vehicles," SAE Paper No. 760185, 1976, 18 p.

15. Gilhaus, A.M., and Renn, V.E., "Drag and Driving-Stability-Related Aerodynamic Forces and Their Interdependence—Results of Measurement on 3/8-Scale Basic Car Shapes," SAE Paper No. 860211, 1986, 15 p.

16. Cole, D., "Elementary Vehicle Dynamics," course notes in Mechanical Engineering, The University of Michigan, Ann Arbor, MI, 1972.

17. Aerodynamics of Road Vehicles, Wolf-Heinrich Hucho, ed., Butterworths, London, 1987, 566 p.

18. Hogue, J.R., "Aerodynamics of Six Passenger Vehicles Obtained from Full Scale Wind Tunnel Tests," SAE Paper No. 800142, 1980, 17 p.

19. MacAdam, C.C., Sayers, M.W., Pointer, J.D., and Gleason, M., "Cross-wind Sensitivity of Passenger Cars and the Influence of Chassis and Aerodynamic Properties on Driver Performance," *Vehicle Systems Dynamics,* Vol. 19, 1990, 36 p.

20. Willumeit, H.P., *et al.*, "Method to Correlate Vehicular Behavior and Driver's Judgment under Side Wind Disturbances," Dynamics of Vehicles on Roads and Tracks, Proceedings, Swets and Zeitlinger B. V. - Lisse, 1988, pp. 509-524.

21. Uffelmann, F., "Influence of Aerodynamics and Suspension on the Cross-Wind Behaviour of Passenger Cars - Theoretical Investigation under Consideration of the Driver's Response," Dynamics of Vehicles on Roads and Tracks, O. Nordstrom, ed., Swets and Zeitlinger B. V. - Lisse, 1986, pp. 568-581.

22. van den Hemel, H., *et al.*, "The Cross-Wind Stability of Passenger Cars: Development of an Objective Measuring Method," Fourth IAVSD Congress, 1987.

23. Clark, S.K., *et al.*, "Rolling Resistance of Pneumatic Tires," The University of Michigan, Interim Report No. UM-010654-3-1, July 1974, 65 p.

24. Fancher, P.S., and Winkler, C.B., "Retarders for Heavy Vehicles: Phase III Experimentation and Analysis; Performance, Brake Savings, and Vehicle Stability," U. S. Department of Transportation, Report No. DOT HS 806 672, Jan. 1984, 144 p.

25. Buchheim, R., "Contributions of Aerodynamics to Fuel Economy Improvements of Future Cars," Fuel Economy Research Conference, Section 2: Technical Presentations.

CHAPTER 5
RIDE

Last segment of Cinturato P3 endurance test. (Photo courtesy of Pirelli Tire Corp.)

Automobiles travel at high speed, and as a consequence experience a broad spectrum of vibrations. These are transmitted to the passengers either by tactile, visual, or aural paths. The term "ride" is commonly used in reference to tactile and visual vibrations, while the aural vibrations are categorized as "noise." Alternatively, the spectrum of vibrations may be divided up according to frequency and classified as ride (0-25 Hz) and noise (25-20,000 IIz). The 25 Hz boundary point is approximately the lower frequency threshold of hearing, as well as the upper frequency limit of the simpler vibrations common to all motor vehicles. The different types of vibrations are usually so interrelated that it may be difficult to consider each separately; i.e., noise is usually present when lower-frequency vibrations are excited.

The vibration environment is one of the more important criteria by which people judge the design and construction "quality" of a car. Being a judgment, it is subjective in nature, from which arises one of the greatest difficulties in developing objective engineering methods for dealing with ride as a performance mode of the vehicle.

The lower-frequency ride vibrations are manifestations of dynamic behavior common to all rubber-tired motor vehicles. Thus, the study of these modes is an important area of vehicle dynamics. As an aid in developing a systematic picture of ride behavior, it is helpful to think of the overall dynamic system as shown in Figure 5.1. The vehicle is a dynamic system, but only exhibits vibration in response to excitation inputs. The response properties determine the magnitude and direction of vibrations imposed on the passenger compartment, and ultimately determine the passenger's perception of the vehicle. Thus, understanding ride involves the study of three main topics:

- Ride excitation sources
- Basic mechanics of vehicle vibration response
- Human perception and tolerance of vibrations

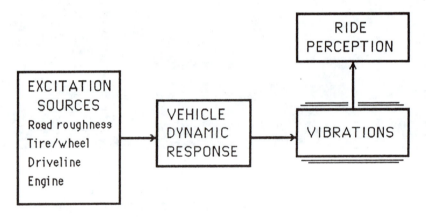

Fig. 5.1 The ride dynamic system.

EXCITATION SOURCES

There are multiple sources from which vehicle ride vibrations may be excited. These generally fall into two classes—road roughness and on-board sources. The on-board sources arise from rotating components and thus include the tire/wheel assemblies, the driveline, and the engine.

Road Roughness

Road roughness encompasses everything from potholes resulting from localized pavement failures to the ever-present random deviations reflecting the practical limits of precision to which the road surface can be constructed and maintained. Roughness is described by the elevation profile along the wheel tracks over which the vehicle passes. Road profiles fit the general category of "broad-band random signals" and, hence, can be described either by the profile itself or its statistical properties. One of the most useful representations is the Power Spectral Density (PSD) function.

Like any random signal, the elevation profile measured over a length of road can be decomposed by the Fourier Transform process [1] into a series of sine waves varying in their amplitudes and phase relationships. A plot of the amplitudes versus spatial frequency is the PSD. Spatial frequency is expressed as the "wavenumber" with units of cycles/foot (or cycles/meter), and is the inverse of the wavelength of the sine wave on which it is based.

Road elevation profiles can be measured either by performing close interval rod and level surveys [2], or by high-speed profilometers [3]. When the PSDs are determined, plots such as those shown in Figure 5.2 are typically obtained [4,5,6]. Although the PSD of every road section is unique, all roads show the characteristic drop in amplitude with wavenumber. This simply reflects the fact that deviations in the road surface on the order of hundreds of feet in length may have amplitudes of inches, whereas those only a few feet in

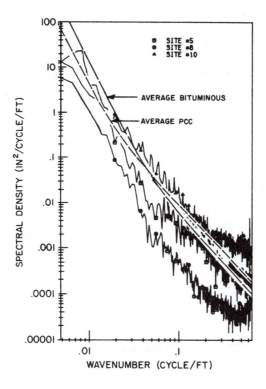

Fig. 5.2 Typical spectral densities of road elevation profiles.

127

length are normally only fractions of an inch in amplitude. The general amplitude level of the plot is indicative of the roughness level—higher amplitudes implying rougher roads. The wavenumber range in the figure corresponds to wavelengths of 200 feet (61 m) on the left at 0.005 cycle/foot (0.016 cycle/meter), to about 2 feet (0.6 m) on the right at 0.5 cycle/foot (1.6 cycles/meter).

The upper PSD in the figure is a deteriorating Portland Cement concrete (often called "rigid pavement") road surface. Note a marked periodicity in the range of wavenumber between 0.01 cycle/foot (0.03 cycle/meter) and 0.1 cycle/foot (0.3 cycle/meter), which is related to the fixed slab length used in construction of the road. The lowest PSD is a similar road overlaid with a bituminous asphalt surface layer yielding a much smoother surface (the PSD amplitude is reduced by an order of magnitude). The reduction is especially pronounced in the high-wavenumber range. The intermediate PSD is a typical asphalt road (often called "flexible pavement").

Although many ride problems are peculiar to a specific road, or road type, the notion of "average" road properties can often be helpful in understanding the response of a vehicle to road roughness. The general similarity in the spectral content of the roads seen in Figure 5.2 (that elevation amplitude diminishes systematically with increasing wavenumber) has long been recognized as true of most roads [7,8,9]. Consequently, road inputs to a vehicle are often modeled with an amplitude that diminishes with frequency to the second or fourth power approximating the two linear segments of the curve shown in the figure. The average properties shown in the figure are derived from recent studies of a large number of roads [4]. The spectral contents are slightly different for bituminous and Portland Cement concrete (PCC) roads. Other less common road types, such as surface treatment and gravel roads, will have slightly differing spectral qualities [6]. The general level of the elevation of the curve may be raised or lowered to represent different roughness levels, but the characteristic slopes and inflection points are constant. The difference between the bituminous and PCC average curves is the relative magnitude of high- versus low-wavenumber content. For a given overall roughness, more is concentrated in the high-wavenumber (short-wavelength) range with PCC surfaces, causing high-frequency vehicle vibrations, whereas it is in the low- wavenumber range on bituminous surfaces, causing greater excitation in the low-frequency range.

The PSD for average road properties shown in the figure can be represented by the equation as follows:

$$G_z(v) = G_0[1+(v_0/v)^2]/(2\pi v)^2 \qquad (5\text{-}1)$$

where:

$G_z(v) =$ PSD amplitude (feet^2/cycle/foot)

v $\quad =$ Wavenumber (cycles/ft)

G_o $\quad =$ Roughness magnitude parameter (roughness level)

$\quad\quad = 1.25 \times 10^5$ for rough roads

$\quad\quad = 1.25 \times 10^6$ for smooth roads

v_o $\quad =$ Cutoff wavenumber

$\quad\quad = .05$ cycle/foot for bituminous roads

$\quad\quad = .02$ cycle/foot for PCC roads

The above equation in combination with a random number sequence provides a very useful method to generate road profiles with random roughness having the spectral qualities of typical roads [4] for study of vehicle ride dynamic behavior.

As described above, the roughness in a road is the deviation in elevation seen by a vehicle as it moves along the road. That is, the roughness acts as a vertical displacement input to the wheels, thus exciting ride vibrations. Yet the most common and meaningful measure of ride vibration is the acceleration produced. Therefore, for the purpose of understanding the dynamics of ride, the roughness should be viewed as an acceleration input at the wheels, in which case a much different picture emerges. Two steps are involved. First a speed of travel must be assumed such that the elevation profile is transformed to displacement as a function of time. Thence, it may be differentiated once to obtain the velocity of the input at the wheels, and a second time to obtain an acceleration. Figure 5.3 shows the transformation of road profile elevation to a velocity and then acceleration input to a vehicle. A vehicle speed of 50 mph has been assumed. The conversion from spatial frequency (cycles/foot) to temporal frequency (cycles/second or Hz) is obtained by multiplying the wavenumber by the vehicle speed in feet/second.

Note that the acceleration spectrum has a relatively constant amplitude at low frequency, but begins increasing rapidly above 1 Hz such that it is an order of magnitude greater at 10 Hz. Viewed as an acceleration input, road roughness presents its largest inputs to the vehicle at high frequency, and thus has the greatest potential to excite high-frequency ride vibrations unless attenuated accordingly by the dynamic properties of the vehicle. As will be seen, the vehicle's attenuation of this high-frequency input is an important aspect of the "ride isolation" behavior obtained via the primary suspension commonly used on highway vehicles today.

129

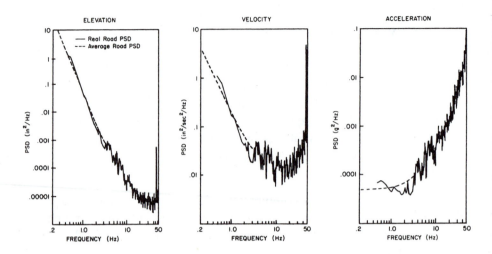

Fig. 5.3 Elevation, velocity and acceleration PSDs of road roughness input to a vehicle traveling at 50 mph on a real and average road.

By viewing road roughness as an acceleration input the primary effect of travel speed can be readily seen. At any given temporal frequency the amplitude of the acceleration input will increase with the square of the speed. This is illustrated by considering a simple sine wave representation of roughness. Then:

$$Z_r = A \sin (2\pi v X) \tag{5-2}$$

where:

Z_r = Profile elevation
A = Sine wave amplitude
v = Wavenumber (cycles/foot)
X = Distance along the road

Now, because the distance, X, equals the velocity, V, times the time of travel, t:

$$Z_r = A \sin (2\pi v V t) \tag{5-3}$$

Differentiating twice to obtain acceleration produces:

$$\ddot{Z_r} = - (2\pi v V)^2 A \sin (2\pi v V t) \tag{5-4}$$

Thus as an acceleration the amplitude coefficient contains the velocity squared. In general, increasing the assumed speed in Figure 5.3 will cause the acceleration plot to shift upward due to the speed-squared effect just described. It may also be noted that the curve would shift to the left slightly because of the corresponding change in the temporal frequency represented by each roughness wavenumber. This also adds to the acceleration amplitude, although not as strongly as the speed-squared effect.

Thus far the road roughness input has been considered only as a vertical input to the vehicle that would excite bounce and pitch motions. For this purpose the road profile points in the left and right wheeltracks are usually averaged before processing to obtain the PSD, although the PSDs of either wheeltrack will usually appear quite similar to that of the average. The difference in elevation between the left and right road profile points represents a roll excitation input to the vehicle. The PSD for the roll displacement input to a vehicle is typically similar to that for the elevation as was shown in Figure 5.2, although its amplitude is attenuated at wavenumbers below 0.02 to 0.03 cycle/foot. Typical roll excitation characteristics of road roughness are more readily seen by normalizing the roll amplitude (difference between the wheeltracks) by the vertical amplitude (average of the wheeltracks) in each wavenumber band of the PSD, so that roll excitation is seen in relationship to the vertical excitation present in the road.

The PSDs obtained have the characteristics as shown in Figure 5.4. At low wavenumbers (long wavelengths) the roll input from a roadway is much lower in its relative magnitude than that of the vertical input to the vehicle, because the difference in elevation is constrained by requirements to maintain the superelevation of the road.

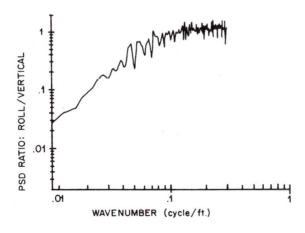

Fig. 5.4 Spectral density of normalized roll input for a typical road.

131

However, the normalized roll input magnitude grows with wavenumber because of the natural tendency for bumps in the left and right wheeltracks to become less and less correlated at high wavenumbers (short wavelengths). For most vehicles, resonance in roll occurs at a lower frequency (0.5 to 1.0 Hz) than resonance in bounce. Thus of the two, bounce is the more dominant response. At higher frequency, where the bounce and roll inputs are more nearly equal in magnitude, vehicles are less responsive to roll.

As an illustration, consider a vehicle with a roll natural frequency of 1.0 Hz traveling at 60 mph (88 ft/sec). Roll excitation in the road at the 88 ft wavelength (0.011 cycle/ft) will therefore directly excite roll motions. However, the roll amplitude at this wavenumber is only 10 percent of the vertical input, so the vehicle passengers will be more conscious of bounce vibrations rather than roll.

At a low speed, for example at 6 mph, a 1.0 Hz roll resonant frequency would be excited by input from wavenumbers on the order of 0.1 cycle/ft at which the roll and vertical inputs are essentially equal in magnitude. Thus roll and bounce motions would be approximately equal as well. The common case where this is observed is in off-road operation of 4x4 vehicles where the exaggerated ride vibrations are often composed of roll as well as bounce motions.

Tire/Wheel Assembly

Ideally, the tire/wheel assembly is soft and compliant in order to absorb road bumps as part of the ride isolation system. At the same time, it ideally runs true without contributing any excitation to the vehicle. In practice, the imperfections in the manufacture of tires, wheels, hubs, brakes and other parts of the rotating assembly may result in nonuniformities of three major types:

1) Mass imbalance

2) Dimensional variations

3) Stiffness variations

These nonuniformities all combine in a tire/wheel assembly causing it to experience variations in the forces and moments at the ground as it rolls [10]. These in turn are transmitted to the axle of the vehicle and act as excitation sources for ride vibrations [11]. The force variations may be in the vertical (radial) direction, longitudinal (tractive) direction, or the lateral direction [12].

The moment variations in the directions of the overturning moment, aligning torque, and rolling resistance moment generally are not significant as sources of ride excitation, although they can contribute to steering system vibrations.

Imbalance derives from a nonuniform distribution of mass in the individual components of the assembly along or about the axis of rotation [14]. Asymmetry about the axis of rotation is observed as static imbalance. The resultant effect is a force rotating in the wheel plane with a magnitude proportional to the imbalance mass, the radius from the center of rotation, and the square of the rotational speed. Because it is rotating in the wheel plane, this force produces both radial as well as longitudinal excitations. The imbalance force is given by the equation:

$$F_i = (m\ r)\ \omega^2 \qquad\qquad (5\text{-}5)$$

where:

F_i = Imbalance force

m r = The imbalance magnitude (mass times radius)

ω = the rotational speed (radians/second)

A nonuniform and asymmetric mass distribution along the axis of rotation causes a dynamic imbalance [14]. Dynamic imbalance creates a rotating torque on the wheel, appearing as variations in overturning moment and aligning torque at the wheel rotational frequency. Dynamic imbalance is most important on steered wheels which may experience steering vibrations as a result of the excitation. Static imbalance can exist in the absence of dynamic imbalance, and vice versa. The tires, wheels, hubs and brake drums may all contribute to imbalance effects.

The tire, being an elastic body analogous to an array of radial springs, may exhibit variations in stiffness about its circumference. Figure 5.5 illustrates that tire model. The free length of the springs establishes the dimensional nonuniformities (free radial runout), yet the variations in their compressed length at a nominal load determines the rolling nonuniformities (loaded radial runout).

Dimensional runouts in the wheel or hub on which the tire is mounted do not produce stiffness variations directly, but may contribute to the free- or loaded-radial runouts that are observed.

The significant effect of the nonuniformities in a tire/wheel assembly is the generation of excitation forces and displacements at the axle of the vehicle as

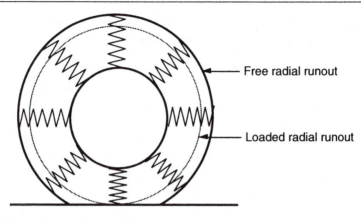

Fig. 5.5 Tire radial spring model.

the wheel rotates. The excitation force observed when the wheel is rolled at constant radius and speed repeats with each revolution of the wheel [15].

Radial force variations measured at constant radius typically take the form illustrated in Figure 5.6. The peak-to-peak magnitude of force variation is called "composite force variation." The force signature may be described in more detail by the amplitude of the harmonics of which it is composed. That is, by a Fourier transform [1], the composition of the signal as a series of sine waves at the fundamental and each multiple frequency can be determined. The amplitude of each harmonic is usually the parameter of primary interest. Although the phase angle of each must also be known to reconstruct the original signal, phase angle information appears to have little relationship to ride phenomena [16].

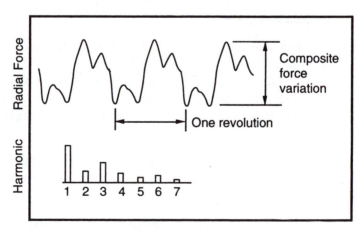

Fig. 5.6 Tire radial force variations.

The first harmonic of the radial force variation tends to be less than that of the composite, and the higher-order harmonics tend to be of diminishing magnitude. For passenger-car tires a decrease of about 30 percent per order has been observed [15], with less of an effect at high speed. Runout of the hub and wheel may also contribute to the radial force variations. The runout may be quantified by finding the point-by-point average radius of the two bead seats around the circumference of the assembled wheel. The first harmonic force variation arising from this source is closely linked to the runout. At first thought, the force variation would be expected to be the runout times the stiffness of the tire mounted on the wheel. Experiments, however, have shown that the force variation is only about 70 percent of that magnitude, indicating that the tire partially masks runout of the wheel [17]. Higher harmonic runouts in the wheel are not as closely related to radial force variations in the overall assembly.

The various harmonics of radial nonuniformities in a tire/wheel assembly are functionally equivalent to imperfections in the shape as shown in Figure 5.7.

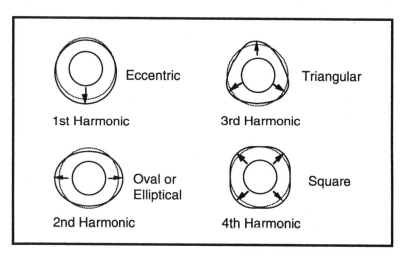

Fig. 5.7 Radial nonuniformities in a tire.

1) Eccentricity—The tires, wheels and hubs individually may exhibit radial eccentricity, resulting in a first-harmonic nonuniformity which produces both radial and tractive excitation on the axle. Since there is one high point and one low point on the assembly, the excitation occurs at the rotational speed of the wheel (10-15 Hz at normal highway speeds). The overall magnitude of the nonuniformity for the assembly depends on the magnitudes in the individual

components and their relative positions when assembled [11]. That is, eccentricity in one component may partially compensate for that in others when the high and low points of the different parts are matched in assembly. The "match-mounting" technique is commonly used in the tire/wheel assemblies for passenger cars to minimize first-harmonic nonuniformities of the assembly. In those cases, wheels may be purposely manufactured with an eccentricity equivalent to the average radial runout of the production tires with which they will be used. The tires and wheels are marked for high/low points to facilitate match mounting.

2) Ovality—Tires and wheels may have elliptical variations that add or subtract depending on the mounting positions [11], although match-mounting is not practical for minimizing this nonuniformity. Because the assembly has two high points and two low points on its circumference, radial and tractive force excitation is produced at twice the wheel rotational frequency (20-30 Hz at normal highway speeds).

3) Higher-order radial variations—Third- and higher-order variations are predominantly of importance in the tire only. Such variations in the wheel are substantially absorbed by the tire [13]. The third harmonic is analogous to a tire with a triangular shape, the fourth harmonic reflects a square shape, and so on. While tires do not purposely have these shapes, the effects may arise from construction methods. For example, in a tire with four plies of fabric material, the overlaps associated with each ply would normally be distributed around the circumference of the tire. The additional stiffness created at each of the overlap positions will then result in a fourth-harmonic stiffness variation and an associated fourth-harmonic force variation. The force variations act in the radial and tractive force directions at the multiple of wheel speed equal to the harmonic number.

Because the magnitude of the radial force variation is relatively independent of speed, low-speed measurements of radial force variations at a constant radius (the method commonly used by tire manufacturers to monitor production) indicate the magnitude of the force tending to excite ride vibrations. Only the frequency is changed with speed. The nonuniformity force can be treated as a direct excitation at the axle. As a point of clarity, it should be noted that the excitation force is not equivalent to the actual force variation experienced at the axle, as the dynamic response of the vehicle can greatly amplify the forces [10]. Nevertheless, the measurement of radial force variations described above is the proper and valid means to characterize the radial excitation potential associated with nonuniformities in the tire and wheel components. Alternatively, measurements of loaded radial runout are also valid and can be transformed to radial force variation by simply multiplying by the radial spring rate of the tire.

<u>Tractive force</u> variations arise from dimensional and stiffness nonuniformities as a result of two effects. The causes are best illustrated by considering a simple eccentric wheel model as shown in Figure 5.8. With eccentricity, even at low speed the axle must roll up and down the "hill" represented by the variation in radius of the wheel assembly. Thus a longitudinal force is involved and a tractive force variation is observed. Its magnitude will be dependent on the load carried and the amount of eccentricity [17]; however, it is independent of speed.

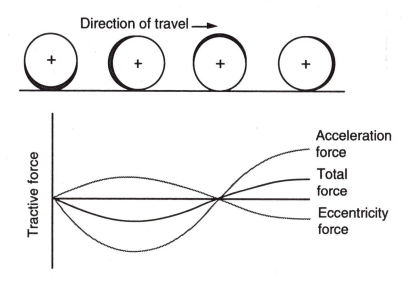

Fig. 5.8 Tractive force variations arising from an eccentric wheel.

On the other hand, at high speed the wheel must accelerate and decelerate in the course of a revolution because of its varying radius. Thus a tractive force at the ground, and accordingly at the hub, must appear in order to produce the acceleration. Quite reasonably, the magnitude of the force will be dependent on the longitudinal stiffness properties of the tire and the rotational moment of inertia of the wheel. Further, the magnitude of the tractive force arising from this mechanism will vary with speed because the acceleration varies with speed, typically increasing by a factor of about 5 over the speed range of 5 to 60 mph. Therefore, the tractive force variation in a tire or tire/wheel assembly can only be measured validly at high speed, and the rotational inertia properties of the wheel assembly must be closely matched to that of actual vehicles.

Lateral force variations may arise from nonuniformities in the tire, but cannot be readily related to lateral runout effects in the wheel or hub components. They tend to be independent of speed, thus measurements of the force magnitudes at low speed are also valid for high speed [11]. First-order lateral variations in the tires or wheels, or in the way in which they are mounted, will cause wobble. These will affect the dynamic balance of the assembly. The wobble in the wheel may contribute a minor lateral force variation, but may also result in radial and tractive force variations comparable to the effect of ovality [11] because the wheel is elliptical in the vertical plane.

Higher-order lateral variations are predominantly important in the tire only. Wheel variations are substantially absorbed by the tire [11,13]. These sources could potentially cause steering vibrations, but have not been identified as the cause of ride problems.

In general, the imperfections in tires and wheels tend to be highly correlated [18,19], such that radial variations are usually accompanied by imbalance and tractive force variations. Thus it may often be difficult to cure a tire-related ride problem simply by correcting one condition, such as imbalance, without consideration of the other nonuniformities likely to be present.

In order to complete the discussion of the tire/wheel assembly as a ride excitation source, it must be recognized that the assembly itself is a dynamic system influencing the excitation seen at the axle (see Chapter 10). The influence derives from the modal resonance properties of the assembly, which at the low-frequency end are dominated by the tire tread band resonances. The capacity to resonate makes the assembly a vibration absorber at certain modal frequencies, while it accentuates the transmissibility properties at the anti-resonance frequencies [20]. Ultimately, the tire can play a significant role in noise, vibration and harshness (NVH) of a motor vehicle, and in the ride development process the vehicle must be properly tuned to avoid various buzzes and booms that can be triggered by tire response.

Driveline Excitation

The third major source of excitation to the vehicle arises from the rotating driveline. While the driveline is often considered to be everything from the engine to the driven wheels, the engine/transmission package will be treated separately in the discussion.

For purposes of discussion, the driveline therefore consists of the driveshaft, gear reduction and differential in the drive axle, and axle shafts connecting to the wheels. Of these various components, the driveshaft with its spline and universal joints has the most potential for exciting ride vibrations. The rear axle gearing and remainder of the driveline are also capable of generating vibrations in the nature of noise as a result of gear mating reactions and torsional vibration along the drivetrain. However, these generally occur at frequencies above those considered as ride.

The most frequent ride excitations arise from the driveshaft. The driveshaft is normally arranged as shown in Figure 5.9 [21]. On rear-drive passenger cars and short-wheelbase trucks, a single-piece shaft is commonly used, whereas, on long-wheelbase trucks and buses a multiple-piece shaft supported by an intermediate bearing is frequently required. Excitations to the vehicle arise directly from two sources—mass imbalance of the driveshaft hardware, and secondary couples, or moments, imposed on the driveshaft due to angulation of the cross-type universal joints [22,23].

WHEELBASE

Short

Intermediate

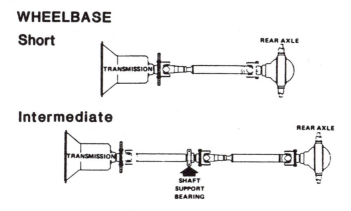

Fig. 5.9 Typical driveline arrangements [39].

Mass Imbalance—Imbalance of the driveshaft may result from the combination of any of the five following factors:

1) Asymmetry of the rotating parts
2) The shaft may be off-center on its supporting flange or end yoke
3) The shaft may not be straight
4) Running clearances may allow the shaft to run off center
5) The shaft is an elastic member and may deflect

An initial imbalance exists as a result of the asymmetry, runouts and looseness in the structure. The imbalance creates a rotating force vector thus imposing forces on the support means in both the vertical and lateral directions. Forces at the front support apply to the transmission. Those at the back exert on the drive axle directly. Where intermediate bearings are employed on trucks, forces may be imposed on the frame via the crossmembers at those points. The force rotates at the speed of the shaft which is always the wheel speed multiplied by the numerical ratio of the final drive, and is equivalent to engine speed when in direct drive. Thus it appears like a harmonic of the wheel of a value equal to the numerical ratio of the rear axle.

In general, the magnitude of the excitation force is equivalent to the product of the imbalance and the square of the rotating speed. However, because the shaft is elastic it may bend in response to the imbalance force allowing additional asymmetry, and an increase in the "dynamic" imbalance. As a result, the apparent magnitude of the imbalance may change with speed, and in theory, the shaft can only be dynamically balanced for one speed [23].

Secondary couples—The use of universal joints in a driveline opens the way for generation of ride excitation forces when they are operated at an angle, due to the secondary couple that is produced. The magnitude and direction of the secondary couple can be determined by a simple vector summation of torques on the universal joint as is illustrated in Figure 5.10.

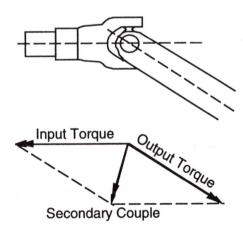

Fig. 5.10 Torque reactions causing a secondary couple.

The magnitude of the secondary torque is proportional to the torque applied to the driveline and the angle of the universal joint. When the torque

140

varies during rotation due to engine torque pulsations and/or nonconstant-velocity joints, the secondary couple will vary accordingly. The secondary couple reacts as forces at the support points of the driveline on the transmission, crossmembers supporting the driveline intermediate bearings, and at the rear axle. Hence, these forces vary with driveline rotation and impose excitation forces on the vehicle [22].

When cross-type (Cardan or Hooke) universal joints are used and operate at an angle, they are a direct source of torque pulsations in the driveline [40]. The joints do not have a constant relationship between input and output velocity when operated at an angle, but must satisfy the equation:

$$\frac{\omega_o}{\omega_i} = \frac{\cos \theta}{1 - \sin^2\beta \, \sin^2\theta} \tag{5-6}$$

where:

ω_o = Output speed
ω_i = Input speed
θ = Angle of the U-joint
β = Angle of rotation of the driving yoke

Because of the "$\sin^2\beta$" term in the denominator, the speed variation reaches a maximum twice per revolution (at 90 and 270 degrees). Thus a second-harmonic speed variation occurs as a result of the symmetry of the cross around each arm. It can be shown from the above equation that the maximum speed variation changes with joint angle as:

$$\left|\frac{\omega_o}{\omega_i}\right|_{max} = \frac{1}{\cos \theta} \tag{5-7}$$

Because of the stiffness of the driveline and the accelerations it experiences, torque variations will necessarily arise from the speed variation. These may cause excitation of torsional vibrations in the driveline, as well as be the source of ride excitation forces on the vehicle. The excitation occurs at the second harmonic of the driveline speed, and will vary with the level of torque applied to the driveline. Excitation from the secondary couple can be minimized by proper design of the driveline—maintaining parallel axes on the transmission output shaft and rear axle input shaft, proper phasing of the joints, and keeping angles within the limits recommended by the manufacturers [21, 22, 23].

The torque variations may also act directly at the transmission and the rear axle. Torque variations at the axle will vary the drive forces at the ground and thus may act directly to generate longitudinal vibrations in the vehicle. The torque variations at the transmission produce excitation in the roll direction on the engine/transmission assembly. In part, these variations must be reacted in the mounting points on the body and thus have a direct path to the interior.

Figure 5.11 illustrates the nature of the vibrations that may be produced as a result of driveline and tire/wheel nonuniformities. In this case the accelerations were measured in a truck cab under carefully controlled conditions (i.e., the vehicle was operated on a smooth road to suppress background vibrations

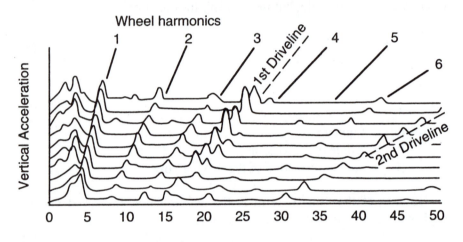

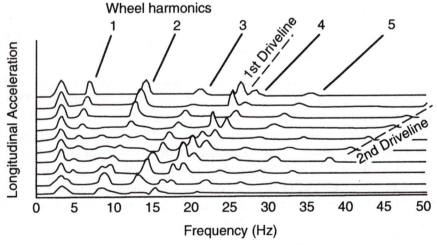

Fig. 5.11 Spectral map of vibrations arising from driveline and tire/wheel nonuniformities.

that would mask the desired effects and all of the tire/wheel assemblies, except for one, were carefully match-mounted to achieve consistency in the tire/wheel excitation).

The figure shows a map of the vibration spectra measured at different vehicle speeds. Excitation from tire/wheel inputs appear as ridges in the spectra moving to higher frequency as the speed increases. The first, second and higher harmonics of the tire/wheel assemblies are evident in the spectra. The ridge at 3.7 times the wheel rotational speed corresponds to first harmonic of the driveline, which is due to imbalance of the driveshaft and other components rotating at this speed. The second harmonic of the driveline at 7.4 times the wheel speed is the result of torque variations in the driveshaft that arise from speed variations caused by the operating angles of the cross-type universal joints.

These or comparable vibrations from the wheels and drivelines will always be present on a vehicle, but are often difficult to recognize in a complex spectrum that includes substantial road roughness excitation. Nevertheless, they constitute one of the factors contributing to the overall ride vibration spectrum and represent one of the areas in which careful design can improve the ride environment of the vehicle.

Engine/Transmission

The engine serves as the primary power source on a vehicle. The fact that it rotates and delivers torque to the driveline opens the possibility that it may be a source for vibration excitation on the vehicle. Further, the mass of the engine in combination with that of the transmission is a substantial part in the chassis, and, if used correctly, can act as a vibration absorber.

Piston engines deliver power by a cyclic process; thus the torque delivered by the engine is not constant in magnitude [24]. At the crankshaft the torque delivered consists of a series of pulses corresponding to each power stroke of a cylinder (see Figure 5.12). The flywheel acts as an inertial damper along with the inertias and compliances in the transmission. Thus the torque output to the driveshaft consists of a steady-state component plus superimposed torque variations. Those torque variations acting through the driveline may result in excitation forces on the vehicle similar to those produced by the secondary couple from the cross-type universal joints explained in the previous sections.

Because of compliance in the engine/transmission mounts, the system will vibrate in six directions—the three translational directions and three rotations around the translational axes. The axis system for a transverse front-engine configuration is shown in Figure 5.13. The figure also shows a three-point mount typical of those used with most transverse engines today.

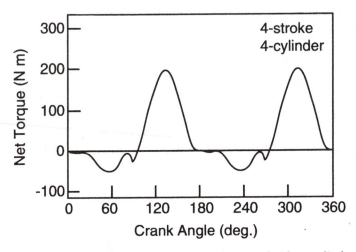

Fig. 5.12 Torque variations at the output of a four-stroke, four-cylinder engine.

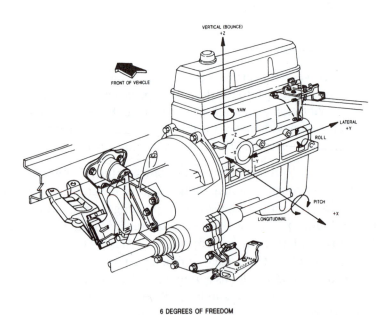

Fig. 5.13 Typical transverse engine and mounting hardware. (Photo courtesy of Ford Motor Company.)

144

Of all the directions of motion, the most important to vibration is the engine roll direction (about the lateral axis of a transverse engine or about the longitudinal axis of an engine mounted in the north-south direction), which is excited by drive torque oscillations. Torque oscillations occur at the engine firing frequency as well as at sub-harmonics of that frequency due to cylinder-to-cylinder variations in the torque.

A key to isolating these excitations from the vehicle body is to design a mounting system with a roll axis that aligns with the engine inertial roll axis, and provide a resonance about this axis at a frequency that is below the lowest firing frequency. By so doing, torque variations which occur above the resonant frequency are attenuated. In effect, the torque is absorbed in the inertial motion of the engine rather than being transmitted to the vehicle body.

The engine inertial axis of four-cylinder engines will generally be inclined downward toward the transmission because of the contribution from the mass of the transmission. Thus the mounting system must be low at the transmission end and high at the front of the engine. On V-type engines (six- and eight-cylinder) the inertial axis is lower in the front permitting a mounting system more closely aligned with the crankshaft.

The worst-case problem is isolation of idle speed torque variations for a four-cylinder engine with the transmission in drive, which may have a firing frequency of 20 Hz or below. Therefore, successful isolation requires a roll axis resonance of 10 Hz or below. Because the system acts like a simple second-order mass-spring dynamic system [14], torque variations at frequencies below the resonant frequency will be felt directly at the mounts, and near the torsional resonant frequency, excitation amplitudes much greater than the torque variations themselves will occur.

Engines may produce forces and moments in directions other than roll as a result of the inherent imbalances in the reciprocating/rotating masses [24]. These take the form of forces or couples at the engine rotational frequency or its second harmonic, and must be isolated in the same manner as for the roll mode described above. For the more commonly used engine configurations the balance conditions are as follows:

1) Four-cylinder inline—Vertical force at twice engine rotational frequency; can be balanced with counter-rotating shafts.

2) Four-cylinder, opposed, flat—Various forces and moments at rotational frequency and twice rotational frequency depending on crankshaft arrangement.

3) Six-cylinder inline—Inherently balanced in all directions.

4) Six-cylinder inline, two-cycle—Vertical couple generating yaw and pitch moments at the engine rotational frequency; can be balanced.

5) Six-cylinder, 60-degree V—Generates a counter-rotating couple at rotational frequency that can be balanced with counter-rotating shaft.

6) Six-cylinder, 90-degree V (uneven firing)—Generates yaw moment of twice rotational frequency; can be balanced with counter-rotating shaft.

7) Six-cylinder, 90-degree V (even firing)—Generates yaw and pitching moments at crankshaft speed, which can be balanced. Also generates complex yaw and pitching moments at twice rotational speed which are difficult to balance.

8) Eight-cylinder inline—Inherently balanced in all directions.

9) Eight-cylinder, 90-degree V—Couple at primary rotational speed; can be counter-balanced.

With proper design of the mounting system the mass of the engine-transmission combination can be utilized as a vibration absorber attenuating other vibrations to which the vehicle is prone. Most often it is used to control vertical shake vibrations arising from the wheel excitations. For this purpose the mounting system is designed to provide a vertical resonance frequency near that of the front wheel hop frequency (12-15 Hz), so that the engine can act as a vibration damper for this mode of vehicle vibration.

VEHICLE RESPONSE PROPERTIES

A systematic treatment of the vehicle as a dynamic system best starts with the basic properties of a vehicle on its suspension system—i.e., the motions of the body and axles. At low frequencies the body, which is considered to be the sprung mass portion of the vehicle, moves as an integral unit on the suspensions. This is rigid-body motion. The axles and associated wheel hardware, which form the unsprung masses, also move as rigid bodies and consequently impose excitation forces on the sprung mass. Beyond this, one must look to structural modes of vibration and resonances of sub-systems on the vehicle. In addition, there are many individual variables of design and operating condition that are known to affect the vibration response on the vehicle.

146

The dynamic behavior of a vehicle can be characterized most meaningfully by considering the input-output relationships. The input may be any of the excitations discussed in the preceding section, or combinations thereof. The output most commonly of interest will be the vibrations on the body. The ratio of output and input amplitudes represents a "gain" for the dynamic system. The term "transmissibility" is often used to denote the gain. Transmissibility is the nondimensional ratio of response amplitude to excitation amplitude for a system in steady-state forced vibration. The ratio may be one of forces, displacements, velocities or accelerations. The magnitude ratio of the transfer function is sometimes used in a similar fashion to denote the gain, although this term is normally reserved for use with linear systems.

Suspension Isolation

At the most basic level, all highway vehicles share the "ride isolation" properties common to a sprung mass supported by primary suspension systems at each wheel. The dynamic behavior of this system is the first level of isolation from the roughness of the road. The essential dynamics can be represented by a quarter-car model, as shown in Figure 5.14.

It consists of a sprung mass supported on a primary suspension, which in turn is connected to the unsprung mass of the axle. The suspension has stiffness and damping properties. The tire is represented as a simple spring, although a damper is often included to represent the small amount of damping inherent to the visco-elastic nature of the tire [25]. A more detailed discussion of this model is provided in the SAE Ride and Vibration Data Manual [26].

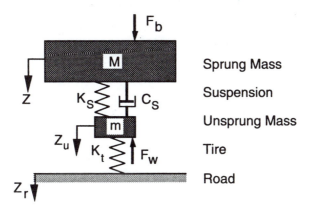

Fig. 5.14 Quarter-car model.

147

The sprung mass resting on the suspension and tire springs is capable of motion in the vertical direction. The effective stiffness of the suspension and tire springs in series is called the "ride rate" determined as follows:

$$RR = \frac{K_s K_t}{K_s + K_t} \qquad (5\text{-}8)$$

where:

> RR = Ride rate
> K_s = Suspension stiffness
> K_t = Tire stiffness

In the absence of damping, the bounce natural frequency at each corner of the vehicle can be determined from:

$$\omega_n = \sqrt{\frac{RR}{M}} \qquad \text{(radians/sec)} \qquad (5\text{-}9a)$$

or:

$$f_n = 0.159 \sqrt{\frac{RR}{W/g}} \qquad \text{(cycles/sec)} \qquad (5\text{-}9b)$$

where:

> M = Sprung mass
> $W = M g$ = Weight of the sprung mass
> g = Acceleration of gravity

When damping is present, as it is in the suspension, the resonance occurs at the "damped natural frequency," ω_d, given by:

$$\omega_d = \omega_n \sqrt{1 - \zeta_s^2} \qquad (5\text{-}10)$$

where:

> ζ_s = Damping ratio
> $= C_s / \sqrt{4 K_s M} \qquad (5\text{-}11)$
> C_s = Suspension damping coefficient

For good ride the suspension damping ratio on modern passenger cars usually falls between 0.2 and 0.4. Because of the way damping influences the resonant frequency in the equation above (i.e., under the square root sign), it is usually quite close to the natural frequency. With a damping ratio of 0.2, the damped natural frequency is 98% of the undamped natural frequency, and even at 0.4 damping, the ratio is about 92%. Because there is so little difference the undamped natural frequency, ω_n, is commonly used to characterize the vehicle.

The ratio of W/K_s represents the static deflection of the suspension due to the weight of the vehicle. Because the "static deflection" predominates in determining the natural frequency, it is a straightforward and simple parameter indicative of the lower bound on the isolation of a system. Figure 5.15 provides a nomograph relating the natural frequency to static deflection.

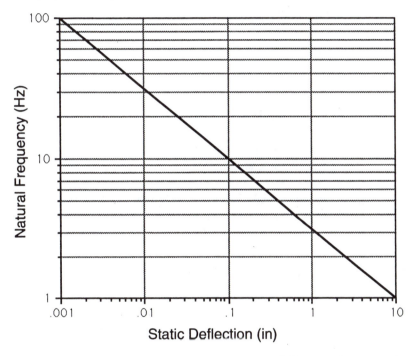

Fig. 5.15 Undamped natural frequency versus static deflection of a suspension.

A static deflection of 10 inches (254 mm) is necessary to achieve a 1 Hz natural frequency—considered to be a design optimum for highway vehicles. A 5-inch (127-mm) deflection results in a 1.4 Hz frequency, and 1 inch (25 mm) equates to a 3.13 Hz frequency. While it is not necessary for the suspension to provide a full 10 inches (254 mm) of travel to achieve the 1 Hz frequency, in general, provisions for larger deflections are necessary with lower frequencies. For example, with a spring rate low enough to yield a 1 Hz frequency, at least

5 inches of stroke must be available in order to absorb a bump acceleration of one-half "g" without hitting the suspension stops. Most large cars have a usable suspension stroke in the range of 7 to 8 inches. On small, compact cars the stroke may be reduced to 5 or 6 inches.

The dynamic behavior for the complete quarter-car model in steady-state vibration can be obtained by writing Newton's Second Law for the sprung and unsprung masses. By considering a free-body diagram for each, the following differential equations are obtained for the sprung and unsprung masses, respectively:

$$M \ddot{Z} + C_s \dot{Z} + K_s Z = C_s \dot{Z}_u + K_s Z_u + F_b \tag{5-12}$$

$$m \ddot{Z}_u + C_s \dot{Z}_u + (K_s + K_t) Z_u = C_s \dot{Z} + K_s Z + K_t Z_r + F_w \tag{5-13}$$

where:

$$
\begin{aligned}
Z &= \text{Sprung mass displacement} \\
Z_u &= \text{Unsprung mass displacement} \\
Z_r &= \text{Road displacement} \\
F_b &= \text{Force on the sprung mass} \\
F_w &= \text{Force on the unsprung mass}
\end{aligned}
$$

While the two equations make solving more complicated, closed-form solutions can be obtained for the steady-state harmonic motion by methods found in classical texts. The solutions of primary interest are those for the sprung mass motion in response to road displacement inputs, forces at the axle, and forces applied directly to the sprung mass. The amplitude ratios for these cases are as follows:

$$\frac{\ddot{Z}}{\ddot{Z}_r} = \frac{K_1 K_2 + j [K_1 C \omega]}{[\chi \omega^4 - (K_1 + K_2 \chi + K_2) \omega^2 + K_1 K_2] + j [K_1 C \omega - (1 + \chi) C \omega^3]} \tag{5-14}$$

$$\frac{\ddot{Z}}{F_w / M} = \frac{K_2 \omega^2 + j [C \omega^3]}{[\chi \omega^4 - (K_1 + K_2 \chi + K_2) \omega^2 + K_1 K_2] + j [K_1 C \omega - (1 + \chi) C \omega^3]} \tag{5-15}$$

$$\frac{\ddot{Z}}{F_b / M} = \frac{[\chi \omega^4 - (K_1 + K_2) \omega^2] + j [C \omega^3]}{[\chi \omega^4 - (K_1 + K_2 \chi + K_2) \omega^2 + K_1 K_2] + j [K_1 C \omega - (1 + \chi) C \omega^3]} \tag{5-16}$$

where:

$$\chi = m/M = \text{Ratio of unsprung to sprung mass}$$
$$C = C_s/M$$
$$K_1 = K_t/M$$
$$K_2 = K_s/M$$
$$j = \text{Complex operator}$$

The equations above are complex in form consisting of real and imaginary components, the latter denoted by the "**j**" operator. To obtain the amplitude ratios, the real and imaginary parts of the numerators and denominators must be evaluated at the frequency of interest. The magnitude of the numerator is then determined by taking the square root of the sum of the squares of the real and imaginary parts. The denominator magnitude is determined similarly, and then the ratio of the two may be taken. With appropriate manipulation, the phase angle of the equations may also be determined.

The quarter-car model is limited to study of dynamic behavior in the vertical direction only. Yet, using equations such as those developed above, it can be used to examine vibrations produced on the sprung mass as a result of inputs from road roughness, radial forces arising from tire/wheel nonuniformities, or vertical forces applied directly to the sprung mass from on-board sources. The response properties can be presented by examining the response gain as a function of frequency, as shown in Figure 5.16. The gain is defined differently for each type of excitation input.

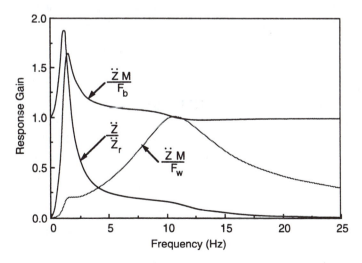

Fig. 5.16 Quarter-car response to road, tire/wheel, and body inputs.

For road roughness input, the gain is the ratio of sprung-mass motion (acceleration, velocity, or displacement) to the equivalent input from the road. At very low frequency the gain is unity (the sprung mass moves in exact duplication of the road input). By classical design of motor vehicles, the sprung mass is chosen to have its natural frequency at or just above 1 Hz. Therefore, at frequencies near 1 Hz the sprung mass is resonating on the suspension and the road inputs are amplified. The amplitude ratio at this peak is very sensitive to damping level, and on typical passenger cars will be in the range of 1.5 to 3. For typical heavy trucks the amplitude ratio is dependent on the road and operating conditions, but in the worst case may reach levels as high as 5 or 6 [27]. Above resonance, the road inputs are increasingly attenuated. In the range of 10 to 12 Hz, the unsprung mass of the tire/wheel assembly goes into a vertical (hop) resonance mode, adding a small bump to the attenuation curve in this region.

The sprung-mass response to tire/wheel excitation is illustrated by choosing an appropriate nondimensional expression for gain of the system. The input is an excitation force at the axle due to the tire/wheel assembly. The output—acceleration of the sprung mass—may be transformed to a force by multiplying by the mass. Thence, the output is the equivalent force on the sprung mass necessary to produce the accelerations. The gain is zero at zero frequency because the force on the axle is absorbed within the tire spring and no sprung-mass acceleration is produced. It rises with frequency through the 1 Hz sprung-mass resonance, but continues to climb until wheel resonance occurs in the 10 to 12 Hz range. Only then does it diminish. This plot tells much about the sensitivity to radial force variations in tires and wheel components that should be expected with conventional motor vehicles. In particular, it illustrates that vehicles will tend to be most responsive to excitation from tire and wheel nonuniformities acting near the resonant frequency of the wheel, and at that frequency the nonuniformity force is transmitted directly to the sprung mass (response gain of unity).

The response gain for direct force excitation on the sprung mass may be expressed nondimensionally by again using the equivalent force on the sprung mass as the output. The response is similar in this case, but shows greater dominance of the sprung-mass resonance. At high frequencies the gain approaches unity because the displacements become so small that suspension forces no longer change and the force is entirely dissipated as acceleration of the sprung mass. By implication, virtually all extraneous forces coming into the body of a vehicle are detrimental to ride vibrations.

The basic isolation properties inherent to a quarter-car model combined with typical spectra of road roughness provide a first picture of the general nature of the ride acceleration spectrum that should be expected on a motor vehicle because of road inputs. The sprung-mass acceleration spectrum can be calculated for a linear model by multiplying the road spectrum by the square of the transfer function. That is:

$$G_{zs}(f) = |H_v(f)|^2 \, G_{zr} \qquad\qquad (5\text{-}17)$$

where:

$G_{zs}(f)$ = Acceleration PSD on the sprung mass
$H_v(f)$ = Response gain for road input
G_{zr} = Acceleration PSD of the road input

The results obtained are illustrated in Figure 5.17. While the road represents an input of acceleration amplitude which grows with frequency, the isolation properties of the suspension system compensate by a decrease in the vehicle's response gain.

The net result is an acceleration spectrum on the vehicle with a high amplitude at the sprung-mass resonant frequency, moderate attenuation out through the resonant frequency of the wheel, and a rapid attenuation thereafter. Note that, even though the road input amplitude increases with frequency, the acceleration response on the vehicle is qualitatively similar to the vehicle's response gain. Thus the acceleration spectrum seen on a vehicle does provide some idea of the response gain of the system even when the exact properties of the road are not known.

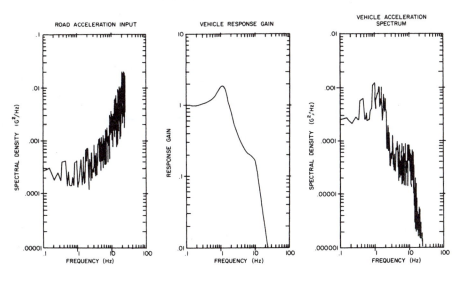

Fig. 5.17 Isolation of road acceleration by a quarter-vehicle model.

Example Problem

Determine the front and rear suspension ride rates for a 5.0 L Mustang given that the tire spring rate is 1198 lb/in. The front suspension rate is 143 lb/in and the rear is 100 lb/in. Also estimate the natural frequencies of the two suspensions when the front tires are loaded to 957 lb and the rear tires are at 730 lb each.

Solution:

The ride rates can be calculated using Eq. (5-8):

$$RR_f = \frac{K_s K_t}{(K_s + K_t)} = \frac{(143)(1198)}{(143 + 1198)} = 127 \frac{lb}{in}$$

$$RR_r = \frac{K_s K_t}{(K_s + K_t)} = \frac{(100)(1198)}{(100 + 1198)} = 92.3 \frac{lb}{in}$$

The natural frequencies for the suspensions can be determined from Eq. (5-9b).

$$f_{nf} = 0.159 \sqrt{\frac{RR_f \, g}{W}} = 0.159 \sqrt{\frac{127 \text{ lb/in} \times 386 \text{ in/sec}^2}{957 \text{ lb}}} = 1.14 \text{ Hz}$$

$$f_{nr} = 0.159 \sqrt{\frac{RR_r \, g}{W}} = 0.159 \sqrt{\frac{92.3 \text{ lb/in} \times 386 \text{ in/sec}^2}{730 \text{ lb}}} = 1.11 \text{ Hz}$$

Suspension Stiffness

Because the suspension spring is in series with a relatively stiff tire spring, the suspension spring predominates in establishing the ride rate and, hence, the natural frequency of the system in the bounce (vertical) mode. Since road acceleration inputs increase in amplitude at higher frequencies, the best isolation is achieved by keeping the natural frequency as low as possible. For a vehicle with a given weight, it is therefore desirable to use the lowest practical suspension spring rate to minimize the natural frequency.

The effect on accelerations transmitted to the sprung mass can be estimated analytically by approximating the road acceleration input as a function that increases with the square of the frequency. Then the mean-square acceleration can be calculated from Eq. (5-17) as a function of frequency. Figure 5.18 shows the acceleration spectra thus calculated for a quarter-car model in which the suspension spring rate has been varied to achieve natural frequencies in the range of 1 to 2 Hz. Because it is plotted on a linear scale, the area under the curves indicates the relative level of mean-square acceleration over the frequency range shown.

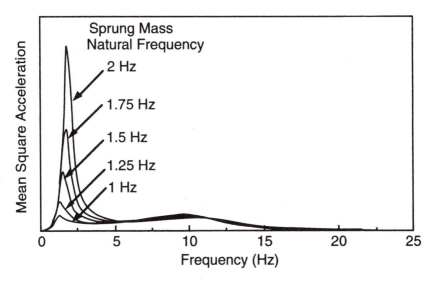

Fig. 5.18 On-road acceleration spectra with different sprung mass natural frequencies.

The lowest acceleration occurs at the natural frequency of 1 Hz. At higher values of natural frequency (stiffer suspension springs), the acceleration peak in the 1 to 5 Hz range increases, reflecting a greater transmission of road acceleration inputs, and the mean square acceleration increases by several hundred percent. In addition, the stiffer springs elevate the natural frequency of the wheel hop mode near 10 Hz, allowing more acceleration transmission in the high-frequency range.

While this analysis clearly shows the benefits of keeping the suspension soft for ride isolation, the practical limits of stroke that can be accommodated within a given vehicle size and suspension envelope constrain the natural frequency for most cars to a minimum in the 1 to 1.5 Hz range. Performance cars on which ride is sacrificed for the handling benefits of a stiff suspension will have natural frequencies up to 2 or 2.5 Hz.

155

Suspension Damping

Damping in suspensions comes primarily from the action of hydraulic shock absorbers. Contrary to their name, they do not absorb the shock from road bumps. Rather the suspension absorbs the shock and the shock absorber's function then is to dissipate the energy put into the system by the bump.

The nominal effect of damping is illustrated for the quarter-car model by the response gains shown in Figure 5.19. The percent damping is determined from the damping ratio given in Eq. (5-11). At very light damping (10%) the response is dominated by a very high response at 1 Hz. This type of response, often referred to as "float," causes the sprung mass to amplify long undulations in the roadway. While this is undesirable, benefit is obtained at all frequencies above the resonant point as a result of the high attenuation achieved.

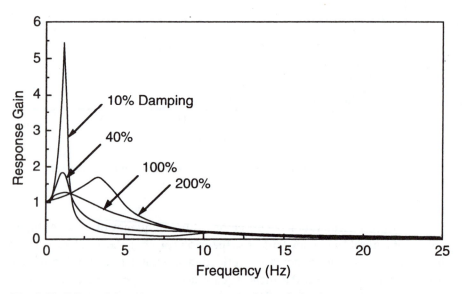

Fig. 5.19 Effect of damping on suspension isolation behavior.

The 40% damping ratio curve is reasonably representative of most cars, recognizable by the fact that the amplification at the resonant frequency is in the range of 1.5 to 2.0. At 100% (critical) damping, the 1 Hz bounce motions of the sprung mass are well controlled, but with penalties in the isolation at higher frequencies. If damping is pushed beyond the critical, for example to 200%, the damper becomes so stiff that the suspension no longer moves and the vehicle bounces on its tires, resonating in the 3 to 4 Hz range.

While this analytical treatment provides a simplified illustration of the ride effect of damping in the suspension, the tailoring of shock absorbers to achieve optimum performance is much more complicated in the modern automobile. Shock absorbers must be tailored not only to achieve the desired ride characteristics, but also play a key role in keeping good tire-to-road contact essential for handling and safety. In general, this is achieved by choosing the "valving" in the shock so that it is not a simple linear element (with force proportional to velocity) as has been assumed so far.

First, the damping in suspension jounce (compression) and rebound (extension) directions is not equal. Damping in the jounce direction adds to the force transmitted to the sprung mass when a wheel encounters a bump, thus it is undesirable to have high damping in this direction. On the other hand, damping in the rebound direction is desirable to dissipate the energy stored in the spring from the encounter with the bump. Consequently, typical shock absorbers are dual-rate with approximately a three-to-one ratio between rebound and jounce damping. Aside from this, the tuning characteristics described below are used for damping control in both directions.

Since the mid-1950s, telescopic shock absorbers have been used almost exclusively for damping in automotive suspensions. Telescopic shocks are a piston-in-tube arrangement with one end connected to the sprung mass and the other to the axle or wheel. There are two types of telescopic shock absorbers—the twin tube and the gas-pressurized monotube—as illustrated in Figure 5.20.

Each has its own advantages, but functionally they are similar. During compression and extension, the piston moves through the fluid in its bore. Valves in the piston restrict the flow of fluid through the piston creating the damping force. In the case of the twin tube shock, additional restriction from valving in the base of the tube may be used to further tailor the damping behavior.

Two types of valving may be used in combination to produce the desired characteristics. A simple orifice valve generates a damping force which grows with the square of the velocity as shown by curve A in Figure 5.21. When designed to provide adequate damping to control body motions at low velocities, simple orifice control yields too much damping at the high velocities typical of axle hop motions. A second type of valving is the "blow-off" valve in which the flow passage is blocked by a spring-loaded valve so that it prevents flow until a desired pressure is reached, at which point it allows "blow-off" with a damping force as shown in curve B. By combining orifice and blow-off controls in series and parallel arrangements, typical shock damping behavior such as that shown in curve C is obtained.

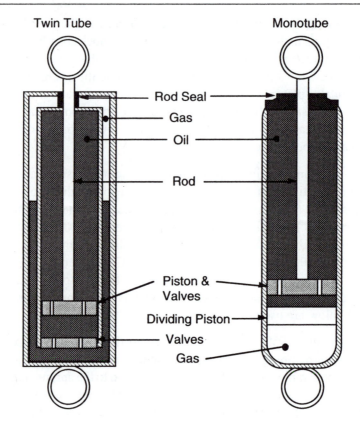

Fig. 5.20 Twin tube and gas-pressurized monotube shock absorbers.

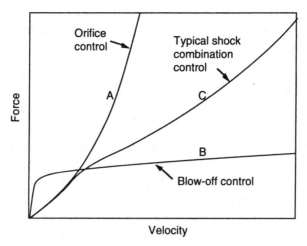

Fig. 5.21 Force-velocity properties of shock absorber valving.

For a comprehensive treatment of shock absorber damping in ride analysis, the shock must be modeled as a nonlinear element. The force-velocity characteristics like those shown in the preceding figure may be represented by polynomials or straight-line segments approximating the curve. Finally, the elastomer bushings in the end attachments of the shocks constitute a significant compliance in the system for small, high-frequency motions typical of axle, and should be taken into account.

Active Control

In the interest of improving the overall ride performance of automotive vehicles in recent years, suspensions incorporating active components have been developed. Most frequently, the active components are hydraulic cylinders that can exert forces in the suspension on command from an electronic controller tailored to produce the desired ride characteristics. The characteristics that are optimal for ride often compromise performance in other modes, most notably in handling.

The quarter-car model can be used to quantify the comparative ride performance of passive and active systems [41]. The two systems are illustrated in Figure 5.22. For a vehicle traveling at a constant forward velocity over a random road surface, the road excitation can be reasonably well approximated by a constant-slope spectrum (road profile displacement inputs are obtained by integrating a white-noise source). Three performance variables are of interest:

1) Vibration isolation—measured by the sprung mass acceleration $(\ddot{Z}_2)$

2) Suspension travel—measured by deflection of the suspension (Z_1)

3) Tire load constancy—measured by deflection of the tire (Z_3)

Since the vehicle is modeled as a linear system subjected to white-noise input, the mean square response of any motion variable of interest can be computed using the relationship:

$$E[y^2] = S_0 \int_{-\infty}^{\infty} |H_y(\omega)|^2 \, d\omega \qquad (5\text{-}18)$$

where:

$E[y^2]$ = Mean square response

S_0 = Spectral density of the white-noise input

$H_y(\omega)$ = Transfer function relating the response 'y' to the white-noise input

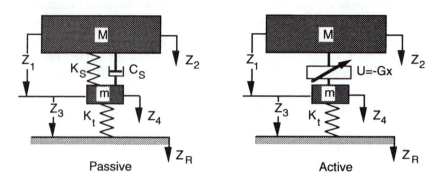

Fig. 5.22 Quarter-car model representations of passive and active suspensions.

The transfer functions may be derived from the governing equations, substituted into Eq. (5-18), and integrated to obtain the rms responses. In the process it is convenient to define certain characteristic parameters as follows:

$$\chi = m / M \qquad \text{Mass ratio} \qquad (5\text{-}19a)$$

$$r_k = K_t / K_s \qquad \text{Stiffness ratio} \qquad (5\text{-}19b)$$

$$\zeta_s = \frac{C_s}{2\sqrt{K_s\,M}} \qquad \text{Damping ratio} \qquad (5\text{-}19c)$$

$$\omega_u = \sqrt{\frac{K_t}{m}} \qquad \text{Natural frequency of the unsprung mass} \quad (5\text{-}19d)$$

Among these vehicle parameters, the suspension designer is only free to select the stiffness and damping values. The influence of these parameters is displayed in Figure 5.23, where the rms vertical acceleration is plotted against the rms suspension travel with different values of stiffness and damping.

The stiffness ratios shown in the figure are typical of the range for most production cars. The stiffest suspension, $r_k = 5$, would be representative of sports and performance cars. The softest, $r_k = 20$, would correspond to luxury cars with air suspensions. For any given stiffness ratio the vertical acceleration varies with damping and has an optimum (the point of lowest rms acceleration). High levels of damping reduce the rms suspension travel, but

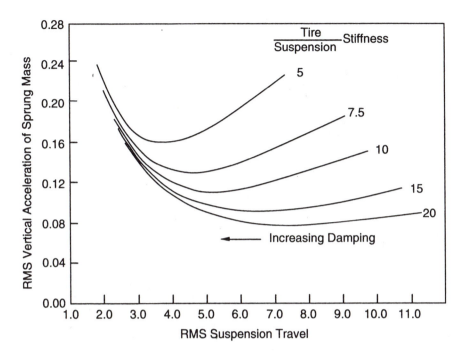

Fig. 5.23 Acceleration versus suspension travel for a passive suspension ($\chi = 0.15$).

at the cost of increased acceleration (the greater damping forces transmit more road input across the suspension into the sprung mass). Low levels of damping allow greater suspension travel, but also increase the rms acceleration because of the uncontrolled motions at the sprung mass resonant frequency.

In practice it is not possible to use the entire range of performance shown in the figure. In particular the low damping levels use suspension strokes that are beyond the range available on most passenger cars, and the low damping is insufficient to control wheel hop oscillations which compromise road-holding behavior. The shaded area in Figure 5.24 indicates the practical range of performance for passive suspensions.

The quarter-car model for an active suspension, shown in Figure 5.22, differs by the presence of a force generator in place of the suspension spring and damper. The force generator corresponds to a hydraulic cylinder controlled by an electronic system. The electronics can potentially sense

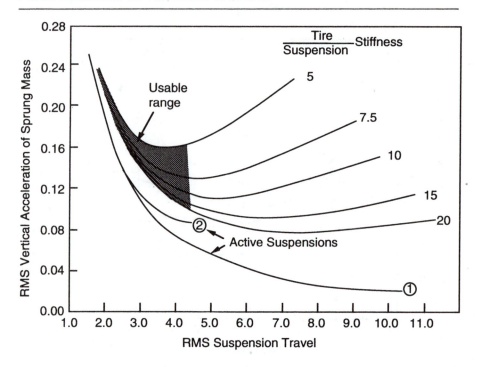

Fig. 5.24 Acceleration versus suspension travel for an active suspension.

accelerations on the sprung and unsprung masses, the suspension displacement and tire radius (or load), and vary the force linearly in proportion to any combination of these state variables.

The optimal control algorithm for an active suspension can be determined analytically [41]. Optimization to minimize vertical acceleration and suspension travel results in the performance shown by curve 1 in Figure 5.24. For any given limit of suspension travel the active suspension always yields better performance. At the upper limit of the usable range for the passive suspension (defined by the shaded area), a ride improvement equivalent to a 30% reduction in rms acceleration is possible with an active suspension.

The optimization for ride (indicated by curve 1) is less than optimal for road-holding because of insufficient damping of the wheel hop resonances. The 30% reduction in vertical acceleration is achieved at the expense of unsprung mass damping, which is only about 5%. If the optimization is constrained to a more reasonable (20%) unsprung mass damping ratio, the best performance for an active suspension becomes curve 2 in Figure 5.24. With

162

this constraint the ride improvement of the active suspension over the best passive suspension is reduced to only about 10 percent.

The benefits of an active suspension are largely obtained by control of the sprung mass motions near its resonant frequency. The vertical acceleration response of the two systems is compared in Figure 5.25.

At the sprung mass resonant frequency the active suspension reduces the response amplitude over that which occurs with the passive suspension. Since the active suspension can sense sprung mass accelerations and exert forces to minimize their amplitude, the system achieves very effective damping of this mode. At the unsprung mass resonant frequency the active suspension behaves just like the passive suspension. This occurs because the suspension forces necessary to control wheel hop motions react against the sprung mass, and would necessarily increase sprung mass accelerations. In effect, optimization of ride accelerations on the sprung mass is not achieved when the sprung mass is used as a reaction point for forces to control unsprung mass motions.

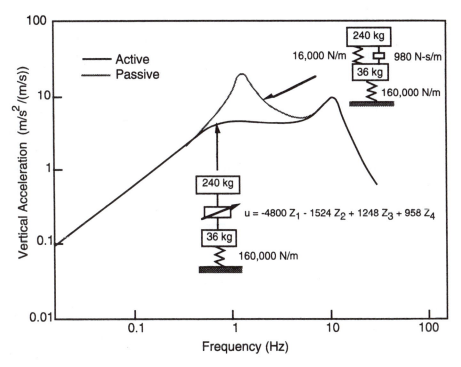

Fig. 5.25 Comparison of vertical acceleration response of active and passive systems [42].

Wheel Hop Resonances

Next to the sprung mass of a vehicle, the axles and wheels (which make up the unsprung mass) are the second largest masses capable of separate resonances as rigid bodies. All individually sprung wheels have a vertical bounce (hop) mode which is excited by road and wheel nonuniformity inputs adding to the vibrations present on the vehicle. The influence of the wheel vertical resonance on the sprung mass vibrations was seen in the basic response of the quarter-car model given earlier in Figure 5.16. The response gains for inputs from the road, or at the wheel, would normally attenuate quickly and continuously in the absence of a wheel resonant mode. However, in both cases, the response is accentuated at frequencies above the body resonance point as a result of the wheel motion, with the effect greatest at the resonant frequency of the wheel.

The resonant frequency is determined by the wheel/axle mass suspended against the springs in the suspension, acting along with that of the tires. Characteristically, the unsprung mass will correspond to a weight that is proportional to the gross axle weight rating (GAWR), which in turn is indicative of the loads normally carried by the axle. For nondriven axles, the weight, W_a, is typically about 10 percent of the GAWR, whereas for drive axles it will be about 15 percent of the GAWR. Inasmuch as the tires and suspension springs are normally sized in proportion to the GAWR, and the resonant frequency depends on the ratio of the mass to the total spring rate of the tires and suspension springs, the resonant frequencies of most wheels, at least theoretically, would fall in a limited range.

Wheel hop frequencies are much higher than the sprung mass resonance, so the sprung mass remains stationary during wheel hop. Thus both the tire and the suspension springs act in parallel to resist wheel hop motions, and the total spring rate controlling the axle mass is the sum of the two. The resonant frequency may then be calculated as:

$$f_a = 0.159 \sqrt{(K_t + K_s) \, g/W_a} \qquad\qquad (5\text{-}20)$$

where:

f_a = Wheel hop resonant frequency (Hz)
K_t = Tire spring rate
K_s = Suspension spring rate
W_a = Axle weight

For passenger cars the typical unsprung weight at a wheel is on the order of 100 lb, with a tire stiffness of 1000 lb/in and a suspension stiffness of 100 lb/in. With these nominal values the calculated resonant frequency will be approximately 10 Hz. Friction in the suspension will increase the effective spring rate for small ride motions which in turn will increase the resonant frequency to 12-15 Hz.

The magnitude of the unsprung mass arising from the wheels, axle/spindle, brakes and suspension components influences the transmission of road inputs to the sprung mass. The quarter-car model may be used to examine the road-to-body transmissibility with variations in unsprung mass. Figure 5.26 compares the response gains as the unsprung mass is changed from the typical value (nominally 10% of the sprung mass value) to a magnitude that is twice the value (heavy) and only one-half the value (light). The body resonant behavior near 1 Hz is unaffected by the unsprung mass changes, but above that frequency changes are evident. A heavier mass pulls the wheel hop resonant frequency down near 7 Hz, greatly increasing transmission of road inputs in this range. Inasmuch as these are more objectionable vibrations and are harder to isolate by other means, ride degradation results. With a lighter unsprung mass the wheel-hop resonant frequency is pushed higher, providing better isolation in the mid-frequency range, although there is some penalty above resonance. Since it is easier to isolate high-frequency vibrations elsewhere in the chassis, the lighter unsprung mass will generally produce better ride performance.

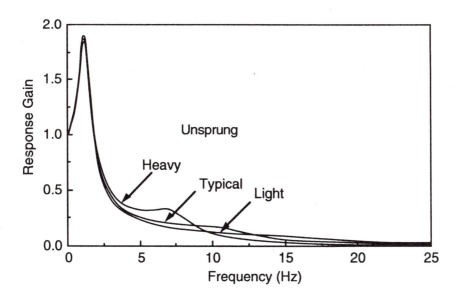

Fig. 5.26 Effect of unsprung mass on suspension isolation behavior.

Suspension Nonlinearities

In practice, the suspension systems on many vehicles are not really linear, as assumed above, due to friction (sometimes termed "stiction") in the struts and bushings, or the interleaf friction in a leaf spring suspension. Rather than a simple linear relationship between force and displacement, the suspension exhibits hysteretic behavior similar to that of the leaf spring shown in Figure 5.27.

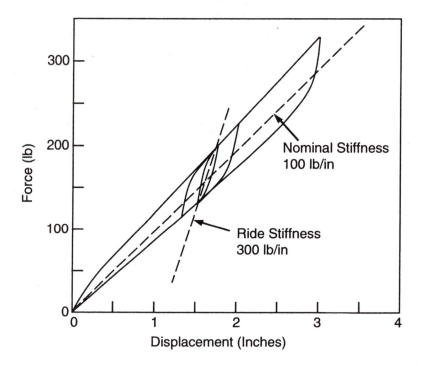

Fig. 5.27 Load-deflection characteristics of a hysteretic leaf spring.

Ride motions are typically quite small in amplitude, involving only a fraction of an inch of suspension travel. Thus in the figure, ride motions are represented by excursions on the small inner loops of the spring hysteresis curves. The area enclosed by a hysteresis loop for any given excursion represents the damping energy dissipated by the suspension. The more important effect is the much higher effective stiffness of suspensions undergoing small deflections.

For the leaf spring shown, the effective stiffness for small ride motions is approximately three times greater than the nominal stiffness of the spring. In more extreme cases of suspension friction, the ride stiffness may be nearly an order of magnitude larger. For this reason, it is important to minimize the friction levels in suspension struts and bushings to achieve good ride.

With hysteresis in the suspension, the vehicle response changes with amplitude and spectral content of the excitation. Because of the nonlinearities, a time-based simulation of its response to random road roughness input must be performed to establish the transmissibility properties. Using a quarter-car model with a hysteretic spring, the response gain (ratio of sprung mass acceleration to road acceleration) across the frequency spectrum can be calculated. Figure 5.28 shows calculations of this type for a truck leaf spring suspension [28]. Using the same road at a high (rough) and low (smooth) amplitude illustrates the way in which suspension nonlinearities affect dynamic behavior of a vehicle.

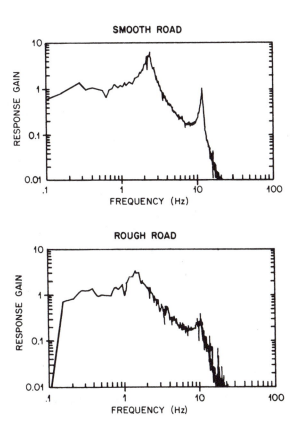

Fig. 5.28 Response of a quarter-vehicle model with a hysteretic suspension.

On smooth roads the suspension deflections are small, resulting in a high effective stiffness and very little damping. Hence, the sprung-mass resonance moves up in frequency to 2.5 Hz and, because of low damping, the resonance becomes a high, narrow peak reaching a gain in excess of 5. Likewise, the axle resonance, which occurs at just above 10 Hz, becomes very accentuated. On the rough road the greater suspension deflections produce a lower effective stiffness and greater damping. Hence, the resonances diminish in both their frequency and amplitude. Though the road is rougher, the reduced response of the vehicle compensates somewhat for the greater input. This type of behavior is very characteristic of heavy trucks; thus it is often observed that some heavy trucks ride better on rough roads than on smooth roads.

Rigid Body Bounce/Pitch Motions

The simple mechanics of the quarter-car model do not fully represent the rigid-body motions that may occur on a motor vehicle. Because of the longitudinal distance between the axles, it is a multi-input system that responds with pitch motions [12] as well as vertical bounce. Depending on the road and speed conditions, one or the other of the motions may be largely absent, or they may not necessarily be observed at the point on the vehicle where the vibration measurements are made. The pitch motions are important because they are generally considered objectionable and are the primary source of longitudinal vibrations at locations above the center of gravity. Understanding the pitch and bounce motions is essential because it is their combination that determines the vertical and longitudinal vibrations at any point on the vehicle.

As the vehicle traverses a road, the roughness excitation at the different axles is not independent. The rear wheels see nearly the same input profile as the front wheels, only delayed in time. The time delay is equal to the wheelbase divided by the speed of travel. The time delay acts to filter the bounce and pitch excitation amplitude, and has been called "wheelbase filtering" [29].

In order to understand the influence of wheelbase filtering, it is convenient to think of the vehicle as having independent pitch and bounce modes. Consider a two-axle vehicle as illustrated in Figure 5.29. As the vehicle moves along the road, the roughness input at the front wheels acts subsequently on the rear wheels, delayed in time by the interval equal to the wheelbase divided by speed. Inasmuch as the road contains roughness at all wavelengths, one can examine the response of the vehicle to individual wavelengths.

Only bounce motion input occurs at a wavelength equal to the wheelbase of the vehicle. The same is true for wavelengths much, much longer than the wheelbase, and also for short wavelengths which have an integer multiple equal to the wheelbase. In a similar fashion, only pitch motion input will be seen on a wavelength that is twice the wheelbase, or on any shorter wavelength that has an odd integer multiple equal to twice the wheelbase. As a consequence the vehicle will be unresponsive in either bounce or pitch to certain wavelengths in the road with the filtering qualities shown in Figure 5.29.

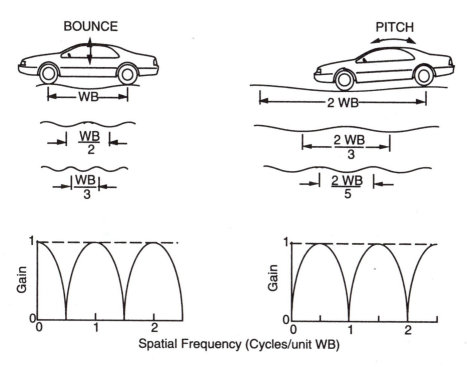

Fig. 5.29 The wheelbase filtering mechanism.

The influence may be seen better in the response gain for a simple vehicle with a 1.25 Hz natural frequency at both front and rear axles (the bounce and pitch vibration modes are uncoupled). The gray line in the top plot of Figure 5.30 shows the vertical response gain that would result at each axle as calculated from the quarter-car model. When the road input is applied with that at the rear axle delayed from that of the front, the response at the passenger positions midway between the wheels will be altered by wheelbase filtering. Assuming a 9-foot (2.7-meter) wheelbase and a speed of 50 mph (80 km/h), response nulls occur at approximately 4, 12, 20 Hz, etc. The null points are equal to the velocity divided by twice the wheelbase and odd multiples thereof.

Thus at high speeds a passenger car tends to experience vertical bounce vibrations more or less as predicted by a quarter-car model with the exception of the null points indicated. At lower speeds the null points shift proportionately to lower frequencies.

The pitch response of the vehicle has similar qualities as given in the bottom plot of the figure. The pitch resonance frequency will typically be close to the bounce frequency, so the same resonant frequency has been assumed in

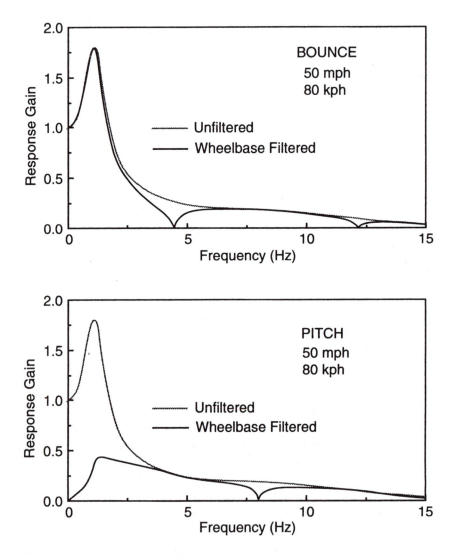

Fig. 5.30 Effect of wheelbase filtering on bounce and pitch response.

the figure. The null frequencies in pitch for this same vehicle will be at 8, 16, 24 Hz, etc. Consequently, there will be little pitch of the vehicle at normal highway speeds. Only at low speed is the pitch mode readily excited by road inputs.

The bounce and pitch vibration behavior of heavy trucks presents a somewhat different picture because of the higher resonant frequencies and longer wheelbases of these vehicles. Because of the stiffer suspensions the natural frequencies of bounce and pitch tend to be closer to 2.5-3 Hz. With wheelbases in the range of 12-15 feet (3.7-4.6 meters), just the reverse behavior occurs at high speed. Namely, the bounce response will have a null at the resonant frequency and the pitch response will be at full amplitude. Consequently, wheelbase filtering will affect the vibration response on a heavy truck as illustrated in Figure 5.31.

FORE/AFT VIBRATION RESPONSE GAIN

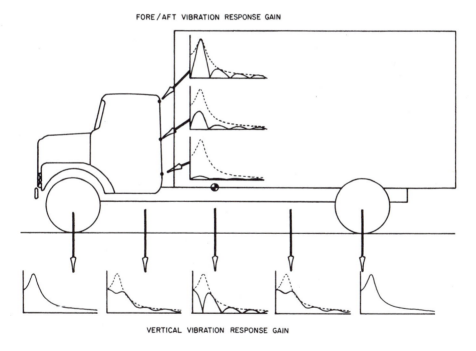

VERTICAL VIBRATION RESPONSE GAIN

Fig. 5.31 Effect of wheelbase filtering on vertical and longitudinal response gain of a truck [44].

The vertical vibration response due to rigid-body motions will vary with location along the length of the vehicle, depending on the relative actions of the bounce and pitch motions. Near the midpoint of the vehicle the vertical vibrations are affected only by bounce, hence, the response properties at this point reflect the wheelbase filtering phenomenon directly. The vertical acceleration spectra measured at this point will show the characteristic decreases at the bounce null frequencies. At points toward the extremes of the vehicle both pitch and bounce will contribute to the vertical accelerations and the effect of wheelbase filtering will become less evident. Over the axles some combination of bounce and pitch will be present at each frequency, such that wheelbase filtering will have no effect on the response; rather, the vertical response will be equivalent to that seen in the quarter-vehicle model earlier.

The pitch action is the predominant source for fore/aft vibrations that are seen on trucks at locations above the center of gravity. Thus the spectrum of fore/aft vibrations will be affected by the wheelbase filtering influence on pitch response, and the amplitude of the vibrations will be dependent on the elevation of the driver's seat above the center of gravity. The fact that fore/aft vibration spectra show periodic reductions at the pitch null frequencies should not be confused as an indication of a multi-resonant system.

Bounce/Pitch Frequencies

The tuning of the bounce and pitch vibration modes on a vehicle has a direct impact on the acceptability of the ride. On most vehicles there is a coupling of motions in the vertical and pitch directions, such that there are no "pure" bounce and pitch modes. The behavior, in terms of natural frequencies and motion centers, for a vehicle with coupled motions in the vertical and pitch directions can be readily determined analytically from the differential equations of motion. Consider a vehicle as shown in Figure 5.32. For simplicity in analysis, the tire and suspension will be considered as a single stiffness (the ride rate), and damping and unsprung masses will be neglected.

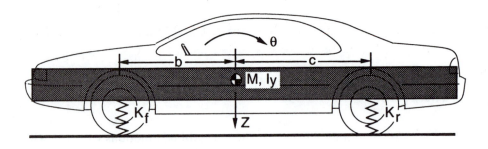

Fig. 5.32 Pitch plane model for a motor vehicle.

For convenience in the analysis the following parameters are defined:

$$\alpha = (K_f + K_r)/M \qquad (5-21)$$

$$\beta = (K_r c - K_f b)/M \qquad (5-22)$$

$$\gamma = (K_f b^2 + K_r c^2)/M k^2 \qquad (5-23)$$

where:

K_f = Front ride rate
K_r = Rear ride rate
b = Distance from the front axle to the CG
c = Distance from the rear axle to the CG
I_y = Pitch moment of inertia
k = Radius of gyration = $\sqrt{I_y/M}$

Then the differential equations for the bounce, Z, and pitch, θ, motions of a simple vehicle can now be written as:

$$\ddot{Z} + \alpha Z + \beta \theta = 0 \qquad (5-24)$$

$$\ddot{\theta} + \beta Z/k^2 + \gamma \theta = 0 \qquad (5-25)$$

Of the several coefficients in these equations, only β appears in both and is appropriately called the coupling coefficient. When $\beta = 0$ no coupling occurs, and the spring center is at the center of gravity. For this condition, a vertical force at the CG produces only bounce motion, and a pure torque applied to the chassis will produce only pitch motion.

Without damping, the solutions to the differential equations will be sinusoidal in form. The vertical motion will be:

$$Z = \mathbf{Z} \sin \omega t \qquad (5-26)$$

and the pitch motion will be:

$$\theta = \mathbf{\theta} \sin \omega t \qquad (5-27)$$

When these are differentiated twice and substituted into Eq. (5-22), we obtain:

$$- \mathbf{Z} \omega^2 \sin \omega t + \alpha \mathbf{Z} \sin \omega t + \beta \mathbf{\theta} \sin \omega t = 0 \qquad (5-28)$$

Since the terms must always equal zero regardless of the instantaneous value of the sine function:

$$(\alpha - \omega^2)\, Z + \beta\, \theta = 0 \tag{5-29}$$

or,

$$Z/\theta = -\beta/(\alpha - \omega^2) \tag{5-30}$$

The same analysis applied to Eq. (5-25) yields:

$$Z/\theta = -k^2\,(\gamma - \omega^2)/\beta \tag{5-31}$$

The above equations define conditions under which the motions can occur. The constraints are that the ratio of amplitudes in bounce and pitch must satisfy Eqs. (5-30) and (5-31).

Equating the right sides of Eqs. (5-30) and (5-31) yields the expressions for the natural frequencies of the two modes of vibration.

$$(\alpha - \omega^2)\,(\gamma - \omega^2) = \beta\,(\beta/k^2) \tag{5-32}$$

Then

$$\omega^4 - (\alpha + \gamma)\,\omega^2 + \alpha\,\gamma - \beta^2/k^2 = 0 \tag{5-33}$$

The values of ω satisfying this equation are the roots representing the frequency of the vibration modes. Two of the roots will be imaginary and can be ignored. The others are obtained from the equations as follows:

$$
\begin{aligned}
(\omega_{1,\,2})^2 &= \frac{(\alpha + \gamma)}{2} \pm \sqrt{\frac{(\alpha + \gamma)^2}{4} - \left(\alpha\,\gamma - \beta^2/k^2\right)} \\[2mm]
&= \frac{(\alpha + \gamma)}{2} \pm \sqrt{\frac{(\alpha - \gamma)^2}{4} + \beta^2/k^2}
\end{aligned}
\tag{5-34}
$$

$$\omega_1 = \sqrt{\frac{(\alpha + \gamma)}{2} + \sqrt{(\alpha - \gamma)^2/4 + \beta^2/k^2}} \tag{5-35}$$

$$\omega_2 = \sqrt{\frac{(\alpha + \gamma)}{2} - \sqrt{(\alpha - \gamma)^2/4 + \beta^2/k^2}} \tag{5-36}$$

These frequencies always lie outside the uncoupled natural frequencies.

The oscillation centers can be found using the amplitude ratios of Eqs. (5-30) and (5-31) with the two frequencies ω_1 and ω_2 in Eqs. (5-35) and (5-36). When substituted it will be found that $Z/\theta(\omega_1)$ and $Z/\theta(\omega_2)$ will have opposite signs.

When Z/θ is positive, both Z and θ must be both positive or negative. Thus the oscillation center will be ahead of the CG by a distance $x = Z/\theta$. Similarly, for the root with a negative value for Z/θ, the oscillation center will be behind the CG by a distance x equal to Z/θ. Likewise, one distance will be large enough that the oscillation center will fall outside the wheelbase, and the other will be small enough that the center falls within the wheelbase. When the center is outside the wheelbase, the motion is predominantly bounce, and the associated frequency will be the bounce frequency. For the center within the wheelbase, the motion will be predominantly pitch, and the associated frequency is the pitch frequency. These cases are illustrated in Figure 5.33.

Predominant Modes

Bounce   Pitch

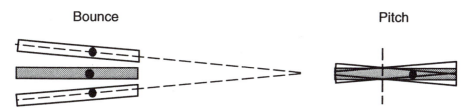

Fig. 5.33 The two vibration modes of a vehicle in the pitch plane.

The locations of the motion centers are dependent on the relative values of the natural frequencies of the front and rear suspensions, where those frequencies are defined by the square root of the ride rate divided by the mass. That is:

$$f_f = \frac{1}{2\pi}\sqrt{\frac{K_f\, g}{W_f}} \qquad (5\text{-}37)$$

$$f_r = \frac{1}{2\pi}\sqrt{\frac{K_r\, g}{W_r}} \qquad (5\text{-}38)$$

Figure 5.34 shows the locus of motion centers as a function of front/rear natural frequency. With equal frequencies, one center is at the CG location and the other is at infinity. Equal frequencies correspond to decoupled vertical and pitch modes, and "pure" bounce and pitch motions result. With a higher front frequency the motion is coupled with the bounce center ahead of the front axle and the pitch center toward the rear axle. A lower front frequency puts the bounce center behind the rear axle and the pitch center forward near the front axle. This latter case was recognized by Maurice Olley in the 1930s as best for achieving good ride.

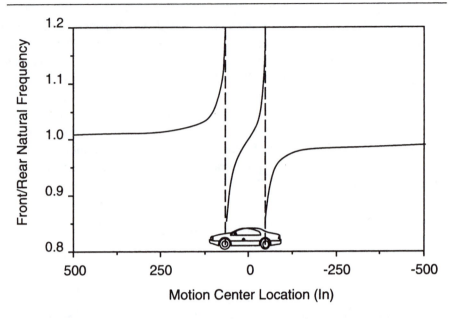

Fig. 5.34 Effect of natural frequency ratio on position of motion centers.

Maurice Olley, one of the founders of modern vehicle dynamics, established guidelines back in the 1930s for designing vehicles with good ride (at least for the low-frequency, rigid-body modes of vibration). These were derived from experiments with a car modified to allow variation of the pitch moment of inertia (his famous "k^2" rig) [43]. Although the measure of ride was strictly subjective, those guidelines are considered valid rules of thumb even for modern cars. The Olley Criteria are:

1) *The front suspension should have a 30% lower ride rate than the rear suspension, or the spring center should be at least 6.5% of the wheelbase behind the CG.* Although this does not explicitly determine the front and rear natural frequencies, since the front-rear weight distribution on passenger cars is close to 50-50, it will generally assure that the rear frequency is greater than the front.

2) *The pitch and bounce frequencies should be close together: the bounce frequency should be less than 1.2 times the pitch frequency. For higher ratios, "interference kicks" resulting from the superposition of the two motions are likely.* In general, this condition will be met for modern cars because the dynamic index is near unity with the wheels located near the forward and rearward extremes of the chassis.

3) *Neither frequency should be greater than 1.3 Hz, which means that the*

effective static deflection of the vehicle should exceed roughly 6 inches. The value of keeping natural frequencies below 1.3 Hz was demonstrated in Figure 5.18.

4) *The roll frequency should be approximately equal to the pitch and bounce frequencies.* In order to minimize roll vibrations the natural frequency in roll needs to be low just as for the bounce and pitch modes.

The rule that rear suspensions should have a higher spring rate (higher natural frequency) is rationalized by the observation that vehicle bounce is less annoying as a ride motion than pitch. Since excitation inputs from the road to a car affect the front wheels first, the higher rear to front ratio of frequencies will tend to induce bounce.

To illustrate this concept, consider a vehicle encountering a bump in the road. The time lag between the front and rear wheel road inputs at a forward speed, V, and a car wheelbase, L, will be:

$$t = L/V \qquad\qquad (5\text{-}39)$$

The oscillations at the front and rear of the car for an input of this type are illustrated in Figure 5.35. Note that soon after the rear wheels have passed over the bump the vehicle is at the worst condition of pitching, indicated by the points A and B in the figure. Point A corresponds to the front end of the car being in a maximum upward position, whereas the rear end (point B) is just beginning to move. Therefore, the car is pitching quite heavily.

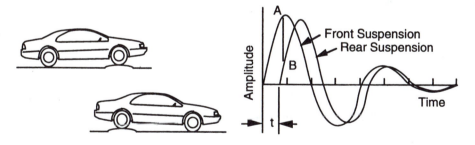

Fig. 5.35 Oscillations of a vehicle passing over a road bump.

With a higher rear frequency, after about one and one-half oscillations of the rear suspensions, both ends of the car are moving in phase. That is, the body is now merely bouncing up and down until the motion is almost fully damped. At different speeds and for different road geometries, the vehicle response will change. Thus the optimum frequency ratio of the front and rear ends of the car has to be determined experimentally.

Special Cases

1) Most modern vehicles with substantial front and rear overhang exhibit a dynamic index close to unity. That is:

$$DI = k^2 / bc = 1 \qquad (5\text{-}40)$$

When the equality holds, the front and rear suspensions are located at conjugate centers of percussion (an input at one suspension causes no reaction at the other). In this case the oscillation centers are located at the front and rear axles. This is a desirable condition for good ride if Olley's ride criteria are also satisfied. There is no interaction between the front and rear suspensions [43].

2) Spring center at the CG—This condition corresponds to the case of uncoupled pitch and bounce motions ($\beta = 0$). The pitch and bounce oscillations are totally independent. Poor ride results because the motions can be very irregular. Coupling tends to even out the ride [43].

3) Dynamic index greater than unity—This occurs when there is substantial overhang at the front and/or rear of the car. The bounce center is in front of the CG (beyond the front axle) and the pitch center is between the CG and the rear axle. The natural frequency of pitch is now less than the bounce frequency and flat ride will still result if the spring center is located far enough behind the CG (front ride rate less than rear ride rate) [43].

4) Uncoupled motion ($\beta = 0$) and dynamic index equal to one—This condition results in equal bounce and pitch frequencies. The ride is inferior because there is essentially no pattern to the road-generated motion; it is quite unpredictable [43].

Example Problem

1) Calculate the pitch and bounce centers and their frequencies for a car with the following characteristics:

Front ride rate = 127 lb/in/wheel   Front tire load = 957 lb/wheel

Rear ride rate = 92.3 lb/in/wheel   Rear tire load = 730 lb/wheel

Wheelbase = 100.6 in   Dynamic index (DI) = 1.1

Solution:

The starting point is to find the values for the three parameters given in Eqs. (5-21), (5-22) and (5-23), but first we must determine the mass, b, c and k.

$$M = W/g = (957 + 957 + 730 + 730) \text{ lb}/386 \text{ in/sec}^2 = 8.74 \text{ lb-sec}^2/\text{in}$$

We can find "b" and "c" from the weight distribution:

$$b = L \, W_r/W = (100.6 \text{ in}) \, 1460 \text{ lb}/3374 \text{ lb} = 43.53 \text{ in}$$

$$c = 100.6 - 43.53 = 57.07 \text{ in}$$

Then:

$$k^2 = b \, c \, DI = (43.53 \text{ in}) \, (57.07 \text{ in}) \, 1.1 = 2732 \text{ in}^2$$

Now we can solve for each of the parameters.

$$\alpha = (K_f + K_r)/M = [2 \, (127) + 2 \, (92.3)] \text{ lb/in}/(8.74 \text{ lb-sec}^2/\text{in}) = 50.18 \text{ sec}^{-2}$$

$$\beta = (K_r \, c - K_f \, b)/M = [184.6 \, (57.07) - 254 \, (43.53)] \text{ lb}/(8.74 \text{ lb-sec}^2/\text{in})$$

$$= -59.67 \text{ in/sec}^2$$

$$\gamma = (K_f \, b^2 + K_r \, c^2)/M \, k^2$$

$$= [254 \, (43.53^2) + 184.6 \, (57.07^2)] \text{ lb-in}/[8.74 \text{ lb-sec}^2/\text{in} \, (2732 \text{ in}^2)] = 45.34 \text{ sec}^{-2}$$

Then from Eqs. (5-35) and (5-36) we can solve for the two frequencies.

$$\omega_1 = \sqrt{\frac{(50.18 + 45.34)}{2 \text{ sec}^2} + \sqrt{\frac{(50.18 - 45.34)^2}{4 \text{ sec}^4} + \frac{(59.67 \text{ in/sec}^2)^2}{2732 \text{ in}^2}}}$$

$$= 7.10 \text{ radians/sec} = 1.13 \text{ Hz}$$

$$\omega_2 = \sqrt{\frac{(50.18 + 45.34)}{2 \text{ sec}^2} - \sqrt{\frac{(50.18 - 45.34)^2}{4 \text{ sec}^4} + \frac{(59.67 \text{ in/sec}^2)^2}{2732 \text{ in}^2}}}$$

$$= 6.71 \text{ radians/sec} = 1.07 \text{ Hz}$$

Now from Eq. (5-30) we can solve for **Z/θ** for each frequency.

$$\mathbf{Z/\theta}_1 = -\beta/(\alpha - \omega_1^2) = 59.67 \text{ in/sec}^2 /(50.18 \text{ sec}^{-2} - 50.41 \text{ sec}^{-2}) = -259 \text{ in/rad}$$

$$\mathbf{Z/\theta}_2 = -\beta/(\alpha - \omega_2^2) = 59.67 \text{ in/sec}^2 /(50.18 \text{ sec}^{-2} - 45.02 \text{ sec}^{-2}) = 11.57 \text{ in/rad}$$

Thus the car has a motion center 259 inches behind the CG with a frequency of 7.1 radians/sec (1.13 Hz). Since it is the most distant from the CG, it is predominantly vertical and would be the bounce center. The second motion center at 6.71 radians/sec (1.07 Hz) is 11.57 inches forward of the CG and therefore will be the pitch center.

2) Find the pitch and bounce centers and their frequencies for a car with the following characteristics:

Front ride rate = 132 lb/in	Front tire load = 1035 lb
Rear ride rate = 93 lb/in	Rear tire load = Varying from 567 to 1000 lb
Wheelbase = 112 inches	Dynamic Index = 1.05

Solution:

Set up a spreadsheet to perform the calculations while the rear axle load is varied over the specified range. The relevant parameters are shown in the table below for rear wheel load of 567 lb and increments above that value.

W_r	b	c	k^2	α	β	γ	$\omega 1$	Z/θ_1	$\omega 2$	Z/θ_2
567	39.64	72.36	3011.8	54.21	360.7	55.55	1.25	49.6	1.11	-60.7
600	41.10	70.90	3059.7	53.12	275.8	53.28	1.21	54.5	1.11	-56.2
625	42.17	69.83	3091.9	52.32	215.8	51.76	1.19	59.8	1.10	-51.7
650	43.20	68.80	3120.9	51.54	159.2	50.39	1.17	68.3	1.10	-45.7
675	44.21	67.79	3146.9	50.79	105.8	49.16	1.15	85.3	1.10	-36.9
700	45.19	66.81	3170	50.06	55.4	48.05	1.13	137.9	1.10	-23.0
725	46.14	65.86	3190.6	49.35	7.7	47.04	1.12	957.3	1.09	-3.4
750	47.06	64.94	3208.9	48.66	-37.2	46.13	1.11	-231.4	1.08	13.9
775	47.96	64.04	3224.9	47.98	-79.8	45.30	1.11	-132.8	1.06	24.3
800	48.83	63.17	3238.8	47.33	-120.0	44.54	1.11	-105.8	1.05	30.6
825	49.68	62.32	3250.8	46.69	-158.0	43.86	1.11	-93.3	1.03	34.9
850	50.50	61.50	3261.1	46.07	-194.0	43.23	1.11	-85.9	1.02	38.0
875	51.31	60.69	3269.7	45.47	-228.1	42.65	1.11	-80.9	1.00	40.4
900	52.09	59.91	3276.8	44.88	-260.3	42.13	1.11	-77.2	0.99	42.5
925	52.86	59.14	3282.4	44.31	-290.9	41.64	1.11	-74.3	0.98	44.2
950	53.60	58.40	3286.8	43.75	-319.8	41.20	1.10	-71.9	0.96	45.7
975	54.33	57.67	3289.9	43.21	-347.2	40.80	1.10	-69.9	0.95	47.1
1000	55.04	56.96	3291.8	42.68	-373.2	40.43	1.10	-68.2	0.94	48.3

- For the base vehicle (567 lb rear wheel load) there is no clear pitch or bounce frequency because both centers are relatively near the wheels (ω_1 has a center 49.6 inches ahead of the CG which is just in front of the front wheels, and ω_2 has its center 60.7 inches behind the CG which is just ahead of the rear wheels).

- At 725 lb the ω_1 becomes almost pure bounce (the center is 957 inches ahead of the CG) and ω_2 is almost pure pitch as its center is only 3.4 inches behind the CG.

- As the rear load continues to increase, the bounce center (ω_1) moves behind the CG and the pitch center continues to move forward. At the highest loads the centers are in close proximity to the wheel positions and it is no longer possible to associate a pitch or bounce with each.

PERCEPTION OF RIDE

The final assessment of ride vibrations must deal with the issue of how ride is perceived. For that purpose, one must first attempt to define ride. Ride is a subjective perception, normally associated with the level of comfort experienced when traveling in a vehicle. Therefore, in its broadest sense, the perceived ride is the cumulative product of many factors. The tactile vibrations transmitted to the passenger's body through the seat, and at the hands and feet, are the factors most commonly associated with ride. Yet it is often difficult to separate the influences of acoustic vibrations (noise) in the perception of ride, especially since noise types and levels are usually highly correlated with other vehicle vibrations. Additionally, the general comfort level can be influenced by seat design and its fit to the passenger, temperature, ventilation, interior space, hand holds, and many other factors. These factors may all contribute to what might be termed the "ride quality" of a vehicle.

Some of the above factors, such as vibrations, can be measured objectively, while others, such as seat comfort, are still heavily dependent on subjective evaluation methods. To further complicate matters, the interactions between factors are not well established. For example, it is the author's experience that tolerance for vibration in a truck often can be drastically reduced if the passenger space provisions do not allow room for body movement without contacting hard points on the vehicle interior.

Tolerance to Seat Vibrations

The judgment of ride vibration in a vehicle is still an area of controversy in the automotive community. As a starting point it is instructive to look at research findings from the scientific community relating to human tolerance for vibration. A brief state-of-the-art review of the vibration limits for human comfort covering work back to the 1920s is presented in the SAE Ride and Vibration Data Manual [26] published in 1965. Major works by Lee and Pradko [30], the International Organization for Standardization [31], Oborne [32], Miwa [33], Parsons [34], Fothergill [35], Leatherwood [36], and others, have made substantial contributions to the data base of information related to vibration tolerance. These studies, in general, tend to focus on tolerance as it relates to discomfort in a seated position in an effort to sort out the frequency sensitivity of the human body. Pure sinusoidal inputs are often used in attempts to establish quantified levels of discomfort, or equal levels of sensation, as a function of frequency. Yet no universally accepted standard exists for judgment of ride vibrations due to variables such as:

- seating position
- influence of hand and foot vibration input
- single- versus multiple-frequency input
- multi-direction input
- comfort scaling
- duration of exposure
- sound and visual vibration inputs

Despite the controversy, certain common denominators can be seen in the results from much of the recent work. When examining tolerance for vertical and fore/aft vibration on seated passengers, the researchers usually observe comparable sensitivity curves.

Figure 5.36 shows lines of constant comfort as determined by a number of researchers. Because of the different interpretations of comfort in each study, the nominal level of one curve is not comparable to the others, nor is it especially meaningful. Nevertheless, note that the majority show a minimum tolerance (maximum sensitivity) of the human body to vertical vibration in the frequency range between 4 and 8 Hz. This sensitivity is well recognized as the result of vertical resonances of the abdominal cavity. At frequencies above and below this range the tolerance increases in proportion to frequency. The actual shape of the boundaries will often show small inflections in the 10 to 20 Hz range due to other organ resonances, especially head resonance near 10 Hz.

As indicated by the plots of the ISO curves in the figure, the duration of the vibration exposure also affects the maximum tolerable level. Hence, two boundaries are shown, one for one minute of exposure, and the second for one hour of exposure. General rules for determining boundaries appropriate to arbitrary exposure levels are available in the ISO Standard [31], and in the work of Lee and Pradko [30].

Very interesting findings were obtained by NASA [37] in research on comfort in mass transport vehicles, notably airplanes. The constant comfort lines for vertical vibration taken from that work are shown in Figure 5.37. The significant point observed is that the sensitivity as a function of frequency is dependent on the acceleration level. At high levels of vibration, the tolerance

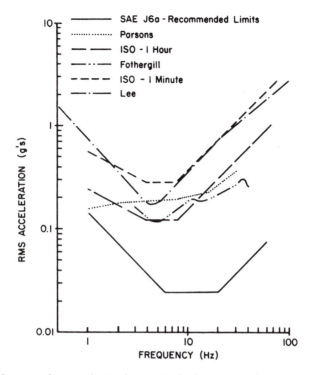

Fig. 5.36 Human tolerance limits for vertical vibration.

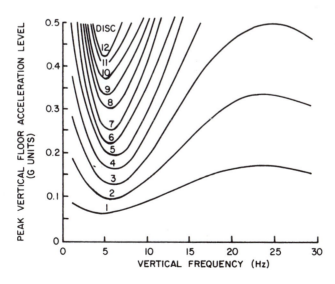

Fig. 5.37 NASA Discomfort Curves for vibration in transport vehicles.

183

curves generally match those of other researchers. But at low amplitudes the horizontal nature of the curves implies that the discomfort is rather independent of frequency. Therefore, low levels of vibration are equally objectionable regardless of their frequency over the indicated range.

Human sensitivity to fore/aft vibration is somewhat different from that of the vertical. Figure 5.38 shows tolerance limits for fore/aft vibration as determined from a number of sources. Again the nominal level of each curve is not especially meaningful, but similar sensitivities are indicated. The most remarkable difference seen is that the region of maximum sensitivity occurs in the 1 to 2 Hz range. This sensitivity is generally recognized to result from the fore/aft resonance of the upper torso. Note also that when the vertical and fore/aft boundaries from individual researchers are compared, the minimum tolerance is observed in the fore/aft direction.

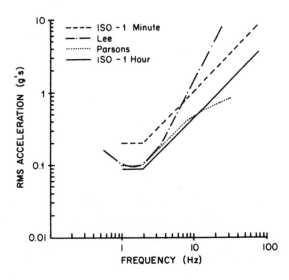

Fig. 5.38 Human tolerance limits for fore/aft vibrations.

The tolerance curves shown in the figures are generally derived from pure sinusoidal inputs to the subjects, whereas the ride environment in a motor vehicle contains all frequencies over a broad spectrum. Thus to apply this information to objective measurements of ride vibration on the seat of a car or truck, it is necessary to resolve this incompatibility. One method commonly used is to filter the acceleration data in inverse proportion to the amplitude of the selected tolerance curve. The inverse filtering then allows the resultant acceleration spectrum to be viewed as if all frequencies were equally important. With this method the vertical and fore/aft vibrations must be evaluated separately. To overcome this problem, the weighted root-mean-square (rms) accelerations in each direction are then sometimes combined by various formulas to obtain an overall rms value.

A more fundamental method for combining vertical and fore/aft vibrations emerged from the work of Lee and Pradko [30]. The level of discomfort was related to the level of vibration power being dissipated in the subject's body, whether from vertical, fore/aft, or lateral inputs. By this method the tolerance curves could be used to weight accelerations so as to arrive at an absorbed power for each direction, and the power quantities are simply added.

The ISO tolerance curves are one of the popular weighting functions used to assess the significance of an acceleration spectrum. However, it should be recognized that the proper method for application of ISO tolerance curves is to evaluate the exposure in one-third octave bands and critique the vibration based on the worst-case band in the spectrum.

When all is said and done, the tolerance curves determined by researchers are instructive to the ride engineer as background in evaluating the vibrations that are imposed on a passenger through the seat. Yet it has been found by many engineers that measurements of these vibrations, even weighted in accordance with selected tolerance curves, bear little correlation to the subjective ratings that will be obtained by a jury in road tests. For example, Healey [38] concluded that a simple measure of rms acceleration in a passenger car is as closely correlated to subjective ratings as any combination of weighted accelerations he could devise.

Whether or not a frequency weighting function is used to adjust the relative importance of specific vibration frequencies, there are formats in which the acceleration spectra can be presented that are more meaningful for ride purposes. In the science of dynamics, it is common practice to present frequency domain information in log-log format as shown at the top of Figure 5.39. In this format, the modal response of a system asymptotically approaches straight lines, and the behavior of complex systems (with multiple modes) can be combined as shown in the bottom of the figure. Thus it has great utility in analysis of dynamic systems. For ride purposes, however, this greatly distorts the relative importance of vibrations at the various frequencies. Presentation of ride acceleration spectra in linear-linear format is more meaningful because the area under the plot is indicative of mean-square or root-mean-square accelerations, depending on the units used on the ordinate axis. (Units of acceleration^2/Hertz correspond to mean-square values, whereas its square root corresponds to the root-mean-square value.)

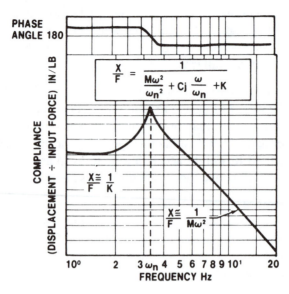

Frequency response plot for one degree of freedom

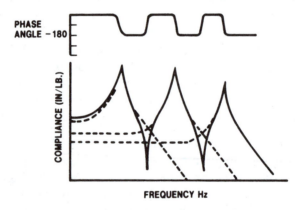

Frequency response plot showing three modes

Fig. 5.39 Representing the response of a dynamic system using log-log plots.

Figure 5.40 contrasts these two means of data presentation. Although the log-log format provides more information for understanding the dynamic systems involved, the linear-linear format allows the engineer to see the relative importance of vibrations in any frequency range by the area involved. Log-log format creates the impression that the vibrations are generally equally important across the entire spectrum. Yet in the linear-linear format it becomes clear that the major portion of the mean-square vibration in this case is concentrated in the low-frequency range of 5 Hz and below.

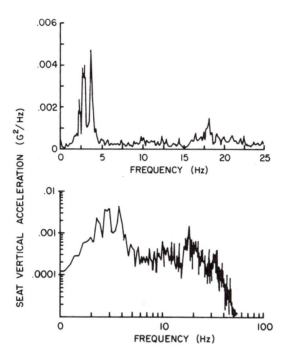

Fig. 5.40 Seat vibrations displayed in linear-linear and log-log format.

Other Vibration Forms

One reason why seat vibration measurements are inadequate as objective measures of ride is that the driver's judgment of the vibration in the vehicle includes far more than what comes through the seat. The point was well demonstrated in studies of the influence of tire/wheel nonuniformities on the ride perception on a highway tractor [17], in which a jury of ten industry engineers rated the ride acceptability of the tractor under varying conditions of tire nonuniformity excitation. The ten-interval rating scale was used for rating seat vertical, seat fore/aft, seat lateral, steering wheel, and cab shake vibrations. Sample results from these tests for left front nonuniformity inputs are shown in Figure 5.41.

A significant point observed in the results was the tendency to degrade the ride due to steering-wheel and cab-shake vibrations. Note that the ratings of seat vibrations showed little sensitivity to nonuniformity input; yet the ratings for cab shake and steering wheel were profoundly affected, especially for the second and third harmonics. The steering-wheel rating reflects vibration inputs to the hands of the driver, while the cab-shake category represents inputs at the feet, as well as visual effects from shake of the A-pillar, rear view mirrors, sun visors, etc.

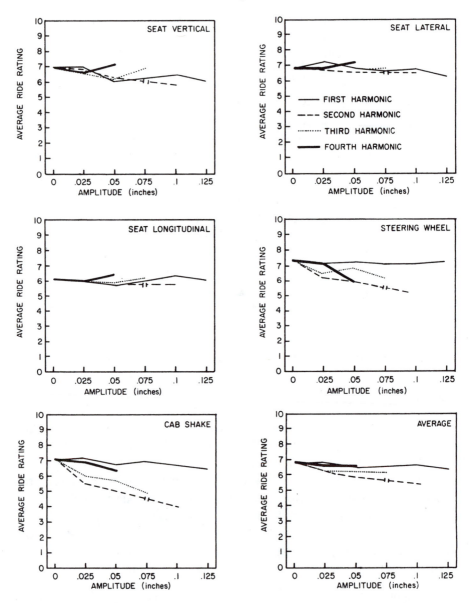

Fig. 5.41 Effect on ride rating of nonuniformities on the left front wheel of a highway tractor.

Note that in these tests the jury was asked to rate the acceptability of the vehicle. Consequently, the ratings reflect the judgment of industry engineers with regard to the acceptability of this vehicle as a product. There is significance in the statement of the question used for a rating experiment. For example, highway engineers, in studies to rate roads for their roughness conditions, have observed that the question "How is the ride?" versus "How is the road?" elicits a much different response from juries. The question "How is the ride?" produces ratings that are dependent on the type of vehicle used to transport the jury during a rating experiment—higher ratings when the jury is transported in a luxury car, and lower ratings when transported in an economy car. Whereas the question "How is the road?" prompts the jury to look beyond the vehicle and judge the road, with the result that the rating process is unaffected by the choice of the vehicle used in the rating study. The ride development engineer should be sensitive to this issue, and is advised to formulate the instructive question carefully before executing ride rating experiments.

CONCLUSION

As a final note, one might hypothesize that the ride engineer's goal should generally be to eliminate all vibrations in a vehicle. Even though this will never be possible in a motor vehicle, it does give direction to development effort. Yet there are two contrary phenomena that must be dealt with. First, the elimination of one vibration will always expose another lesser annoyance. This has been illustrated in past stories of making cars ride so well that the sound of the clock became annoying. Second, in the limit, elimination of all vibration is also undesirable, inasmuch as vibrations are the source of road feel considered to be essential feedback to the driver of a motor vehicle.

REFERENCES

1. Bendat, J.S., and Piersol, A.G., <u>Random Data: Analysis and Measurement Procedures</u>, John Wiley & Sons, New York, 1971, 407 p.

2. Sayers, M., and Gillespie, T.D., "Guidelines for Conducting and Calibrating Road Roughness Measurements," World Bank Technical Paper, ISSN 0253-7494, No. 46, 1986, 87 p.

3. Spangler, E.B., et al. "GMR Profilometer Method for Measuring Road Profile," General Motors Research Publication GMR-452, 1964, 44 p.

4. Gillespie, T.D., Sayers, M., and Segel, L., "Calibration of Response-Type Road Roughness Measuring Systems," Final Report, NCHRP Project 1-18, NCHRP Rept. No. 228, December 1980, 70 p.

5. LaBarre, R.P., *et al.* "The Measurement and Analysis of Road Surface Roughness," Report 1970/5, Motor Industry Research Association, December 1969, 31 p.

6. Sayers, M., Gillespie, T.D., and Queiroz, C.A., "The International Road Roughness Experiment: Establishing Correlation and a Calibration Standard for Measurements," World Bank Technical Paper, ISSN 0253-7494, No. 45, 1986, 453 p.

7. Van Dusen, B.D., "Truck Suspension Optimization," SAE Paper No. 710222, 1971, 12 p.

8. Kropac, D., and Sprinc, J., "Identification of the System Vehicle-Road Parameters," Vehicle Systems Dynamics, Volume II, No. 4, September 1982, pp. 241-249.

9. Ribarits, J.L., *et al.* "Ride Comfort Aspects of Heavy Truck Design," SAE Paper No. 781067, 1978, 24 p.

10. Gillespie, T.D., "The Dynamic Behavior of Nonuniform Tire/Wheel Assemblies," Special Rept., MVMA Project #1163, Transportation Res. Inst., Univ. of Mich., Rept. No. UMTRI-83-8, November 1983, 53 p.

11. Gillespie, T.D., "Relationship of Truck Tire/Wheel Nonuniformities to Cyclic Force Generation," Final Report, MVMA Project No. 1162, Transportation Res. Inst., Univ. of Mich., Rept. No. UMTRI-84-18, April 1984, 136 p.

12. "Vehicle Dynamics Terminology," SAE J670e, Society of Automotive Engineers, Warrendale, PA, 1978 (see Appendix A).

13. Luders, A., *et al.* "Contributions to the Problem of Irregular Running of Vehicle Wheels," *ATZ*, Vol. 73, January 1971, pp. 1-8 (in German).

14. Thomson, W.T., Mechanical Vibrations, 2nd Edition, Prentice-Hall, Englewood Cliffs, N.J., June 1959, 252 p.

15. Klamp, W.K., *et al.* "Higher Orders of Tire Force Variations and Their Significance," SAE Paper No. 720463, 1972, 8 p.

16. Gillespie, T.D., "Influence of Tire/Wheel Nonuniformities on Heavy Truck Ride Quality," Final Rept., MVMA Project #1163, Highway Safety Res. Inst., Univ. of Mich., Rept. No. UM-HSRI-82-30, September 1982, 109 p.

17. Gillespie, T.D., "Tire and Wheel Nonuniformities: Their Impact on Heavy Truck Ride," presentation at the Meeting of the American Chemical Society, Denver, CO, October 1984, 21 p.

18. Lippmann, S.A., "Forces and Torques Associated with Roughness in Tires," SAE Paper No. 610544 (322d), March 1961, 10 p.

19. Morrish, L.M., et al. "The Effect of Loaded Radial Runout on Tire Roughness and Shake," SAE Paper No. 610545 (322e), 1961, 13 p.

20. Potts, G.R., et al. "Tire Vibrations," *Tire Science and Technology,* Vol. 5, No. 4, November 1977, pp. 202-225.

21. Wagner, E.R., "Driveline and Driveshaft Arrangements and Constructions," Universal Joint and Driveshaft Design Manual, Chapter 1, SAE AE-7, 1979, pp. 3-10.

22. Joyner, R.G., "The Truck Driveline as a Source of Vibration," SAE Paper No. 760843, November 1976, 13 p.

23. Mazziotti, P.J., "Dynamic Characteristics of Truck Driveline Systems," 11th L. Ray Buckendale Lecture, SAE SP-262, January 1965.

24. Patterson, D., "Engine Torque and Balance Characteristics," SAE Paper No. 821375, 1982.

25. Dahlberg, T., "Optimization Criteria for Vehicles Traveling on a Randomly Profiled Road—A Survey," *Vehicle Systems Dynamics,* Vol. 8, No. 4, September 1979, pp. 239-252.

26. "Ride and Vibration Data Manual," SAE J6a, Society of Automotive Engineers, Warrendale, PA, December 1965 (see Appendix B).

27. Gillespie, T.D., et al. "Truck Cab Vibrations and Highway Safety," Final Report, FHWA Contract No. DTFH-61-81-C-00083, Highway Safety Research Institute, University of Michigan, Report No. UM-HSRI-82-9-1/2, March 1982, 203 p.

28. Sayers, M., and Gillespie, T.D., "The Effect of Suspension System Nonlinearities on Heavy Truck Vibration," The Dynamics of Vehicles on Roads and Tracks, Proceedings, A. H. Wickens, Ed., Swets

and Zeitlander, Lisse, 1982, pp. 154-166.

29. Butkunas, A.A., "Power Spectral Density and Ride Evaluation," SAE Paper No. 660138, 1966, 8 p.

30. Lee, R.A., and Pradko, F., "Analytical Analysis of Human Vibration," SAE Paper No. 680091, January 1968, 15 p.

31. "Guide for the Evaluation of Human Exposure to Whole-Body Vibration," Second Edition, International Standard ISO 2631-1978(E), International Organization for Standardization, 1978, 15 p.

32. Oborne, D.J., "Techniques Available for the Assessment of Passenger Comfort," *Applied Ergonomics*, Vol. 9, No. 1, March 1978, pp. 45-49.

33. Miwa, T., "Evaluation Methods for Vibration Effect. Part 8—the Vibration Greatness of Random Waves," *Industrial Health*, Vol. 87, 1969, pp. 89-115.

34. Parsons, K.C., Whitham, E.M., and Griffin, M.J., "Six Axis Vehicle Vibration and Its Effects on Comfort," *Ergonomics*, Vol. 22, No. 2, 1979, pp. 211-225.

35. Fothergill, L.C., *et al.*, "The Use of an Intensity Matching Technique to Evaluate Human Response to Whole-Body Vibration," *Ergonomics*, Vol. 20, No. 3, May 1977, pp. 249-261.

36. Leatherwood, J.D., and Dempsey, T.K., "Psychophysical Relationships Characterizing Human Response to Whole-Body Sinusoidal Vertical Vibrations," NASA TN D-8188, NASA Langley Research Center, June 1976, 34 p.

37. Leatherwood, J.D., Dempsey, T.K., and Clevenson, S.A., "A Design Tool for Estimating Passenger Ride Comfort within Complex Ride Environments," *Human Factors*, Vol. 22, No. 3, June 1980, pp. 291-312.

38. Healey, A.J., *et al.*, "An Analytical and Experimental Study of Automobile Dynamics with Random Roadway Inputs," Transactions of the ASME, December 1977, pp. 284-292.

39. Wagner, E.R., "Driveline and Driveshaft Arrangements and Constructions," Universal Joint and Driveshaft Design Manual, Chapter 1, SAE AE-7, 1979, 440 p.

40. Wagner, E.R., and Cooney, C.E., "Cardan or Hooke Universal Joint," Universal Joint and Driveshaft Design Manual, Section 3.1.1, SAE AE-7, 1979, 440 p.

41. Chalasani, R.M., "Ride Performance Potential of Active Suspension Systems — Part I: Simplified Analysis Based on a Quarter-Car Model," Proceedings, Symposium on Simulation and Control of Ground Vehicles and Transportation Systems, AMD-Vol. 80, DSC-Vol 2, American Society of Mechanical Engineers, 1986, pp. 187-204.

42. Chalasani, R.M., "Ride Performance Potential of Active Suspension Systems — Part II: Comprehensive Analysis Based on a Full-Car Model," Proceedings, Symposium on Simulation and Control of Ground Vehicles and Transportation Systems, AMD-Vol. 80, DSC-Vol 2, American Society of Mechanical Engineers, pp. 205-226.

43. Cole, D., "Elementary Vehicle Dynamics," course notes in Mechanical Engineering, The University of Michigan, Ann Arbor, MI, 1972.

44. Gillespie, T.D., Heavy Truck Ride, SP-607, Society of Automotive Engineers, Inc., 1985, 68 p.

CHAPTER 6
STEADY-STATE CORNERING

Stability in handling test track. (Photo courtesy of Volvo of America Corp.)

INTRODUCTION

The cornering behavior of a motor vehicle is an important performance mode often equated with handling. "Handling" is a loosely used term meant to imply the responsiveness of a vehicle to driver input, or the ease of control. As such, handling is an overall measure of the vehicle-driver combination. The driver and vehicle is a "closed-loop" system—meaning that the driver observes the vehicle direction or position, and corrects his/her input to achieve the desired motion. For purposes of characterizing only the vehicle, "open-loop" behavior is used. Open loop refers to vehicle response to specific steering inputs, and is more precisely defined as "directional response" behavior [1].

The most commonly used measure of open-loop response is the understeer gradient [2]. Understeer gradient is a measure of performance under steady-state conditions, although the measure can be used to infer performance properties under conditions that are not quite steady-state (quasi-steady-state conditions).

195

Open-loop cornering, or directional response behavior, will be examined in this section. The approach is to first analyze turning behavior at low speed, and then consider the differences that arise under high-speed conditions. The importance of tire properties will appear in the high-speed cornering case and provide a natural point for systematic study of the suspension properties influential to turning.

LOW-SPEED TURNING

The first step to understanding cornering is to analyze the low-speed turning behavior of a motor vehicle. At low speed (parking lot maneuvers) the tires need not develop lateral forces. Thus they roll with no slip angle, and the vehicle must negotiate a turn as illustrated in Figure 6.1. If the rear wheels have no slip angle, the center of turn must lie on the projection of the rear axle. Likewise, the perpendicular from each of the front wheels should pass through the same point (the center of turn). If they do not pass through the same point, the front tires will "fight" each other in the turn, with each experiencing some scrub (sideslip) in the turn. The ideal turning angles on the front wheels are established by the geometry seen in the figure, and define the steering angles for the turn.

For proper geometry in the turn (assuming small angles), the steer angles are given by:

$$\delta_o \cong \frac{L}{(R + t/2)} \tag{6-1}$$

$$\delta_i \cong \frac{L}{(R - t/2)} \tag{6-2}$$

The average angle of the front wheels (again assuming small angles) is defined [2] as the Ackerman Angle:

$$\delta = L/R \tag{6-3}$$

The terms "Ackerman Steering" or "Ackerman Geometry" are often used to denote the exact geometry of the front wheels shown in Figure 6.1. The correct angles are dependent on the wheelbase of the vehicle and the angle of turn [3]. Errors, or deviations, from the Ackerman in the left-right steer angles can have a significant influence on front tire wear. Errors do not have a significant influence on directional response [4]; however, they do affect the centering torques in the steering system [5]. With correct Ackerman geometry,

196

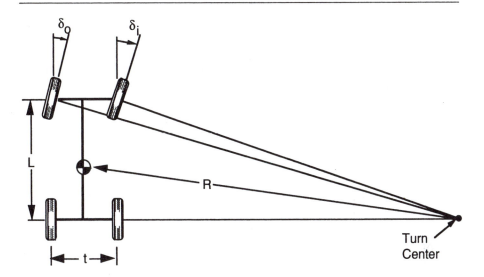

Fig 6.1 Geometry of a turning vehicle.

the steering torques tend to increase consistently with steer angle, thus providing the driver with a natural feel in the feedback through the steering wheel. With the other extreme of parallel steer, the steering torques grow with angle initially, but may diminish beyond a certain point, and even become negative (tending to steer more deeply into the turn). This type of behavior in the steering system is undesirable.

The other significant aspect of low-speed turning is the off-tracking that occurs at the rear wheels. The off-tracking distance, Δ, may be calculated from simple geometry relationships as:

$$\Delta = R[1 - \cos(L/R)] \qquad (6\text{-}4a)$$

Using the expression for a series expansion of the cosine, namely:

$$\cos z = 1 - \frac{z^2}{2!} + \frac{z^4}{4!} - \frac{z^6}{6!} \cdots$$

Then

$$\Delta \cong \frac{L^2}{2\,R} \qquad (6\text{-}4b)$$

For obvious reasons, off-tracking is primarily of concern with long-wheelbase vehicles such as trucks and buses. For articulated trucks, the geometric equations become more complicated and are known as "tractrix" equations.

197

HIGH-SPEED CORNERING

At high speed, the turning equations differ because lateral acceleration will be present. To counteract the lateral acceleration the tires must develop lateral forces, and slip angles will be present at each wheel.

Tire Cornering Forces

Under cornering conditions, in which the tire must develop a lateral force, the tire will also experience lateral slip as it rolls. The angle between its direction of heading and its direction of travel is known as slip angle, α [2]. These are illustrated in Figure 6.2.

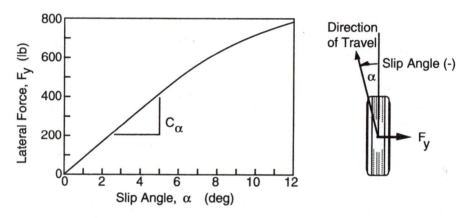

Fig. 6.2 Tire cornering force properties.

The lateral force, denoted by F_y, is called the "cornering force" when the camber angle is zero. At a given tire load, the cornering force grows with slip angle. At low slip angles (5 degrees or less) the relationship is linear, hence, the cornering force is described by:

$$F_y = C_\alpha \, \alpha \qquad\qquad (6\text{-}5)$$

The proportionality constant, C_α, is known as the "cornering stiffness," and is defined as the slope of the curve for F_y versus α at $\alpha = 0$. A positive slip angle produces a negative force (to the left) on the tire, implying that C_α must be negative; however, SAE defines cornering stiffness as the negative of the slope, such that C_α takes on a positive value [2].

The cornering stiffness is dependent on many variables [6]. Tire size and type (radial- versus bias-ply construction), number of plies, cord angles, wheel width, and tread are significant variables. For a given tire, the load and inflation pressure are the main variables. Speed does not strongly influence the cornering forces produced by a tire. The plots in Figure 6.3 illustrate the influence of many of these variables.

Because of the strong dependence of cornering force on load, tire cornering properties may also be described by the "cornering coefficient" which is the cornering stiffness divided by the load. Thus the cornering coefficient, CC_α, is given by:

$$CC_\alpha = C_\alpha/F_z \qquad (lb_y/lb_z/deg) \qquad (6-6)$$

Cornering coefficient is usually largest at light loads, diminishing continuously as the load reaches its rated value (Tire & Rim Association rated load [7]). At 100% load, the cornering coefficient is typically in the range of 0.2 (lb cornering force per lb load per degree of slip angle).

Cornering Equations

The steady-state cornering equations are derived from the application of Newton's Second Law along with the equation describing the geometry in turns (modified by the slip angle conditions necessary on the tires). For purposes of analysis, it is convenient to represent the vehicle by the bicycle model shown in Figure 6.4. At high speeds the radius of turn is much larger than the wheelbase of the vehicle. Then small angles can be assumed, and the difference between steer angles on the outside and inside front wheels is negligible. Thus, for convenience, the two front wheels can be represented by one wheel at a steer angle, δ, with a cornering force equivalent to both wheels. The same assumption is made for the rear wheels.

For a vehicle traveling forward with a speed of V, the sum of the forces in the lateral direction from the tires must equal the mass times the centripetal acceleration.

$$\Sigma F_y = F_{yf} + F_{yr} = M \, V^2/R \qquad (6-7)$$

where:

F_{yf} = Lateral (cornering) force at the front axle
F_{yr} = Lateral (cornering) force at the rear axle
M = Mass of the vehicle
V = Forward velocity
R = Radius of the turn

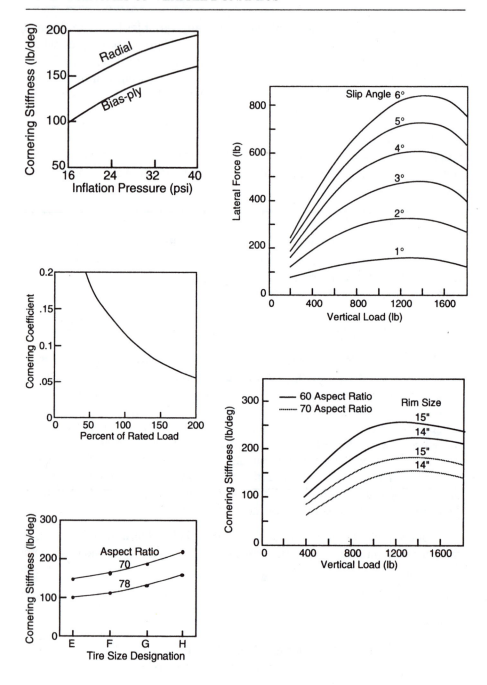

Fig. 6.3 Variables affecting tire cornering stiffness.

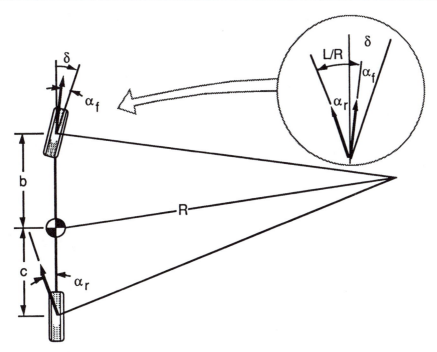

Fig. 6.4 Cornering of a bicycle model.

Also, for the vehicle to be in a moment equilibrium about the center of gravity, the sum of the moments from the front and rear lateral forces must be zero.

$$F_{yf} \, b - F_{yr} \, c = 0 \qquad\qquad (6\text{-}8)$$

Thus

$$F_{yf} = F_{yr} \, c/b \qquad\qquad (6\text{-}9)$$

Substituting back into Eq. (6-7) yields:

$$M \, V^2/R = F_{yr} \, (c/b + 1) = F_{yr} \, (b + c)/b \; = F_{yr} \, L/b \qquad\qquad (6\text{-}10)$$

$$F_{yr} = M \, b/L \, (V^2/R) \qquad\qquad (6\text{-}11)$$

But M b/L is simply the portion of the vehicle mass carried on the rear axle (i.e., W_r/g); thus Eq. (6-11) simply tells us that the lateral force developed at the rear axle must be W_r/g times the lateral acceleration at that point. Solving for F_{yf} in the same fashion will indicate that the lateral force at the front axle must be W_f/g times the lateral acceleration.

With the required lateral forces known, the slip angles at the front and rear wheels are also established from Eq. (6-5). That is:

$$\alpha_f = W_f V^2/(C_{\alpha f} \, g \, R) \tag{6-12}$$

and

$$\alpha_r = W_r V^2/(C_{\alpha r} \, g \, R) \tag{6-13}$$

We must now look to the geometry of the vehicle in the turn to complete the analysis. With a little study of Figure 6.4, it can be seen that:

$$\delta = 57.3 \, L/R + \alpha_f - \alpha_r \tag{6-14}$$

Now substituting for α_f and α_r from Eqs. (6-12) and (6-13) gives:

$$\delta = 57.3 \frac{L}{R} + \frac{W_f V^2}{C_{\alpha f} g \, R} - \frac{W_r V^2}{C_{\alpha r} g \, R}$$

$$\delta = 57.3 \frac{L}{R} + (\frac{W_f}{C_{\alpha f}} - \frac{W_r}{C_{\alpha r}}) \frac{V^2}{g \, R} \tag{6-15}$$

where:

δ = Steer angle at the front wheels (deg)
L = Wheelbase (ft)
R = Radius of turn (ft)
V = Forward speed (ft/sec)
g = Gravitational acceleration constant = 32.2 ft/sec^2
W_f = Load on the front axle (lb)
W_r = Load on the rear axle (lb)
$C_{\alpha f}$ = Cornering stiffness of the front tires (lb$_y$/deg)
$C_{\alpha r}$ = Cornering stiffness of the rear tires (lb$_y$/deg)

Understeer Gradient

The equation is often written in a shorthand form as follows:

$$\delta = 57.3 \, L/R + K \, a_y \tag{6-16}$$

where:

K = Understeer gradient (deg/g)
a_y = Lateral acceleration (g)

Equation (6-15) is very important to the turning response properties of a motor vehicle. It describes how the steer angle of the vehicle must be changed with the radius of turn, R, or the lateral acceleration, $V^2/(g\,R)$. The term $[W_f/\,C_{\alpha f} - W_r/\,C_{\alpha r}]$ determines the magnitude and direction of the steering inputs required. It consists of two terms, each of which is the ratio of the load on the axle (front or rear) to the cornering stiffness of the tires on the axle. It is called the "Understeer Gradient," and will be denoted by the symbol, K, which has the units of degrees/g. Three possibilities exist:

1) **Neutral Steer**: $\qquad\qquad W_f/C_{\alpha f} = W_r/C_{\alpha r} \rightarrow K = 0 \rightarrow \alpha_f = \alpha_r$

On a constant-radius turn, no change in steer angle will be required as the speed is varied. Specifically, the steer angle required to make the turn will be equivalent to the Ackerman Angle, 57.3 L/R. Physically, the neutral steer case corresponds to a balance on the vehicle such that the "force" of the lateral acceleration at the CG causes an identical increase in slip angle at both the front and rear wheels.

2) **Understeer**: $\qquad\qquad W_f/C_{\alpha f} > W_r/C_{\alpha r} \rightarrow K > 0 \rightarrow \alpha_f > \alpha_r$

On a constant-radius turn, the steer angle will have to increase with speed in proportion to K (deg/g) times the lateral acceleration in g's. Thus it increases linearly with the lateral acceleration and with the square of the speed. In the understeer case, the lateral acceleration at the CG causes the front wheels to slip sideways to a greater extent than at the rear wheels. Thus to develop the lateral force at the front wheels necessary to maintain the radius of turn, the front wheels must be steered to a greater angle.

3) **Oversteer**: $\qquad\qquad W_f/C_{\alpha f} < W_r/C_{\alpha r} \rightarrow K < 0 \rightarrow \alpha_f < \alpha_r$

On a constant-radius turn, the steer angle will have to decrease as the speed (and lateral acceleration) is increased. In this case, the lateral acceleration at the CG causes the slip angle on the rear wheels to increase more than at the front. The outward drift at the rear of the vehicle turns the front wheels inward, thus diminishing the radius of turn. The increase in lateral acceleration that follows causes the rear to drift out even further and the process continues unless the steer angle is reduced to maintain the radius of turn.

The way in which steer angle changes with speed on a constant-radius turn for each of these cases is illustrated in Figure 6.5. With a neutral steer vehicle, the steer angle to follow the curve at any speed is simply the Ackerman Angle.

With understeer the angle increases with the square of the speed, reaching twice the initial angle at the characteristic speed. In the oversteer case, the steer angle decreases with the square of the speed and becomes zero at the critical speed value.

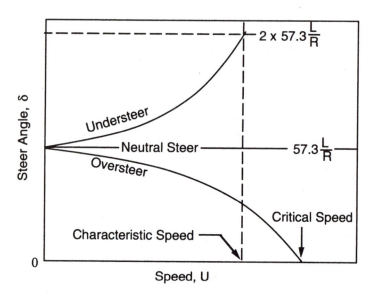

Fig. 6.5 Change of steer angle with speed.

Characteristic Speed

For an understeer vehicle, the understeer level may be quantified by a parameter known as the characteristic speed [8]. Characteristic speed is simply the speed at which the steer angle required to negotiate any turn is twice the Ackerman Angle. This can be seen in Eq. (6-16) when:

$$K \, a_y = 57.3 \, L/R \qquad (6\text{-}17)$$

Since a_y is a function of speed squared, the characteristic speed is:

$$V_{char} = \sqrt{57.3 \, L \, g/K} \qquad (6\text{-}18)$$

Critical Speed

In the oversteer case, a critical speed will exist above which the vehicle will be unstable. The critical speed is given by the expression:

$$V_{crit} = \sqrt{-57.3 \, L \, g / K} \qquad\qquad (6\text{-}19)$$

where it must be remembered that K is negative in value, such that the expression under the square root is positive and has a real value. Note that the critical speed is dependent on the wheelbase of the vehicle; for a given level of oversteer, long-wheelbase vehicles have a higher critical speed than short-wheelbase vehicles. An oversteer vehicle can be driven at speeds less than the critical, but becomes directionally unstable at and above the critical speed. The significance of critical speed becomes more apparent through its influence on the lateral acceleration gain and yaw rate gain as discussed in the next sections.

Lateral Acceleration Gain

Inasmuch as one of the purposes for steering a vehicle is to produce lateral acceleration, the turning equation can be used to examine performance from this perspective. Equation (6-16) can be solved for the ratio of lateral acceleration, a_y, to the steering angle, δ. The ratio is the lateral acceleration gain, and is given by:

$$\frac{a_y}{\delta} = \frac{\dfrac{V^2}{57.3 \, L \, g}}{1 + \dfrac{K V^2}{57.3 \, L \, g}} \qquad (\text{g/sec}) \qquad\qquad (6\text{-}20)$$

Note that when K is zero (neutral steer), the lateral acceleration gain is determined only by the numerator and is directly proportional to speed squared. When K is positive (understeer), the gain is diminished by the second term in the denominator, and is always less than that of a neutral steer vehicle. Finally, when K is negative (oversteer), the second term in the denominator subtracts from 1, increasing the lateral acceleration gain. The magnitude of the term is dependent on the square of the speed, and goes to the value of 1 when the speed reaches the critical speed. Thus the critical speed of Eq. (6-19) corresponds to the denominator becoming zero (infinite gain) in the above equation.

Yaw Velocity Gain

A second reason for steering a vehicle is to change the heading angle by developing a yaw velocity (sometimes called "yaw rate"). The yaw velocity, **r**, is the rate of rotation in heading angle and is given by:

$$\mathbf{r} = 57.3 \ V/R \qquad\qquad (\text{deg/sec}) \qquad\qquad (6\text{-}21)$$

Substituting this expression into Eq. (6-16) and solving for the ratio of yaw velocity to steering angle produces:

$$\frac{\mathbf{r}}{\delta} = \frac{V/L}{1 + \dfrac{K \ V^2}{57.3 \ L \ g}} \qquad\qquad (6\text{-}22)$$

The ratio represents a "gain" which is proportional to velocity in the case of a neutral steer vehicle. This is illustrated in Figure 6.6. It is readily shown that in the oversteer case the yaw velocity gain becomes infinite when the speed reaches the critical speed in accordance with Eq. (6-19). In the case of the understeer vehicle, the yaw velocity increases with speed up to the characteristic speed, then begins to decrease thereafter. Thus the characteristic speed has significance as the speed at which the vehicle is most responsive in yaw.

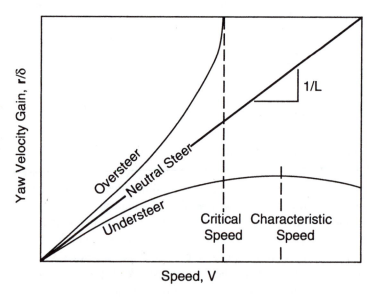

Fig. 6.6 Yaw velocity gain as a function of speed.

Sideslip Angle

From the discussion of turning behavior, it is evident that when the lateral acceleration is negligible, the rear wheel tracks inboard of the front wheel. But as lateral acceleration increases, the rear of the vehicle must drift outboard to develop the necessary slip angles on the rear tires. At any point on the vehicle a sideslip angle may be defined as the angle between the longitudinal axis and the local direction of travel. In general, the sideslip angle will be different at every point on a car during cornering.

Taking the center of gravity as a case in point, the sideslip angle is defined as shown in Figure 6.7. The sideslip angle is defined as positive for this case because the direction of travel (the local velocity vector) is oriented clockwise from the longitudinal axis (clockwise angles viewed from above are positive in SAE convention). At high speed the slip angle on the rear wheels causes the sideslip angle at the CG to become negative as in Figure 6.8.

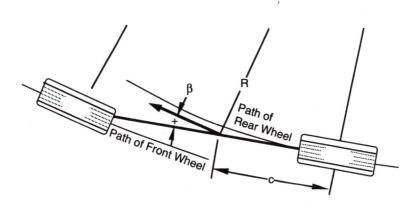

Fig 6.7 Sideslip angle in a low-speed turn.

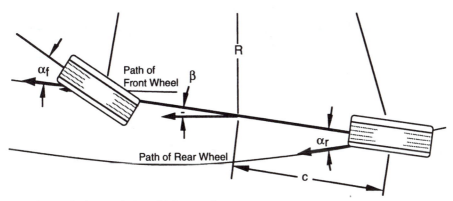

Fig. 6.8 Sideslip angle in a high-speed turn.

For any speed the sideslip angle, β, at the CG will be:

$$\beta = 57.3 \; c/R - \alpha_r$$
$$= 57.3 \; c/R - W_r \; V^2/(C_{\alpha r} \; g \; R) \tag{6-23}$$

Note that the speed at which the sideslip angle becomes zero is:

$$V_{\beta=0} = \sqrt{57.3 \; g \; c \; C_{\alpha r}/W_r} \tag{6-24}$$

and is independent of the radius of turn.

Static Margin

A term often used in discussions of handling is the static margin and, like understeer coefficient or characteristic speed, provides a measure of the steady-state handling behavior.

Static margin is determined by the point on the vehicle where a side force will produce no steady-state yaw velocity (i.e., the neutral steer point). We may go one step further and define a neutral steer line as shown in Figure 6.9. The neutral steer line is the locus of points in the x-z plane along which external lateral forces produce no steady-state yaw velocity.

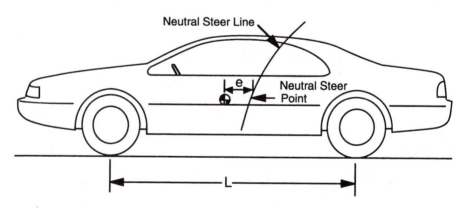

Fig. 6.9 Neutral steer line on a vehicle.

The static margin is defined as the distance the neutral steer point falls behind the CG, normalized by the wheelbase. That is:

Static Margin = e/L $\qquad$ (6-25)

When the point is behind the CG the static margin is positive and the vehicle is understeer. At the CG the margin is zero and the vehicle is neutral steer. When ahead of the CG, the vehicle is oversteer. On typical vehicles the static margin falls in the range of 0.05 to 0.07 behind the CG.

SUSPENSION EFFECTS ON CORNERING

The analysis of turning, thus far, has shown that the behavior is dependent on the ratios of load/cornering coefficient on the front and rear axles ($W_f/C_{\alpha f}$ and $W_r/C_{\alpha r}$). The ratios have the engineering units of deg/g, and have been called the "cornering compliance" [9]. The name arises from the fact that the ratio indicates the number of degrees of slip angle at an axle per "g" of lateral force imposed at that point. Inasmuch as the lateral force in a turn is actually a "D'Alembert" force at the CG, it is distributed at the axles in exact proportions to the weight (as the gravitational force is distributed).

Although the understeer gradient was derived for the case of a vehicle in a turn, it can be shown that the gradient determines vehicle response to disturbances in straight-ahead driving. In particular, an analysis by Rocard [10] demonstrates that oversteer vehicles have a stability limit at the critical speed due to normal disturbances in straight-ahead travel.

When the front axle is more compliant than the rear (understeer vehicle), a lateral disturbance produces more sideslip at the front axle; hence, the vehicle turns away from the disturbance. This is illustrated in Olley's definitions for understeer and oversteer [11] shown in Figure 6.10. If the rear axle exhibits more cornering compliance (oversteer), the rear of the vehicle drifts out, and it turns into the disturbance. The lateral acceleration acting at the CG adds to the disturbance force further increasing the turning response and precipitating instability.

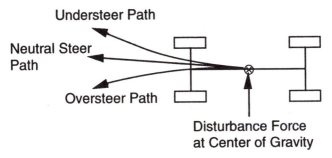

Fig. 6.10 Olley's definitions for understeer/oversteer.

Although tire cornering stiffness was used as the basis for developing the equations for understeer/oversteer, there are multiple factors in vehicle design that may influence the cornering forces developed in the presence of a lateral acceleration. Any design factor that influences the cornering force developed at a wheel will have a direct effect on directional response. The suspensions and steering system are the primary sources of these influences. In this section the suspension factors affecting handling will be discussed.

Roll Moment Distribution

For virtually all pneumatic tires the cornering forces are dependent on, and nonlinear with, load. This is important because load is transferred in the lateral direction in cornering due to the elevation of the vehicle CG above the ground plane. Figure 6.11 shows a typical example of how lateral force varies with vertical load.

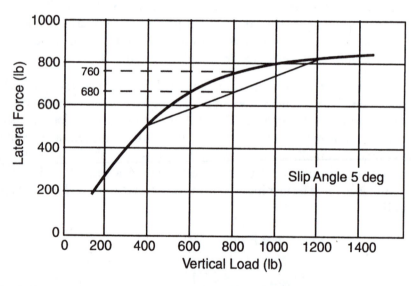

Fig. 6.11 Lateral force-vertical load characteristics of tires.

For a vehicle at 800 lb load on each wheel, about 760 lb of lateral force will be developed by each wheel at the 5-degree slip angle. In hard cornering, the loads might typically change to 400 lb on the inside wheel and 1200 lb on the outside. Then the average lateral force from both tires will be reduced to about

210

680 lb. Consequently, the tires will have to assume a greater slip angle to maintain the lateral force necessary for the turn. If these are front tires, the front will plough out and the vehicle will understeer. If on the rear, the rear will slip out and the vehicle will oversteer.

Actually, this mechanism is at work on both axles of all vehicles. Whether it contributes to understeer or oversteer depends on the balance of roll moments distributed on the front and rear axles. More roll moment on the front axle contributes to understeer, whereas more roll moment on the rear axle contributes to oversteer. Auxiliary roll stiffeners (stabilizer bars) alter handling performance primarily through this mechanism—applied to the front axle for understeer, and to the rear for oversteer.

The mechanics governing the roll moment applied to an axle are shown in the model of Figure 6.12. All suspensions are functionally equivalent to the two springs. The lateral separation of the springs causes them to develop a roll resisting moment proportional to the difference in roll angle between the body and axle. The stiffness is given by:

$$K_\phi = 0.5 \ K_s \ s^2 \tag{6-26}$$

where:

K_ϕ = Roll stiffness of the suspension
K_s = Vertical rate of each of the left and right springs
s = Lateral separation between the springs

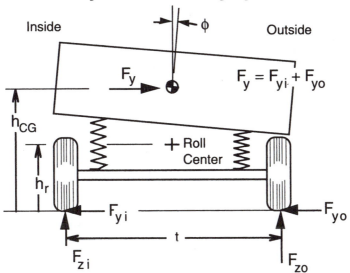

Fig. 6.12 Force analysis of a simple vehicle in cornering.

(In the case of an independent suspension, the above expression can be used by substituting the rate at the wheel for K_s and using the tread as the separation distance. When a stabilizer bar is present, the roll stiffness of the bar must be added to the stiffness calculated above.)

The suspension is further characterized by a "roll center," the point at which the lateral forces are transferred from the axle to the sprung mass. The roll center can also be thought of as the point on the body at which a lateral force application will produce no roll angle, and it is the point around which the axle rolls when subjected to a pure roll moment.

By writing Newton's Second Law for moments on the axle, we can determine the relationship between wheel loads and the lateral force and roll angle. In addition to the vertical forces imposed at the tires there is a net lateral force, F_y (the sum of the lateral forces on the inside and outside wheels), acting to the right on the axle at its roll center. The body roll acting through the springs imposes a torque on the axle proportional to the roll stiffness, K_ϕ, times the roll angle, ϕ. This results in an equation for the load difference from side to side of the form:

$$F_{zo} - F_{zi} = 2\, F_y\, h_r/t + 2\, K_\phi\, \phi/t = 2\, \Delta F_z \tag{6-27}$$

where:

F_{zo} = Load on the outside wheel in the turn
F_{zi} = Load on the inside wheel in the turn
F_y = Lateral force = $F_{yi} + F_{yo}$
h_r = Roll center height
t = Tread (track width)
K_ϕ = Roll stiffness of the suspension
ϕ = Roll angle of the body

Note that lateral load transfer arises from two mechanisms:

1) $2\, F_y\, h_r/t$—Lateral load transfer due to cornering forces. This mechanism arises from the lateral force imposed on the axle, and is thus an instantaneous effect. It is independent of roll angle of the body and the roll moment distribution.

2) $2\, K_\phi\, \phi/t$—Lateral load transfer due to vehicle roll. The effect depends on the roll dynamics, and thus may lag the changes in cornering conditions. It is directly dependent on front/rear roll moment distribution.

The total vehicle must be considered to obtain the expression for the roll moment distribution on the front and rear axles. In this case, we define a roll axis as the line connecting the roll centers of the front and rear suspensions, as shown in Figure 6.13. Now the moment about the roll axis in this case is:

$$M_\phi = [W\, h_1 \sin \phi + W\, V^2/(R\, g)\, h_1 \cos \phi] \cos \varepsilon \qquad (6\text{-}28)$$

For small angles, $\cos \phi$ and $\cos \varepsilon$ may be assumed as unity, and $\sin \phi = \phi$. Then:

$$M_\phi = W\, h_1\, [V^2/(R\, g) + \phi] \qquad (6\text{-}29)$$

But:

$$M_\phi = M_{\phi f} + M_{\phi r} = (K_{\phi f} + K_{\phi r})\, \phi \qquad (6\text{-}30)$$

Equations (6-28) and (6-29) can be solved for the roll angle, ϕ:

$$\phi = \frac{W\, h_1\, V^2/(R\, g)}{K_{\phi f} + K_{\phi r} - W\, h_1} \qquad (6\text{-}31)$$

The derivative of this expression with respect to the lateral acceleration produces an expression for the roll rate of the vehicle:

$$R_\phi = d\phi/da_y = W\, h_1/[K_{\phi f} + K_{\phi r} - W\, h_1] \qquad (6\text{-}32)$$

The roll rate is usually in the range of 3 to 7 degrees/g on typical passenger cars.

Combining the expression for ϕ from Eq. (6-31) with Eq. (6-29) allows solution for the roll moments on the front and rear axles:

$$M'_{\phi f} = K_{\phi f} \frac{W\, h_1\, V^2/(R\, g)}{K_{\phi f} + K_{\phi r} - W\, h_1} + W_f h_f V^2/(R\, g) = \Delta F_{zf}\, t_f \qquad (6\text{-}33)$$

$$M'_{\phi r} = K_{\phi r} \frac{W\, h_1\, V^2/(R\, g)}{K_{\phi f} + K_{\phi r} - W\, h_1} + W_r h_r V^2/(R\, g) = \Delta F_{zr}\, t_r \qquad (6\text{-}34)$$

where:

$$\Delta F_{zf} = F_{zfo} - W_f/2 = -(F_{zfi} - W_f/2)$$
$$\Delta F_{zr} = F_{zro} - W_r/2 = -(F_{zri} - W_r/2)$$

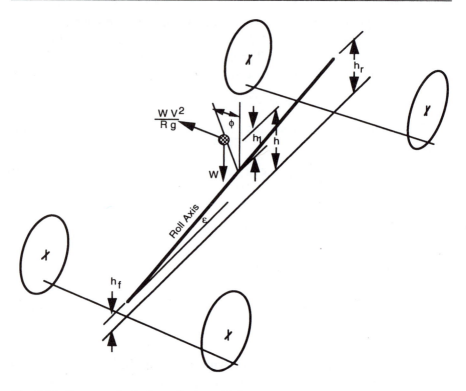

Fig. 6.13 Force analysis for roll of a vehicle.

In general, the roll moment distribution on vehicles tends to be biased toward the front wheels due to a number of factors:

1) Relative to load, the front spring rate is usually slightly lower than that at the rear (for flat ride), which produces a bias toward higher roll stiffness at the rear. However, independent front suspensions used on virtually all cars enhance front roll stiffness because of the effectively greater spread on the front suspension springs.

2) Designers usually strive for higher front roll stiffness to ensure understeer in the limit of cornering.

3) Stabilizer bars are often used on the front axle to obtain higher front roll stiffness.

4) If stabilizer bars are needed to reduce body lean, they may be installed on the front or the front and rear. Caution should be used when adding a stabilizer bar only to the rear because of the potential to induce unwanted oversteer.

We now have the solution for roll moments front and rear, and can calculate the difference in load between the left and right wheels on the axle. To translate the lateral load transfer into an effect on understeer gradient, it is necessary to have data which relates the tire cornering force to slip angle and load. At the given conditions, the slip angle on each axle will change when the load transfer is taken into account. The difference between the change on the front and rear (normalized by the lateral acceleration) represents the understeer effect. The effect can be modeled by expressing the tire load sensitivity as a polynomial. In the first analysis the cornering characteristics of the tires on an axle were described simply by a constant called the cornering stiffness, C_α. The cornering force developed on the axle was given by:

$$F_y = C_\alpha\, \alpha \qquad\qquad (6\text{-}35)$$

where:

F_y = Lateral force developed on the axle
C_α = Cornering stiffness of two tires, each at one-half the axle load
α = Slip angle

To represent load sensitivity effect, the two tires (inside and outside) must be treated separately. The cornering stiffness of each tire can be represented by a second- or higher-order polynomial, and the lateral force developed by either will be given by:

$$F_y{}' = C_\alpha{}'\, \alpha = (a\, F_z - b\, F_z{}^2)\, \alpha \qquad\qquad (6\text{-}36)$$

where:

$F_y{}'$ = Lateral force of one tire
$C_\alpha{}'$ = Cornering stiffness of one tire
a = First coefficient in the cornering stiffness polynomial ($lb_y/lb_z/deg$)
b = Second coefficient in the cornering stiffness polynomial ($lb_y/lb_z{}^2/deg$)
F_z = Load on one tire (assumed equal on both tires in previous analysis)

For a vehicle cornering as shown in Figure 6.12, the lateral force of both tires, F_y , is given by:

$$F_y = (a\, F_{zo} - b\, F_{zo}{}^2 + a\, F_{zi} - b\, F_{zi}{}^2\,)\, \alpha \qquad\qquad (6\text{-}37)$$

Now, let the load change on each wheel be given by ΔF_z.

215

$$F_{zo} = F_z + \Delta F_z \qquad\qquad F_{zi} = F_z - \Delta F_z \tag{6-38}$$

Then:

$$F_y = [a\,(F_z + \Delta F_z) - b\,(F_z + \Delta F_z)^2 + a\,(F_z - \Delta F_z) - b\,(F_z - \Delta F_z)^2\,]\,\alpha \tag{6-39}$$

This equation reduces to:

$$F_y = [2\,a\,F_z - 2\,b\,F_z{}^2 - 2\,b\,\Delta F_z{}^2\,]\,\alpha \tag{6-40}$$

The equation can be simplified if we recognize that the first two terms in the brackets are equivalent to the cornering stiffness of the tires at their static load conditions (as it has been defined in the previous analysis). Namely:

$$C_\alpha = 2\,a\,F_z - 2\,b\,F_z{}^2 \tag{6-41}$$

or:

$$F_y = [C_\alpha - 2\,b\,\Delta F_z{}^2\,]\,\alpha \tag{6-42}$$

Recall that the steer angle necessary to maintain a turn is given by:

$$\delta = 57.3\,L/R + \alpha_f - \alpha_r \tag{6-43}$$

For the two tires on the front we can write:

$$F_{yf} = [C_{\alpha f} - 2\,b\,\Delta F_{zf}{}^2\,]\,\alpha_f = W_f\,V^2/(R\,g) \tag{6-44}$$

and on the rear:

$$F_{yr} = [C_{\alpha r} - 2\,b\,\Delta F_{zr}{}^2\,]\,\alpha_r = W_r\,V^2/(R\,g) \tag{6-45}$$

Substituting to eliminate the slip angles in Eq. (6-43):

$$\delta = 57.3\,\frac{L}{R} + \frac{W_f\,V^2/(R\,g)}{(C_{\alpha f} - 2\,b\,\Delta F_{zf}^2)} - \frac{W_r\,V^2/(R\,g)}{(C_{\alpha r} - 2\,b\,\Delta F_{zr}^2)} \tag{6-46}$$

This equation can be simplified by utilizing the fact that $C_\alpha \gg 2\,b\,\Delta F_z{}^2$. Then:

$$\frac{1}{(C_\alpha - 2\,b\,\Delta F_z^2)} = \frac{1}{C_\alpha(1 - \dfrac{2\,b\,\Delta F_z^2}{C_\alpha})} \approx \frac{1}{C_\alpha}(1 + \frac{2\,b\,\Delta F_z^2}{C_\alpha}) \tag{6-47}$$

Equation (6-45) can be rewritten in the form:

$$\delta = 57.3 \frac{L}{R} + [\,(\frac{W_f}{C_{\alpha f}} - \frac{W_r}{C_{\alpha r}}) + (\frac{W_f}{C_{\alpha f}} \frac{2\,b\,\Delta F_{zf}^2}{C_{\alpha f}} - \frac{W_r}{C_{\alpha r}} \frac{2\,b\,\Delta F_{zr}^2}{C_{\alpha r}})]\frac{V^2}{R\,g}$$

$$\underleftrightarrow{\;\;\;\;1\;\;\;\;} \underleftrightarrow{\;\;\;\;\;\;\;\;2\;\;\;\;\;\;\;\;}$$

$$(6-48)$$

Term number 1 inside the brackets is simply the understeer gradient arising from the nominal cornering stiffness of the tires, K_{tires}, as was developed earlier. The second term represents the understeer gradient arising from lateral load transfer on the tires; i.e.:

$$K_{llt} = \frac{W_f}{C_{\alpha f}} \frac{2\,b\,\Delta F_{zf}^2}{C_{\alpha f}} - \frac{W_r}{C_{\alpha r}} \frac{2\,b\,\Delta F_{zr}^2}{C_{\alpha r}} \qquad (6-49)$$

The values for ΔF_{zf} and ΔF_{zr} can be obtained from Eqs. (6-33) and (6-34) as a function of lateral acceleration. Since all the variables in the above equation are positive, the contribution from the front axle is always understeer; that from the rear axle is always negative, meaning it is an oversteer effect.

Camber Change

The inclination of a wheel outward from the body is known as the camber angle [2]. Camber on a wheel will produce a lateral force known as "camber thrust." Figure 6.14 shows a typical camber thrust curve.

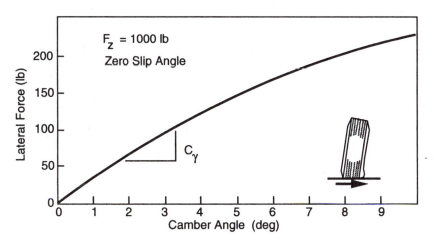

Fig. 6.14 Lateral force caused by camber of a tire.

Camber angle produces much less lateral force than slip angle. About 4 to 6 degrees of camber are required to produce the same lateral force as 1 degree of slip angle on a bias-ply tire. Camber stiffness of radial tires is generally lower than that for bias-ply tires; hence, as much as 10 to 15 degrees are required on a radial. Nevertheless, camber thrust is additive to the cornering force from slip angle, thus affecting understeer. Camber thrust of bias-ply tires is strongly affected by inflation pressure, although not so for radial tires, and it is relatively insensitive to load and speed for both radial and bias tires.

Camber angles are small on solid axles, and at best only change the lateral forces by 10% or less. On independent wheel suspensions, however, camber can play an important role in cornering. Camber changes both as a result of body roll and the normal camber change in jounce/rebound. Figure 6.15 illustrates the mechanisms of camber change as a vehicle rolls in cornering.

The total camber angle during cornering will be:

$$\gamma_g = \gamma_b + \phi \qquad (6\text{-}50)$$

where:

γ_g = Camber angle with respect to the ground
γ_b = Camber angle of the wheel with respect to the body
ϕ = Roll angle of the vehicle

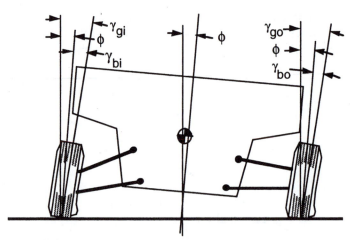

Fig. 6.15 Camber change in cornering of a vehicle.

Now the camber angle arising from the suspension is a function of the roll angle, because the jounce on the inside wheel and the rebound on the outside wheel relate directly to roll angle. Thus we can obtain the derivative of camber angle with roll angle from analysis of the suspension kinematics. The relationship is dependent on the geometry of the suspension, but for every suspension a kinematic analysis can be performed to develop a camber gradient of the form:

$$\partial\gamma/\partial\phi = f_\gamma \text{ (track width, suspension geometry, roll angle)} \qquad (6\text{-}51)$$

In turn, the roll angle can be related to lateral acceleration through Eq. (6-31) obtained earlier.

The influence on cornering comes about from the fact that the lateral force results not only from slip angle of the tire, but also the camber angle. That is:

$$F_y = C_\alpha \, \alpha + C_\gamma \, \gamma \qquad (6\text{-}52)$$

Thus:

$$\alpha = \frac{F_y}{C_\alpha} - \frac{C_\gamma}{C_\alpha}\gamma \qquad (6\text{-}53)$$

Now both F_y and γ are related to the lateral acceleration—F_y through Eq. (6-11) and γ through Eq. (6-52). Thus the equations for α_f and α_r take the forms:

$$\alpha_f = \frac{W_f}{C_\alpha} a_y - \frac{C_\gamma}{C_\alpha}\frac{\partial\gamma_f}{\partial\phi}\frac{\partial\phi}{\partial a_y} a_y \text{ and } \alpha_r = \frac{W_r}{C_\alpha} a_y - \frac{C_\gamma}{C_\alpha}\frac{\partial\gamma_r}{\partial\phi}\frac{\partial\phi}{\partial a_y} a_y \qquad (6\text{-}54)$$

When these are substituted into the turning equation (6-14), it takes the form:

$$\delta = 57.3\frac{L}{R} + [(\frac{W_f}{C_{\alpha f}} - \frac{W_r}{C_{\alpha r}}) + (\frac{C_{\gamma f}}{C_{\alpha f}}\frac{\partial\gamma_f}{\partial\phi} - \frac{C_{\gamma r}}{C_{\alpha r}}\frac{\partial\gamma_r}{\partial\phi})\frac{\partial\phi}{\partial a_y}]\frac{V^2}{R g} \qquad (6\text{-}55)$$

Therefore, the understeer deriving from camber angles on each axle is given by:

$$K_{camber} = (\frac{C_{\gamma f}}{C_{\alpha f}}\frac{\partial\gamma_f}{\partial\phi} - \frac{C_{\gamma r}}{C_{\alpha r}}\frac{\partial\gamma_r}{\partial\phi})\frac{\partial\phi}{\partial a_y} \qquad (6\text{-}56)$$

Roll Steer

When a vehicle rolls in cornering, the suspension kinematics may be such that the wheels steer. Roll steer is defined as the steering motion of the front or rear wheels with respect to the sprung mass that is due to the rolling motion of the sprung mass. Consequently, roll steer effects on handling lag the steer input, awaiting roll of the sprung mass.

The steer angle directly affects handling as it alters the angle of the wheels with respect to the direction of travel. Let "ε" be the roll steer coefficient on an axle (degrees steer/degree roll). Then by reasoning similar to the above, we can derive the understeer gradient contribution from roll steer as:

$$K_{\text{roll steer}} = (\varepsilon_f - \varepsilon_r)\frac{\partial \phi}{\partial a_y} \qquad (6\text{-}57)$$

A positive roll steer coefficient causes the wheels to steer to the right in a right-hand roll. Inasmuch as a right-hand roll occurs when the vehicle is turning to the left, positive roll steer on the front axle steers out of the turn and is understeer. Conversely, positive roll steer on the rear axle is oversteer.

On solid axles the suspension will allow the axle to roll about an imaginary axis which may be inclined with respect to the longitudinal axis of the vehicle. The kinematics of the suspension, regardless of design, may be envisioned as functionally equivalent to leading or trailing arm systems; and the roll axis inclination is equal to that of the arms. Given an initial inclination angle, β, on the arms, as the body rolls, the arm on the inside wheel rotates downward while the arm on the outside wheel rotates upward as illustrated in Figure 6.16.

Fig. 6.16 Roll steer with a solid axle.

If the initial orientation of a rear axle trailing arm is angled downward, as seen in the figure, the effect of the trailing arm angle change is to pull the inside wheel forward while pushing the outside wheel rearward. This produces roll steer of the solid axle contributing to oversteer. The roll steer coefficient is equal to the inclination angle ($\varepsilon = \beta$, in radians) of the trailing arms. On a rear trailing arm system, roll understeer is achieved by keeping the transverse pivots of the trailing arms below the wheel center. Figure 6.17 illustrates the effect of trailing arm angle on understeer.

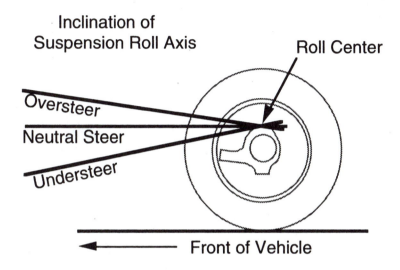

Fig. 6.17 Influence of rear axle trailing arm angle on understeer.

With independent suspensions the roll steer coefficient must be evaluated from the kinematics of the suspension. On steered wheels, the interactions with the steering system must also be taken into account.

Lateral Force Compliance Steer

With the soft bushings used in suspension linkages for NVH reasons, there is the possibility of steer arising from lateral compliance in the suspension. With the simple solid axle, compliance steer can be represented as rotation about a yaw center as illustrated in Figure 6.18.

With a forward yaw center on a rear axle, the compliance allows the axle to steer toward the outside of the turn, thus causing oversteer. Conversely, a rearward yaw center results in understeer. On a front axle, just the opposite is true—a rearward yaw center is oversteer, and a forward yaw center is understeer.

221

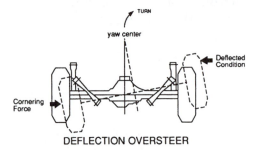

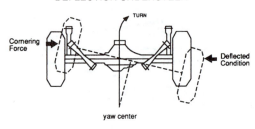

Fig. 6.18 Steer due to lateral compliance in the suspension.

The handling influence of lateral force compliance steer can be quantified by defining an appropriate coefficient as follows:

$$A = \delta_c/F_y \qquad \text{(degrees steer/unit lateral force)} \qquad (6\text{-}58)$$

where:

δ_c = Steer angle
F_y = Lateral force

The lateral force experienced on an axle is simply the load on the axle times the lateral acceleration. Thus on the front axle:

$$\delta_{cf} = A_f W_f a_y \qquad (6\text{-}59)$$

Since the understeer effect is directly related to the steer angles produced on the front and rear axles, the understeer arising from lateral force compliance steer is:

$$K_{lfcs} = A_f W_f - A_r W_r \qquad (6\text{-}60)$$

Of course, the kinematics of linkages must be analyzed and taken into account to determine the coefficients on independent wheel suspensions and on steered wheels.

Aligning Torque

The aligning torque experienced by the tires on a vehicle always resists the attempted turn, thus it is the source of an understeer effect. Aligning torque is the manifestation of the fact that the lateral forces are developed by a tire at a point behind the tire center. The distance is known as the "pneumatic trail (p)."

The direct handling influence can be determined by deriving the turning equations with the assumption that the lateral forces are developed not at the wheels, but at a distance "p" behind each wheel. The understeer term obtained is:

$$K_{at} = W \frac{p}{L} \frac{C_{\alpha f} + C_{\alpha r}}{C_{\alpha f} C_{\alpha r}}$$

(6-61)

Because the C_α values are positive, the aligning torque effect is positive (understeer) and cannot ever be negative (oversteer).

The understeer due to this mechanism is normally less than 0.5 deg/g. However, aligning torque is indirectly responsible for additional, and more significant, understeer mechanisms through its influence on the steering system. These mechanisms will be discussed with the steering system.

Effect of Tractive Forces on Cornering

The turning analysis developed at the outset of this chapter does not consider the potential effects of drive forces present at the wheels. We will now look at the case of drive forces present at front and rear wheels to develop the general equation showing their influence.

With drive forces, the "bicycle" model for turning is as shown in Figure 6.19. The application of Newton's Second Law in the lateral direction takes the form:

$$W_f V^2/(R\,g) = F_{yf} \cos(\alpha_f + \delta) + F_{xf} \sin(\alpha_f + \delta)$$

(6-62)

$$W_r V^2/(R\,g) = F_{yr} \cos\alpha_r + F_{xr} \sin\alpha_r$$

(6-63)

where:

W_f, W_r = Load on the front and rear axles
V = Forward speed
R = Radius of turn
F_{yf}, F_{yr} = Cornering forces on front and rear axles
F_{xf}, F_{xr} = Tractive forces on the front and rear axles
α_f, α_r = Slip angles at front and rear wheels

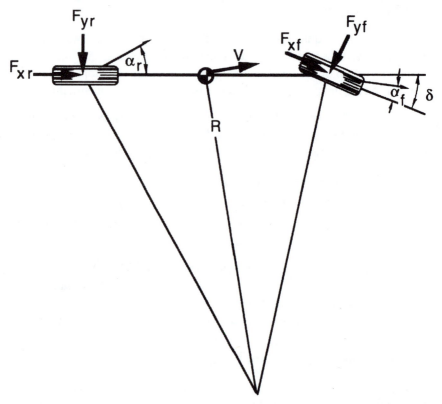

Fig. 6.19 Cornering model with tractive forces.

Now the lateral forces, F_{yf} and F_{yr}, are simply the cornering stiffness on the axle times the slip angle. When that substitution is made in Eqs. (6-62) and (6-63), the right-hand side contains only the tractive forces and the slip angles. Assuming small angles, $\cos \alpha = 1$ and $\sin \alpha = \alpha$, we can then solve for α_f and α_r for substitution into the geometry equation:

$$\delta = 57.3 \, L/R + \alpha_f - \alpha_r \tag{6-64}$$

On substitution, δ occurs on both the left- and right-hand sides of the equation, and it is necessary to manipulate to get it only on the left-hand side. Then:

$$\delta = \frac{57.3 \, L/R}{1 + F_{xf}/C_{\alpha f}} + \frac{\dfrac{W_f \, V^2}{C_{\alpha f} \, R \, g}}{1 + F_{xf}/C_{\alpha f}} - \frac{\dfrac{W_r \, V^2}{C_{\alpha r} \, R \, g}}{1 + F_{xr}/C_{\alpha r}} \tag{6-65}$$

The equation can be gotten into a somewhat more convenient form when we realize that $F_{xf}/C_{\alpha f}$ and $F_{xr}/C_{\alpha r}$ are much less than one. In that case,

$$\frac{1}{1 + F_{xf}/C_{\alpha f}} \approx 1 - F_{xf}/C_{\alpha f} \tag{6-66}$$

The same is true for the rear axle.

Then Eq. (6-65) can be manipulated to the form:

$$\delta = \frac{57.3 \, L/R}{1 + F_{xf}/C_{\alpha f}} + [(\frac{W_f}{C_{\alpha f}} - \frac{W_r}{C_{\alpha r}}) - (\frac{W_f}{C_{\alpha f}}\frac{F_{xf}}{C_{\alpha f}} - \frac{W_r}{C_{\alpha r}}\frac{F_{xr}}{C_{\alpha r}})]\frac{V^2}{R \, g}$$

$$\longleftarrow 1 \longrightarrow \quad \longleftarrow 2 \longrightarrow \longleftarrow \qquad 3 \qquad \longrightarrow \tag{6-67}$$

This is the final turning equation for the case where tractive forces are taken into account. Note that though it appears more complicated than that developed earlier, it contains the same basic terms. The three terms on the right-hand side are as follows:

Term 1—This is the Ackerman steer angle altered by the tractive force on the front axle (rear tractive force does not show up here).

- If F_{xf} is positive (drive force applied in a FWD) it reduces the required steer angle for low-speed maneuvers, and accounts for the sense that FWD "pulls" a vehicle around in low-speed maneuvers.

- If F_{xf} is negative (equivalent to rolling resistance on a RWD or engine drag on a FWD) it tends to increase the required steer angle for turning.

- When front wheels spin on snow or ice, tractive force is still produced but $C_{\alpha f}$ goes to zero. In that case the denominator of the term becomes infinite, suggesting that turns of zero radius can be made with virtually no steer angle. This accounts for the "trick" of turning a FWD vehicle within its own length on an icy surface by turning the wheel sharply and making them spin.

Term 2—This is the understeer gradient, unchanged from its earlier form.

Term 3—This term represents the effect of tractive forces on the understeer behavior of the vehicle.

- If F_{xf} is positive it causes an oversteer influence (pulls the front of the vehicle into the turn). Thus this mechanism is an oversteer influence with a FWD in the throttle-on case.

225

- If F_{xr} is positive it causes an understeer influence by the same reasoning on a RWD.

- On a 4WD these mechanisms would suggest that the rear axle should "over drive" the front axle to ensure understeer behavior.

Anyone familiar with a FWD vehicle may be aware that the throttle-on oversteer mechanism described here is not evident with most vehicles. In discussion of the effects of FWD in the steering section, it will be seen that the modification of tire cornering properties caused by traction forces has a stronger influence on handling than the direct action of the forces on the vehicle.

SUMMARY OF UNDERSTEER EFFECTS

The understeer coefficient, K, for a vehicle is the result of tire, vehicle and steering system parameters. Its total value is computed as the sum of a number of effects as summarized in the following table.

UNDERSTEER COMPONENT	SOURCE
$K_{tires} = \dfrac{W_f}{C_{\alpha f}} - \dfrac{W_r}{C_{\alpha r}}$	Tire cornering stiffness
$K_{camber} = (\dfrac{C_{\gamma f}}{C_{\alpha f}} \dfrac{\partial \gamma_f}{\partial \phi} - \dfrac{C_{\gamma r}}{C_{\alpha r}} \dfrac{\partial \gamma_r}{\partial \phi}) \dfrac{\partial \phi}{\partial a_y}$	Camber thrust
$K_{roll\ steer} = (\varepsilon_f - \varepsilon_r)\, d\phi/da_y$	Roll steer
$K_{lfcs} = A_f W_f - A_r W_r$	Lateral force compliance steer
$K_{at} = W \dfrac{p}{L} \dfrac{C_{\alpha f} + C_{\alpha r}}{C_{\alpha f} C_{\alpha r}}$	Aligning torque
$K_{llt} = \dfrac{W_f}{C_{\alpha f}} \dfrac{2\, b\, \Delta F_{zf}^2}{C_{\alpha f}} - \dfrac{W_r}{C_{\alpha r}} \dfrac{2\, b\, \Delta F_{zr}^2}{C_{\alpha r}}$	Lateral load transfer
$K_{strg} = W_f \dfrac{r\, v + p}{K_{ss}}$	Steering system

EXPERIMENTAL MEASUREMENT OF UNDERSTEER GRADIENT

Understeer gradient is defined by the SAE [2] as "The quantity obtained by subtracting the Ackerman steer angle gradient from the ratio of the steering wheel angle gradient to the overall steering ratio." Methods for experimental measurement of understeer gradient [12, 13, 14] are all based on the definition of the gradient reflected in Eq. (6-16). Namely,

$$\delta = 57.3 \, L/R + K \, a_y \tag{6-16}$$

The derivation of this equation assumes the vehicle to be in a steady-state operating condition; therefore, understeer is defined as a steady-state property. For experimental measurement the vehicle must be placed into a steady-state turn with appropriate measures of the quantities in the above equation so that the value of K can be determined. Four test methods have been suggested as means to measure this property—constant radius, constant speed, constant steer angle and constant throttle. Only the first two reasonably reflect normal driving circumstances, hence, the discussion will be limited to these two.

Constant Radius Method

Understeer can be measured by operating the vehicle around a constant radius turn and observing steering angle versus lateral acceleration. The method closely replicates vehicle operation in many highway situations, such as the constant radius turns in off ramps from limited access highways. At a minimum, instrumentation must be available to measure steering wheel angle and lateral acceleration. Given the radius of turn and some measure of vehicle velocity (from the speedometer, fifth wheel or by lap time), lateral acceleration can be computed using the relationship:

$$a_y = V^2/(Rg) \tag{6-68}$$

The recommended procedure is to drive the vehicle around the circle at very low speed, for which the lateral acceleration is negligible, and note the steer angle (Ackerman steer angle) required to maintain the turn. (The experimenter is challenged to develop good technique for this process as cross-slope on the test surface, bumps, etc., will cause the vehicle to drift in and out as it proceeds, complicating the determination of the average steer angle.) Vehicle speed is then increased in steps that will produce lateral accelerations at reasonable increments (typically 0.1 g), noting the steer angle at each speed. The steer angle (divided by the steering ratio to obtain the road wheel angle) is then plotted as a function of lateral acceleration as illustrated in Figure 6.20.

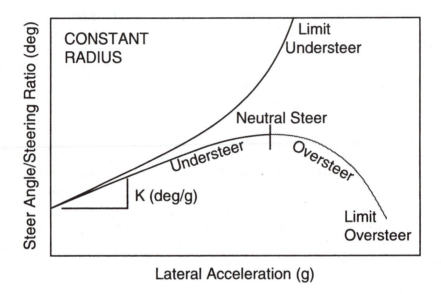

Fig 6.20 Example measurements of understeer gradient by constant radius method.

The meaning of this plot can be seen by taking the derivative of Eq. (6-16):

$$\frac{\partial \delta}{\partial a_y} = \frac{\partial}{\partial a_y} (57.3 \frac{L}{R}) + K \frac{\partial a_y}{\partial a_y} \qquad (6\text{-}69)$$

Since the radius of turn is constant, the Ackerman steer angle is also constant and its derivative is zero. Thus:

$$K = \frac{\partial \delta}{\partial a_y} \qquad (6\text{-}70)$$

The slope of the steer angle curve is the understeer gradient. A positive slope (upward to the right) indicates understeer, zero slope is neutral steer, and a negative slope is oversteer. Typical measurements will take one of the forms shown in Figure 6.20. Some vehicles will be understeer over the entire operating range, remaining so to the limit. Others may be understeer at low lateral acceleration levels but change to oversteer at high lateral acceleration levels and exhibit limit oversteer.

Note that the implied road wheel steer angle (obtained by dividing the steering wheel angle by the steering ratio) is used for characterizing the understeer gradient. While understeer is measured at the steering wheel, it is described by the degrees per g of steer required at the road wheel. As will be seen in the discussion of steering systems, the steering ratio is not a constant because of compliance in the system. This does not invalidate the measurement method, but rather recognizes that those properties in the steering system are a legitimate source of understeer on the vehicle. In cases where the road wheel steer angle is measured directly, a different understeer gradient will be obtained because the steering system effects will not be included. While this method is not incorrect, it fails to fully characterize the understeer properties of the vehicle by excluding contribution from the steering system. Recognizing that the driver must control the vehicle from the steering wheel, the steering system effects should be included in a full characterization of understeer.

The constant radius method has the advantage that minimal instrumentation is required, but has the disadvantage that it is difficult to execute in an objective fashion. Determination of a precise steering wheel angle is difficult because of the deviations necessary to keep the vehicle on the selected radius of turn. This aspect of test technique is not readily controlled.

The minimum radius of turn for this test procedure is normally 30 m ($\approx$100 ft). For two-axle vehicles the understeer gradient is not affected by the radius of the circle. The gradient for multi-axle straight trucks (three axles or more), however, is sensitive to turn radius in this range.

Constant Speed Method

Understeer can be measured at constant speed by varying the steer angle. Measurements by this method closely duplicate many real driving situations since vehicles are normally driven at near constant speed. With this method the radius of turn will vary continuously requiring more extensive data collection to determine the gradient. In addition to measuring speed and steer angle, the radius of turn must be determined for each condition as well. The most practical means to measure radius of turn is either by measuring lateral acceleration or yaw rate. The radius of turn is derived from the measurements using the appropriate form of the relationships below:

$$R = V^2/a_y = V/r \qquad (6\text{-}71)$$

where:

V = Forward speed (ft/sec or m/sec)
a_y = Lateral acceleration (ft/sec^2 or m/sec^2)
r = Yaw rate (radians/sec)

The Ackerman steer angle gradient for this test procedure is obtained by substituting Eq. (6-68) into Eq. (6-16), eliminating the radius. This produces the form:

$$\delta = 57.3 \, L/R + K a_y = 57.3 \, L a_y / V^2 + K a_y \qquad (6\text{-}72)$$

Again taking derivatives with respect to lateral acceleration, we obtain the expression for the understeer gradient:

$$K = \frac{\partial \delta}{\partial a_y} - 57.3 \, \frac{L}{V^2} \qquad (6\text{-}73)$$

Since speed and wheelbase are constant, the Ackerman steer angle gradient (the second term on the right-hand side) is a straight line of constant slope and appears in a data plot as shown in Figure 6.21. The Ackerman steer angle gradient is neutral steer. In regions where the steer angle gradient is greater than that of the Ackerman, the vehicle is understeer. A point where the two have the same slope is neutral steer, and where the steer angle gradient is less than the Ackerman, the vehicle is oversteer. For the oversteer vehicle, the point where the slope of the steer angle curve is zero is the stability boundary corresponding to the critical speed.

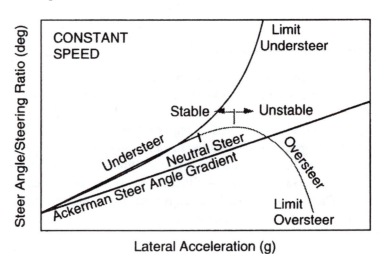

Fig. 6.21 Example measurements of understeer gradient by constant speed method.

EXAMPLE PROBLEMS

1) A car has a weight of 1901 lb front axle and 1552 lb on the rear with a wheelbase of 100.6 inches. The tires have the following cornering stiffness values:

Load	Cornering Stiffness	Cornering Coefficient
225 lb	67 lb/deg	0.298 lb/lb/deg
450	121	0.269
675	171	0.253
900	225	0.250
1125	257	0.228
1350	300	0.222

Determine the following cornering properties for the vehicle:

a) Ackerman steer angles for 500, 200, 100 and 50 ft turn radius
b) Understeer gradient
c) Characteristic speed
d) Lateral acceleration gain at 60 mph
e) Yaw velocity gain at 60 mph
f) Sideslip angle at the CG on an 800 ft radius turn at 60 mph
g) Static margin

Solution:

a) The Ackerman steer angles are only a function of the wheelbase and radius of turn, and can be found easily from Eq. (6-3).

$$\delta_{500} = L/R = 100.6 \text{ in } (1 \text{ ft}/12 \text{ in})/500 \text{ ft} = 0.01677 \text{ radians} = 0.96 \text{ deg}$$

$$\delta_{200} \quad = 100.6 \text{ in } (1 \text{ ft}/12 \text{ in})/200 \text{ ft} = 0.0419 \text{ radians} = 2.4 \text{ deg}$$

$$\delta_{100} \quad = 100.6 \text{ in } (1 \text{ ft}/12 \text{ in})/100 \text{ ft} = 0.0838 \text{ radians} = 4.8 \text{ deg}$$

$$\delta_{50} \quad = 100.6 \text{ in } (1 \text{ ft}/12 \text{ in})/50 \text{ ft} = 0.1677 \text{ radians} = 9.6 \text{ deg}$$

b) In order to find the understeer gradients we must know the cornering stiffness of the tires at the prevailing loads. On the front axle the tire load is 950 lb per tire. Interpolating the cornering stiffness data between the loads of 900 and 1125 lb leads to a stiffness of 232 lb/deg at 950 lb. On the rear axle the load is 776 lb per tire. Again interpolating between the appropriate loads in the tire data, we obtain a cornering stiffness of 195 lb/deg. Now from Eq. (6-15) we get:

$$K = \frac{W_f}{C_{\alpha f}} - \frac{W_r}{C_{\alpha r}} = \frac{950 \text{ lb}}{232 \text{ lb/deg}} - \frac{776 \text{ lb}}{195 \text{ lb/deg}} = (4.09 - 3.98) \text{ deg} = 0.11 \text{ deg(/g)}$$

c) The characteristic speed is determined from Eq. (6-18):

$$V_{char} = \sqrt{57.3 \, L \, g/K} = \sqrt{57.3 \, \frac{\text{deg}}{\text{rad}} \, \frac{110.6 \text{ in}}{12 \text{ in/ft}} \, \frac{32.2 \text{ ft/sec}^2}{0.11 \text{ deg/g}}} = 393 \text{ ft/sec} = 268 \text{ mph}$$

d) The lateral acceleration gain comes from Eq. (6-20):

$$\frac{a_y}{\delta} = \frac{\dfrac{V^2}{57.3 \, L \, g}}{1 + \dfrac{K \, V^2}{57.3 \, L \, g}} = \frac{\dfrac{(88 \text{ ft/sec})^2}{57.3 \text{ deg/rad} \, (8.38 \text{ ft}) \, (32.2 \text{ ft/sec}^2)}}{1 + \dfrac{0.11 \text{ deg/g} \, (88 \text{ ft/sec})^2}{57.3 \text{ deg/rad} \, (8.38 \text{ ft}) \, (32.2 \text{ ft/sec}^2)}}$$

$$= \frac{0.501 \text{ g/deg}}{1.055} = 0.475 \text{ g/deg}$$

e) The yaw rate gain comes from Eq. (6-22):

$$\frac{r}{\delta} = \frac{V/L}{1 + \dfrac{K \, V^2}{57.3 \, L \, g}} = \frac{\dfrac{88 \text{ ft/sec}}{8.38 \text{ ft}}}{1 + \dfrac{0.11 \text{ deg/g} \, (88 \text{ ft/sec})^2}{57.3 \text{ deg/rad} \, (8.38 \text{ ft}) \, (32.2 \text{ ft/sec}^2)}}$$

$$= \frac{10.5 \text{ /sec}}{1.055} = 9.95 \, \frac{\text{deg/sec}}{\text{deg}}$$

f) The sideslip angle is obtained from Eq. (6-23), but first we must find the value for "c," the distance from the CG to the rear axle. This is obtained from a simple moment balance about the front axle.

$$c = \frac{100.6 \text{ in}}{12 \text{ in/ft}} \, \frac{1901 \text{ lb}}{(1901 + 1552) \text{ lb}} = 4.62 \text{ ft}$$

$$\beta = 57.3 \; \frac{c}{R} - \frac{W_r \; V^2}{C_{\alpha r} \; R \, g} = 57.3 \; \text{deg/rad} \; \frac{4.62 \; \text{ft}}{800 \; \text{ft}} - \frac{1552 \; \text{lb} \; (88 \; \text{ft/sec})^2}{390 \; \text{lb/deg} \; (32.2 \; \text{ft/sec}^2)(800 \; \text{ft})}$$

$$= (0.331 - 1.196) \; \text{deg} = -0.865 \; \text{deg}$$

g) To find the static margin it is necessary to find the neutral steer point. The neutral steer point (nsp) is the point on the side of the vehicle where one can push laterally and produce the same slip angle at both the front and rear tires. In the solution for part 'a' we determined that the cornering stiffnesses of the front and rear tires are 232 and 195 lb/deg, respectively. From a moment balance in the plan view, we can show that the distance from the nsp to the rear axle (c') must be:

$$c' = L \frac{C_{\alpha f}}{C_{\alpha f} + C_{\alpha r}} = 8.38 \; \text{ft} \; \frac{232}{232 + 195} = 4.55 \; \text{ft}$$

Now c was 4.62 ft, therefore the neutral steer point is 0.07 ft (0.8% of the wheelbase) behind the CG.

Note:

The calculated cornering properties for this vehicle (understeer gradient and static margin) are very close to neutral steer. However, this is a consequence of only the tire properties. Many other systems on the vehicle, particularly the steering and suspension systems, will contribute to the understeer gradient.

2) A passenger car has an equal arm (parallel) independent front suspension and a conventional solid rear axle with leaf spring suspension. The front suspension has a roll stiffness, $K_{\phi f}$, of 1500 in-lb/deg. The leaf springs have a rate of 115 lb/in and a lateral separation of 40 inches.

a) What is the rear suspension roll stiffness?

b) If the sprung mass is 2750 lb at a CG height of 10 inches above the roll axis, what is the roll rate?

c) Assuming a camber stiffness that is 10 percent of the cornering stiffness, estimate the understeer gradient due to camber effects.

d) The rear leaf springs have an effective trailing arm angle of -7 degrees

233

(the negative sign means that the pivot of the arms is below the wheel center). What is the understeer gradient due to rear roll steer?

Solution:

a) The rear suspension roll stiffness can be computed from Eq. (6-26):

$$K_\phi = 0.5 \, K_s \, s^2 = 0.5 \, (115 \text{ lb/in}) \, (40 \text{ in})^2 = 92,000 \text{ in-lb/rad} = 1606 \text{ in-lb/deg}$$

b) The roll rate can be calculated from Eq. (6-32):

$$d\phi/da_y = W \, h_1/[K_{\phi f} + K_{\phi r} - W \, h_1]$$

$$= \frac{2750 \text{ lb } (10 \text{ in})}{(1500 \, \frac{\text{in-lb}}{\text{deg}} + 1606 \, \frac{\text{in-lb}}{\text{deg}} - \frac{2750 \text{ lb } (10 \text{ in})}{57.3 \text{ deg/rad}})} = \frac{27500 \text{ in-lb}}{(1500 + 1606 - 480) \, \frac{\text{in-lb}}{\text{deg}}} = 10.5 \text{ deg/g}$$

c) The understeer gradient due to camber effects can be estimated from Eq. (6-56):

$$K_{camber} = (\frac{C_{\gamma f}}{C_{\alpha f}} \frac{\partial \gamma_f}{\partial \phi} - \frac{C_{\gamma r}}{C_{\alpha r}} \frac{\partial \gamma_r}{\partial \phi}) \frac{\partial \phi}{\partial a_y}$$

Although we know the ratio of camber stiffness to cornering stiffness (given as 0.1), the camber gradients must be determined. For an independent front suspension with parallel equal arms, the wheel does not incline with jounce and rebound. Therefore, the camber angle will change exactly with the roll angle and the gradient for the front axle is 1. The rear axle is a solid axle which does not roll significantly. Therefore, its gradient is zero. Finally, we know the roll gradient from part b, where it was calculated as 10.5 deg/g, and the equation can now be solved.

$$K_{camber} = (0.1 \times 1.0 - 0.1 \times 0) \, 10.5 \text{ deg/g} = 1.05 \text{ deg/g}$$

d) The understeer gradient due to roll steer on the rear axle comes from Eq. (6-57):

$$K_{roll \, steer} = (\varepsilon_f - \varepsilon_r) \, d\phi/da_y$$

Since we are concerned only with the rear axle, only the second term must be determined. A solid axle will exhibit roll steer dependent on the effective angle of the imaginary trailing arms, which in this case is -7 degrees (- 0.122 radians). Then:

$$K_{roll \, steer \, (rear)} = + 0.122 \text{ radians } (10.5 \text{ deg/g}) = 1.28 \text{ deg/g}$$

By SAE convention, the positive sign means that the wheels steer right as the body rolls to the right (left turn). Since the roll steer turns the rear wheels to the right on a left turn, the rear will swing out more in the turn and oversteer is experienced.

REFERENCES

1. Good, M.C., "Sensitivity of Driver-Vehicle Performance to Vehicle Characteristics Revealed in Open-Loop Tests," Vehicle Systems Dynamics, Vol. 6, No. 4, 1977, pp. 245-277.

2. "Vehicle Dynamics Terminology," SAE J670e, Society of Automotive Engineers, Warrendale, PA (see Appendix A).

3. Durstine, J.W., The Truck Steering System from Hand Wheel to Road Wheel, SAE SP-374, 1973, 76 p.

4. Lugner, P., and Springer, H., "Uber den Enfluss der Lenkgeometrie auf die stationare Kurventfahrt eines LKW," (Influence of Steering Geometry on the Stationary Cornering of a Truck), *Automobil-Industrie,* November 1974, pp. 21-25.

5. Pitts, S., and Wildig, A.W., "Effect of Steering Geometry on Self-Centering Torque and 'Feel' During Low-Speed Maneuvers," *Automotive Engineer,* Institution of Mechanical Engineers, June-July 1978, pp. 45-48.

6. Nordeen, D.L., and Cortese, A.D., "Force and Moment Characteristics of Rolling Tires," SAE Paper No. 640028 (713A), 1963, 13 p.

7. 1991 Yearbook, The Tire & Rim Association Inc., Copley, OH, 1991.

8. Bundorf, R.T., "The Influence of Vehicle Design Parameters on Characteristic Speed and Understeer," SAE Paper No. 670078, 1967, 10 p.

9. Bundorf, R.T., and Leffert, R.L., "The Cornering Compliance Concept for Description of Vehicle Directional Control Properties," SAE Paper No. 760713, 1976, 14 p.

10. Ellis, J.R., Vehicle Dynamics, Business Books Limited, London, 1969, 243 p.

11. Olley, M., "National Influences on American Passenger Car Design," <u>Proceedings of The Institution of Automobile Engineers,</u> Vol. 32, 1938, pp. 509-572.

12. "Passenger Car and Light Truck Directional Control Response Test Procedures," SAE XJ266, Proposed Recommended Practice, Society of Automotive Engineers, Warrendale, PA.

13. "Steady-State Circular Test Procedure for Trucks and Buses," SAE J2181, Proposed Recommended Practice, Society of Automotive Engineers, Warrendale, PA.

14. "Road Vehicles - Steady-State Circular Test Procedure," International Standard ISO 4138, International Organization for Standardization, 1982, 14 p.

CHAPTER 7
SUSPENSIONS

Suspension system of the Beat. (Courtesy of Honda Motor Co., Ltd.)

With a background understanding of suspension properties that affect ride and directional response provided in the previous chapters, it is now appropriate to examine the various types of suspensions used on modern passenger cars.

The primary functions of a suspension system are to:

- Provide vertical compliance so the wheels can follow the uneven road, isolating the chassis from roughness in the road.

- Maintain the wheels in the proper steer and camber attitudes to the road surface.

- React to the control forces produced by the tires—longitudinal (acceleration and braking) forces, lateral (cornering) forces, and braking and driving torques.

- Resist roll of the chassis.

- Keep the tires in contact with the road with minimal load variations.

237

The properties of a suspension important to the dynamics of the vehicle are primarily seen in the kinematic (motion) behavior and its response to the forces and moments that it must transmit from the tires to the chassis [1, 2, 3, 4]. In addition, other characteristics considered in the design process are cost, weight, package space, manufacturability, ease of assembly, and others.

Suspensions generally fall into either of two groups—solid axles and independent suspensions. Each group can be functionally quite different, and so will be divided accordingly for discussion.

SOLID AXLES

A solid axle is one in which wheels are mounted at either end of a rigid beam so that any movement of one wheel is transmitted to the opposite wheel [5] causing them to steer and camber together. Solid <u>drive</u> (sometimes called "live") axles are used on the rear of many cars and most trucks and on the front of many four-wheel-drive trucks. Solid beam axles are commonly used on the front of heavy trucks where high load-carrying capacity is required.

Solid axles have the advantage that wheel camber is not affected by body roll. Thus there is little wheel camber in cornering, except for that which arises from slightly greater compression of the tires on the outside of the turn. In addition, wheel alignment is readily maintained, minimizing tire wear. The major disadvantage of solid steerable axles is their susceptibility to tramp-shimmy steering vibrations.

Hotchkiss

The most familiar form of the solid drive axle is the Hotchkiss drive [5]. The axle is located by semi-elliptic leaf springs as shown in Figure 7.1, and is driven through a longitudinal driveshaft with universal joints at the transmission and axle. The springs, mounted longitudinally, connect to the chassis at their ends with the axle attached near the midpoint.

Leaf springs are perhaps the simplest and least expensive of all suspensions. While compliant in the vertical direction, the leaf is relatively stiff in the lateral and longitudinal directions, thereby reacting the various forces between the sprung and unsprung masses. The Hotchkiss was used widely on the rear axle of passenger cars into the 1960s, and is still used on most light and heavy trucks. The demise of leaf springs on passenger cars was caused by the inherent

238

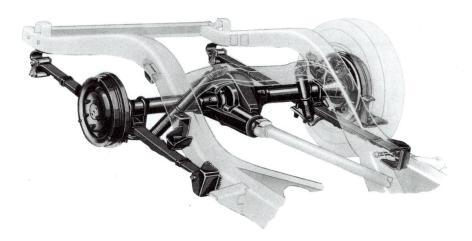

Fig. 7.1 The Hotchkiss rear suspension. (Photo courtesy of Ford Motor Company.)

friction of the springs and the loss in side stability of the springs as they were made longer to achieve lower spring rates [6]. With softer springs, compliance in the windup direction often required the addition of trailing arms to react brake torques and also the greater drive torques coincident with high-power engines popular in the post-war decades.

Four Link

In response to the shortcomings of leaf spring suspensions, the four-link rear suspension, shown in Figure 7.2, evolved as the suspension of choice in recent decades for the larger passenger cars with solid rear-drive axles. The lower control arms provide longitudinal control of the axle while the upper arms absorb braking/driving torques and lateral forces. Occasionally, the two upper arms will be replaced by a single, triangular arm, but it remains functionally similar to the four-link. The ability to use coil springs (or air springs) in lieu of leaf springs provides better ride and NVH by the elimination of the coulomb friction characteristic of leaf springs.

Although more expensive than the leaf spring, the geometric design of the four-link allows better control of roll center location, anti-squat and anti-dive performance, and roll steer properties.

239

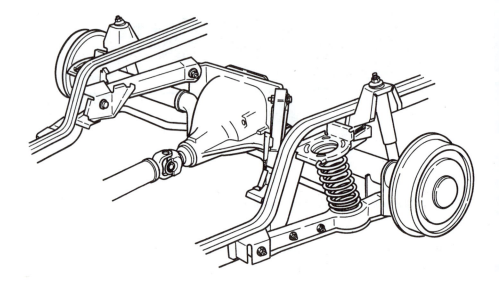

Fig. 7.2 The four-link rear suspension. (Photo courtesy of Ford Motor Company.)

De Dion

A cross between the solid axle and an independent suspension is the classic, but little used, de Dion system (patented in 1894 by Count de Dion and George Bouton), shown in Figure 7.3. It consists of a cross tube between the two driving wheels with a chassis-mounted differential and halfshafts. Like a solid axle, the de Dion keeps the wheels upright while the unsprung weight is reduced by virtue of the differential being removed from the axle. Axle control is provided by any of a variety of linkages from leaf springs to trailing arms. The design also has advantages for interior space because there is no need to provide differential clearance. One of the main disadvantages of the de Dion is the need to have a sliding tube or splined halfshafts, which can add friction to the system.

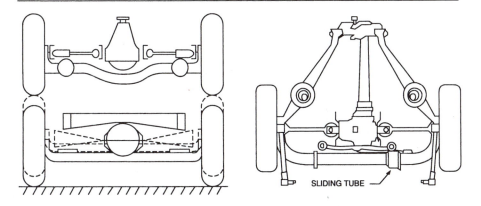

Fig. 7.3 The de Dion rear suspension.

INDEPENDENT SUSPENSIONS

In contrast to solid axles, independent suspensions allow each wheel to move vertically without affecting the opposite wheel. Nearly all passenger cars and light trucks use independent front suspensions, because of the advantages in providing room for the engine, and because of the better resistance to steering (wobble and shimmy) vibrations. The independent suspension also has the advantage that it provides an inherently higher roll stiffness relative to the vertical spring rate.

The first independent suspensions appeared on front axles in the early part of this century. Maurice Olley [7, 8] deserves much of the credit for promoting its virtues, recognizing that it would reduce some of the wobble and shimmy problems characteristic of the solid axles (by decoupling the wheels and interposing the mass of the car between the two wheels). Further advantages included easy control of the roll center by choice of the geometry of the control arms, the ability to control tread change with jounce and rebound, larger suspension deflections, and greater roll stiffness for a given suspension vertical rate.

Trailing Arm Suspension

One of the most simple and economical designs of an independent front suspension is the trailing arm used by Volkswagen and Porsche around the time of World War II. This suspension, shown in Figure 7.4, uses parallel, equal-length trailing arms connected at their front ends to lateral torsion bars, which provide the springing. With this design the wheels remain parallel to the body and camber with body roll.

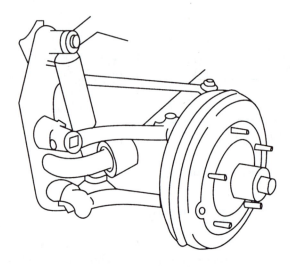

Fig. 7.4 The trailing arm independent front suspension.

SLA Front Suspension

The most common design for the front suspension of American cars following World War II used two lateral control arms to hold the wheel as shown in Figure 7.5. The upper and lower control arms are usually of unequal length from which the acronym SLA (short-long arm) gets its name. The arms are often called "A-arms" in the United States and "wishbones" in Britain. This layout sometimes appears with the upper A-arm replaced by a simple lateral link, or the lower arm replaced by a lateral link and an angled tension strut, but the suspensions are functionally similar.

The SLA is well adapted to front-engine, rear-wheel-drive cars because of the package space it provides for the engine oriented in the longitudinal direction. Additionally, it is best suited to vehicles with a separate frame for mounting the suspension and absorbing the loads.

Design of the geometry for an SLA requires careful refinement to give good performance. The camber geometry of an unequal-arm system can improve camber at the outside wheel by counteracting camber due to body roll, but usually carries with it less-favorable camber at the inside wheel. (Equal-

Fig. 7.5 The A-arm front suspension. (Photo courtesy of Ford Motor Company.)

length parallel arms eliminate the unfavorable condition on the inside wheel but at the loss of camber compensation on the outside wheel.) At the same time, the geometry must be selected to minimize tread change in jounce and rebound to avoid excessive tire wear.

MacPherson Strut

Earle S. MacPherson developed a suspension with geometry similar to the unequal-arm front suspensions using a strut configuration (see Figure 7.6). The strut is a telescopic member incorporating damping with the wheel rigidly attached at its lower end, such that the strut maintains the wheel in the camber direction. The upper end is fixed to the body shell or chassis, and the lower end is located by linkages which pick up the lateral and longitudinal forces. Because of the need to offset the strut inboard of the wheel, the wheel loads the strut with an overturning moment which adds to friction in the strut. This is often counteracted by mounting the coil spring at an angle on the strut.

243

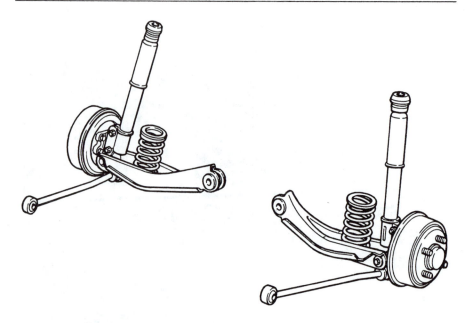

Fig. 7.6 The MacPherson strut suspension. (Photo courtesy of Ford Motor Company.)

The MacPherson strut provides major advantages in package space for transverse engines, and thus is used widely for front-wheel-drive cars. Because of the separation of the connection points on the body, it is well-suited to vehicles with unibody construction. The strut has further advantages of fewer parts and capability to spread the suspension loads to the body structure over a wider area. Among its disadvantages is the high installed height which limits the designer's ability to lower hood height.

Multi-Link Rear Suspension

In recent years, multi-link versions of independent rear suspensions have become quite popular. Figure 7.7 shows that used on the Ford Taurus/Sable cars. The multi-link is characterized by ball-joint connections at the ends of the linkages so that they do not experience bending moments. Generally speaking, four links are required to provide longitudinal and lateral control of the wheels, and react brake torques. Occasionally five links are used, as in the Mercedes Benz rear suspensions. The additional link over-constrains the wheel, but capitalizes on compliances in the bushings to allow more accurate control of toe angles in cornering. The use of linkages provides flexibility for the designer to achieve the wheel motions desired.

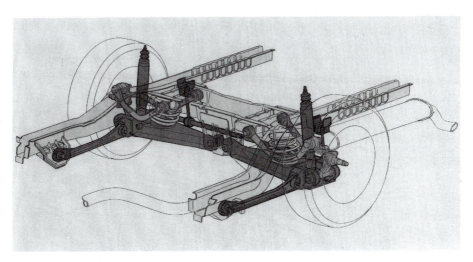

Fig. 7.7 Multi-link rear suspension of the Ford Taurus and Sable. (Photo courtesy of Ford Motor Company.)

Trailing-Arm Rear Suspension

Suspensions of this type are often used on more expensive and high-performance cars. A popular American car example is the Corvette rear suspension shown in Figure 7.8. The control arms (trailing arms) absorb longitudinal forces and braking moments, and control squat and lift. In the Corvette design the U-jointed halfshafts serve as an upper lateral control arm with a simple strut rod serving as the lower lateral arm. The independent suspension has the advantage of reducing unsprung weight by mounting the differential on the body.

Semi-Trailing Arm

The semi-trailing arm rear suspension was popularized by BMW and Mercedes Benz. This design, as shown in Figure 7.9, gives rear wheel camber somewhat between that of a pure trailing arm (no camber change relative to the body) and a swing axle. Its pivot axis is usually about 25 degrees to a line running across the car. The semi-trailing arm produces a steering effect as the wheels move in jounce and rebound. The steer/camber combination on the outside wheel acts against the direction of cornering, thus generating roll understeer on the rear axle, but lateral force compliance steer will contribute oversteer if not controlled.

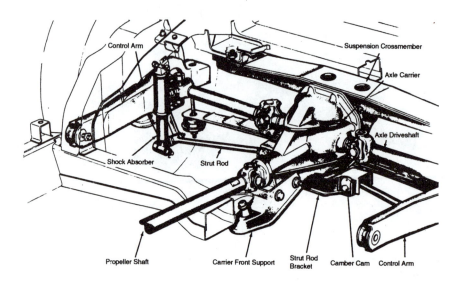

Fig. 7.8 The Corvette independent rear suspension. (Courtesy of Chevrolet Motor Division.)

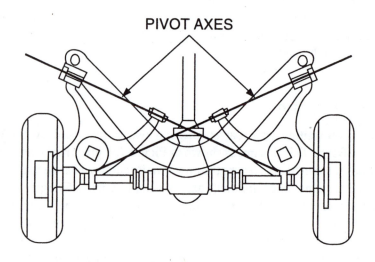

Fig. 7.9 The semi-trailing arm rear suspension.

Swing Axle

The easiest way to get independent rear suspension is by swing axles as shown in Figure 7.10. Edmund Rumpler is credited with inventing this system around the turn of the century, and by 1930 they were used on several European cars, most notably the Volkswagen "Beetle."

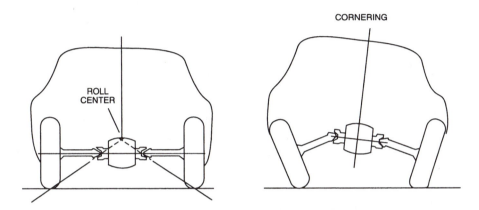

Fig. 7.10 The swing-axle rear suspension.

The camber behavior is established entirely by the axle shafts pivoting at the U-joint adjacent to the differential. The swing radius is small and thus the camber change with jounce and rebound movements can be large. As a result, it is difficult to get consistent cornering performance from swing-axle arrangements.

Critical to any independent suspension, but especially with the swing axle and semi-trailing arm, is a phenomenon known as "jacking." Jacking occurs during cornering when both tires are developing cornering forces but with the outside (more heavily loaded tire) contributing the greater cornering force. The inward direction of the cornering force attempts to lift the vehicle such that the wheels "tuck under." This has the effect of elevating the vehicle body (reducing its rollover resistance) and causing loss of cornering force on the axle due to camber thrust, leading to the possibility that the vehicle will spin out and roll over. Unless an additional control arm is included in the suspension to limit wheel travel under high lateral acceleration conditions, serious directional control problems are likely. This property of swing axle suspensions received much publicity as a result of its use on the Corvair passenger car in the 1960s [9].

ANTI-SQUAT AND ANTI-PITCH SUSPENSION GEOMETRY

In earlier chapters it was observed that under acceleration the load on the rear wheels increases due to longitudinal weight transfer. The load on the rear axle is:

$$W_r = W(\frac{b}{L} + \frac{a_x}{g} \frac{h}{L})$$ (7-1)

The second term on the right side of this equation is the weight transfer effect. The weight is transferred to the axle and wheels principally through the suspension. Therefore, there is an implied compression in the rear suspension which, in the case of rear-drive vehicles, has been called "Power Squat." Concurrently, there is an associated rebound in the front suspension. The combination of rear jounce and front rebound deflections produces vehicle pitch. Suspension systems may be designed to counteract the weight transfer and minimize squat and pitch.

Equivalent Trailing Arm Analysis

Anti-squat forces can be generated on a rear-drive axle by choice of the suspension geometry. The mechanics of the system can be understood most easily by recognizing that all suspensions are functionally equivalent to a trailing arm with regard to the reaction of forces and moments onto the vehicle. For analysis purposes consider a drive axle restrained by upper and lower control arms as shown in Figure 7.11. The horizontal drive force at the ground is F_x. The force F_z is the vertical reaction at the ground caused by the vertical components of the control arm forces. Static vertical loads may be neglected in the analysis. Thus, writing Newton's Second Law for the horizontal and vertical directions and for the moments around the point "o" yields:

$$F_x + P_1 \cos \theta_1 - P_2 \cos \theta_2 = 0$$ (7-2)

$$F_z - P_1 \sin \theta_1 - P_2 \sin \theta_2 = 0$$ (7-3)

$$F_x z_2 - P_1 \cos \theta_1 z_1 = 0$$ (7-4)

From Eq. (7-4) we can solve for P_1:

$$P_1 = \frac{F_x z_2}{z_1 \cos \theta_1}$$ (7-5)

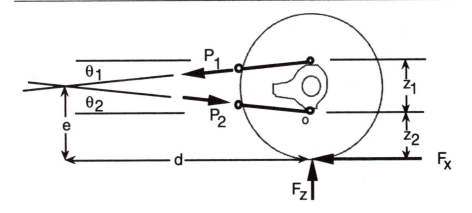

Fig. 7.11 Forces acting on a drive axle suspension system.

Assuming small angles and combining Eqs. (7-2) and (7-5):

$$P_2 = \frac{F_x(1 + z_2/z_1)}{\cos\theta_2} \tag{7-6}$$

Then using Eq. (7-3) to solve for F_z:

$$F_z = P_1 \sin\theta_1 + P_2 \sin\theta_2 = F_x\frac{z_2}{z_1}\tan\theta_1 + F_x(1 + \frac{z_2}{z_1})\tan\theta_2 \tag{7-7}$$

From the geometry it is seen that:

$$\tan\theta_1 = \frac{z_2 + z_1 - e}{d} \quad \text{and} \quad \tan\theta_2 = \frac{e - z_2}{d} \tag{7-8}$$

When this is substituted into Eq. (7-7), the ratio of forces is seen to be:

$$\frac{F_z}{F_x} = \frac{e}{d} \tag{7-9}$$

The expression given in Eq. (7-9) is identical to that which would be obtained if the control arms were replaced with a single (trailing) arm pivoting on the body at the projected intersection of the control arm axes. The intersection represents a "virtual reaction point" where the torque reaction of the suspension control arms can be resolved into equivalent longitudinal and vertical forces imposed on the vehicle body.

Given that any suspension is functionally equivalent to a trailing arm, the anti-squat performance can be quantified by analyzing the free-body diagram of a rear-drive axle as shown in Figure 7.12. In the figure, point "A" is the imaginary pivot on the vehicle body. Since the arm is rigidly fastened to the axle (resisting axle windup), it has the ability to transmit a vertical force to the sprung mass which can be designed to counteract squat.

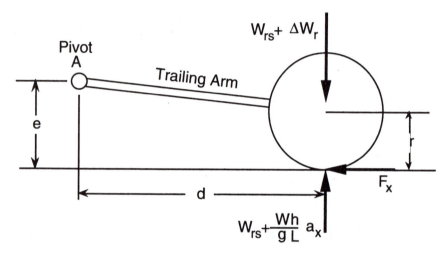

Fig. 7.12 Forces acting on a rear-drive axle during acceleration.

Rear Solid Drive Axle

The system is analyzed by applying NSL for the torques around the pivot point "A." The sum of these torques must be zero when the system is in equilibrium. Note that the wheel load is shown as consisting of a static component plus the dynamic component arising from longitudinal load transfer during acceleration. Also, for simplicity in the analysis, the rear axle weight will be neglected. Taking counterclockwise torques to be positive:

$$\Sigma M_A = W_{rs}\, d + \frac{W}{g}\frac{h}{L}\, a_x\, d - W_{rs}\, d - \Delta W_r\, d - F_x\, e = 0 \qquad (7\text{-}10)$$

where:

W_{rs} = Static load on the axle = Static load in the suspension

ΔW_r = Change in the suspension load under acceleration

This equation can be solved for the change in rear suspension load.

$$\Delta W_r = \frac{W}{g} \frac{h}{L} a_x - F_x \frac{e}{d} = K_r \delta_r \tag{7-11}$$

where:

K_r = Rear suspension spring rate

δ_r = Rear suspension deflection (positive in jounce)

The front suspension is undergoing a rebound deflection because of the longitudinal load transfer, and has a magnitude of:

$$\Delta W_f = - \frac{W}{g} \frac{h}{L} a_x = K_f \delta_f \tag{7-12}$$

The pitch angle of the vehicle, θ_p, during acceleration is simply the sum of the suspension deflections divided by the wheelbase. Thus we can write:

$$\theta_p = \frac{\delta_r - \delta_f}{L} = \frac{1}{L} \frac{W}{g} \frac{h}{L} \frac{a_x}{K_r} - \frac{1}{L} \frac{F_x}{K_r} \frac{e}{d} + \frac{1}{L} \frac{W}{g} \frac{h}{L} \frac{a_x}{K_f} \tag{7-13}$$

Since F_x is simply the mass times the acceleration, $(W/g)a_x$, the equation can be rewritten:

$$\theta_p = \frac{1}{L} \frac{W}{g} \frac{h}{L} \frac{a_x}{K_r} - \frac{1}{L} \frac{W}{g} \frac{a_x}{K_r} \frac{e}{d} + \frac{1}{L} \frac{W}{g} \frac{h}{L} \frac{a_x}{K_f}$$

$$= \frac{1}{L} \frac{W}{g} a_x (\frac{1}{K_r} \frac{h}{L} - \frac{1}{K_r} \frac{e}{d} + \frac{1}{K_f} \frac{h}{L}) \tag{7-14}$$

From this equation it is easy to show that zero pitch angle is achieved when the following condition is satisfied:

$$\frac{e}{d} = \frac{h}{L} + \frac{h}{L} \frac{K_r}{K_f} \tag{7-15}$$

The first term on the right-hand side corresponds to the condition by which anti-squat is achieved on the rear suspension. That is, if $e/d = h/L$, the rear suspension will not deflect (jounce) during acceleration. The degree to which this is achieved is described as the percent anti-squat. For example, if $e/d = 0.5$ h/L, the suspension is said to be 50% anti-squat. Since h/L is in the vicinity of 0.2 for most passenger cars, full anti-squat generally requires an effective trailing arm length of about five times the elevation of "e."

251

The anti-squat equation ($e/d = h/L$) defines a locus of points extending from the tire contact point on the ground to the height of the CG over the front axle. Locating the trailing arm pivot at any point on this line will provide 100% anti-squat.

Satisfying the equation with inclusion of the second term implies that the rear suspension will lift to compensate for rebound of the front suspension, thereby keeping the vehicle level. The complete equation may be interpreted as the full anti-pitch relationship. Because the ratio of suspension stiffnesses is nominally 1, the anti-pitch condition is approximately:

$$\frac{e}{d} \approx \frac{h}{L} + \frac{h}{L} = 2\frac{h}{L} \qquad \text{(Full anti-pitch)} \qquad (7\text{-}16)$$

The locus of points for anti-pitch extends from the tire contact point on the ground to the height of the CG at the mid-wheelbase position. Anti-pitch is achieved when the trailing arm pivot point is located on the line from the center of tire contact on the ground to the CG of the vehicle.

Normally some degree of squat and pitch is expected during vehicle acceleration, so full compensation is unusual. Anti-squat performance cannot be designed without considering other performance modes of the vehicle as well. When the trailing arm is short, the rear axle may experience "power hop" during acceleration near the traction limit. The goals for anti-squat may conflict with those for braking or handling. In this latter case, placing the pivot center above the wheel center can produce roll oversteer.

Independent Rear Drive

In the case of an independent rear-drive configuration, the free-body diagram and the analysis is changed slightly from that above. The difference arises from the fact that there is a drive torque reaction acting on the free-body system with a magnitude $T_d = r F_x$ (where r is the wheel radius). The differential is mounted on the vehicle body imposing a drive torque on the system of Figure 7.12 through the halfshafts. This alters Eq. (7-10) to the form:

$$\Sigma M_A = W_{rs} d + \frac{W}{g}\frac{h}{L} a_x d - W_{rs} d - \Delta W_r d - F_x (e - r) = 0 \quad (7\text{-}17)$$

Carrying this through the analysis alters Eq. (7-14) to the form:

$$\theta_p = \frac{1}{L}\frac{W}{g} a_x \left(\frac{1}{K_r}\frac{h}{L} - \frac{1}{K_r}\frac{e-r}{d} + \frac{1}{K_f}\frac{h}{L} \right) \qquad (7\text{-}18)$$

and it is concluded that full squat compensation is achieved when:

$$\frac{e-r}{d} = \frac{h}{L} + \frac{h}{L}\frac{K_r}{K_f} \tag{7-19}$$

Similarly, 100% anti-squat in the rear suspension corresponds to $(e-r)/d = h/L$.

Front Solid Drive Axle

With a front-drive axle, performing this type of analysis only results in a change of sign on some of the terms. The comparable equations that are obtained are:

$$\theta_p = \frac{1}{L}\frac{W}{g}a_x\left(\frac{1}{K_r}\frac{h}{L} + \frac{1}{K_f}\frac{h}{L} + \frac{1}{K_f}\frac{e}{d}\right) \tag{7-20}$$

and

$$\frac{e}{d} = -\frac{h}{L} - \frac{h}{L}\frac{K_f}{K_r} \tag{7-21}$$

The first term on the right side of Eq. (7-21) now corresponds to an anti-lift on the front axle, rather than an anti-squat on the rear axle. The negative signs on the right-hand side of the equation imply that the pivot must be behind the axle, corresponding to an effective leading arm arrangement to obtain anti-lift behavior.

Independent Front-Drive Axle

The comparable equations for an independent front-drive axle, as is common on most front-drive cars today, are:

$$\theta_p = \frac{1}{L}\frac{W}{g}a_x\left(\frac{1}{K_r}\frac{h}{L} + \frac{1}{K_f}\frac{h}{L} + \frac{1}{K_f}\frac{e-r}{d}\right) \tag{7-22}$$

and

$$\frac{e-r}{d} = -\frac{h}{L} - \frac{h}{L}\frac{K_f}{K_r} \tag{7-23}$$

253

Four-Wheel Drive

The four-wheel-drive case will be considered assuming independent suspensions on both axles. The performance will depend on how the tractive force is distributed between the front and rear axles. Let ξ represent the fraction of the total drive force developed on the front axle. Then:

$$F_{xf} = \xi\,F_x \quad \text{and} \quad F_{xr} = (1 - \xi)\,F_x \tag{7-24}$$

The expressions for the change in load on each axle are:

$$\Delta W_r = \frac{W}{g}\frac{h}{L}\,a_x - (1 - \xi)\,F_x\frac{e_r - r}{d_r} = K_r\delta_r \tag{7-25}$$

$$\Delta W_f = \frac{W}{g}\frac{h}{L}\,a_x + \xi\,F_x\frac{e_f - r}{d_f} = K_f\delta_f \tag{7-26}$$

Then the pitch equation becomes:

$$\theta_p = \frac{1}{L}\frac{W}{g}\,a_x\left(\frac{1}{K_r}\frac{h}{L} - \frac{(1 - \xi)}{K_r}\frac{e_r - r}{d_r} + \frac{1}{K_f}\frac{h}{L} + \frac{\xi}{K_f}\frac{e_f - r}{d_f}\right) \tag{7-27}$$

This equation indicates that zero pitch will be obtained when the terms in parentheses sum to zero. The terms included here indicate that the anti-squat and anti-pitch performance depends on a combination of vehicle properties—suspension geometry, suspension stiffness and tractive force distribution.

In the event one or both of the axles is a solid axle, the pitch expression can be obtained from the above equation by setting "r" equal to zero in the term of the equation applicable to that axle.

ANTI-DIVE SUSPENSION GEOMETRY

The longitudinal load transfer incidental to braking acts to pitch the vehicle forward producing "brake dive." Just as a suspension can be designed to resist acceleration squat, the same principles apply to generation of anti-dive forces during braking. Because virtually all brakes are mounted on the suspended wheel (the only exception is in-board brakes on independent suspensions), the brake torque acts on the suspension and by proper design can create forces which resist dive.

Using an analysis similar to that developed for the four-wheel-drive anti-squat example given previously, it can be shown that the anti-dive is accomplished when the following relationships hold:

Front suspension:

$$\frac{e_f}{d_f} = \tan \beta_f = -\frac{h}{\xi L} \qquad (7\text{-}28)$$

Rear suspension:

$$\frac{e_r}{d_r} = \tan \beta_r = \frac{h}{(1 - \xi) L} \qquad (7\text{-}29)$$

where:

ξ = Fraction of the brake force developed on the front axle

Therefore, to obtain 100% anti-dive on the front and 100% anti-lift on the rear, the pivot for the effective trailing arm must fall on the locus of points defined by these ratios. Figure 7.13 illustrates these conditions. If the pivots are located below the locus, less than 100% anti-dive will be obtained; if above the locus the front will lift and the rear will squat during braking.

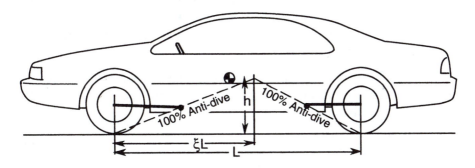

Fig. 7.13 Illustration of the conditions for anti-dive.

In practice, 100% anti-dive is rarely used. The maximum anti-dive seldom exceeds 50%. A number of reasons for this have been cited:

1) Full anti-dive requires that the pivot be located above the point required for full anti-squat. Thus acceleration lift would be produced on solid drive axles.

2) Flat stops are subjectively undesirable.

3) With full anti-dive, front suspension caster angle changes may increase the steering effort substantially during braking.

4) The required steering system geometry may be quite complex.

5) Excessive variation in rotational speed can occur in the drivetrain as the wheels move in jounce and rebound causing rattling and noise in the drive gears.

6) In the rear suspension, oversteer problems may be created by the high location of the pivot.

7) Brake hop may be induced if the effective trailing arm is too short. The propensity for brake hop is reduced by a suspension design with a long effective arm.

8) NVH performance may be compromised.

EXAMPLE PROBLEMS

1) Find the geometry that would be necessary to achieve 100% anti-squat in the rear suspension, and the geometry to achieve full anti-pitch for the solid-axle, rear-wheel-drive vehicle described below. Also, find the pitch rate (degrees pitch/g acceleration) when the geometry is set for 100% anti-squat in the rear suspension.

The front and rear suspension spring rates are 285 and 169 lb/in, respectively (rates are combination of left and right sides). The CG height is 20.5 inches and wheelbase is 108.5.

Solution:

Since the vehicle has a rear-drive solid axle, Eq. (7-15) applies.

$$\frac{e}{d} = \frac{h}{L} + \frac{h}{L}\frac{K_r}{K_f} \qquad (7\text{-}15)$$

$$\frac{e}{d} = \frac{20.5}{108.5} + \frac{20.5}{108.5}\frac{169}{285} = 0.189 + 0.112 = 0.301$$

If the suspension is to achieve 100% anti-squat, then e/d must equal 0.189. Full pitch compensation would be achieved with e/d = 0.301.

The acceleration pitch rate can be calculated using Eq. (7-18).

$$\theta_p = \frac{1}{L}\frac{W}{g}\, a_x \left(\frac{1}{K_r}\frac{h}{L} - \frac{1}{K_r}\frac{e}{d} + \frac{1}{K_f}\frac{h}{L}\right)$$

$$\frac{\theta_p}{a_x} = \frac{1}{108.5 \text{ in}}\frac{4074 \text{ lb}}{386 \text{ in/sec}^2}\left(\frac{1}{169 \text{ lb/in}}\frac{20.5}{108.5} - \frac{0.189}{169 \text{ lb/in}} + \frac{1}{285 \text{ lb/in}}\frac{20.5}{108.5}\right)$$

$$= 0.0000645 \text{ rad/(in/sec}^2) = 0.0249 \text{ rad/g} = 1.43 \text{ deg/g}$$

2) Determine the acceleration pitch rate for the following front-drive vehicle with no anti-lift in the front suspension, and its value if full anti-lift was designed into the suspension. Essential data are—CG height of 20.5", wheelbase of 108.5", design weight of 4549 lb, and front and rear spring rates of 287 and 174 lb/in, respectively.

Solution:

The pitch equation for a front-wheel drive comes from Eq. (7-22). With no anti-lift the third term on the right side is zero. Thus:

$$\frac{\theta_p}{a_x} = \frac{1}{L}\frac{W}{g}\left(\frac{1}{K_r}\frac{h}{L} + \frac{1}{K_f}\frac{h}{L}\right)$$

$$= \frac{1}{108.5"}\frac{4549 \text{ lb}}{386 \text{ in/sec}^2}\left(\frac{1}{174 \text{ lb/in}}\frac{20.5"}{108.5"} + \frac{1}{287 \text{ lb/in}}\frac{20.5"}{108.5"}\right)$$

$$= 0.0455 + 0.0276 \text{ rad/g} = 4.2 \text{ deg/g}$$

If anti-lift is designed into the suspension it would cancel the second term in this equation, in which case the acceleration pitch rate would be:

$$\frac{\theta_p}{a_x} = 0.0455 \text{ rad/g} = 2.61 \text{ deg/g}$$

ROLL CENTER ANALYSIS

One very important property of a suspension relates to the location at which lateral forces developed by the wheels are transmitted to the sprung mass. This point, which has been referred to as the roll center, affects the behavior of both the sprung and unsprung masses, and thus directly influences cornering.

Each suspension has a <u>suspension roll center</u>, defined as the point in the transverse vertical plane through the wheel centers at which lateral forces may be applied to the sprung mass without producing suspension roll [10]. It derives from the fact that all suspensions possess a <u>roll axis,</u> which is the instantaneous axis about which the unsprung mass rotates with respect to the sprung mass when a pure couple is applied to the unsprung mass. The roll center is the intersection of the suspension roll axis with the vertical plane through the centers of the two wheels. These definitions are illustrated in Figure 7.14. The roll center height is the distance from the ground to the roll center. Once the front and rear suspension roll centers are located, the <u>vehicle roll axis</u> is defined by the line connecting the centers. This axis is the instantaneous axis about which the total vehicle rolls with respect to the ground.

The reference to "instantaneous" axes in these definitions is used to alert the reader to the fact that the location of the axis is only accurate in the absence of roll. As body roll occurs the change in geometry of most suspensions will cause the "center" to migrate, and thus it is not a true center. Nevertheless, the concept is valid for purposes of establishing where the forces are reacted on the sprung mass, which is necessary for analyzing behavior in the lateral plane.

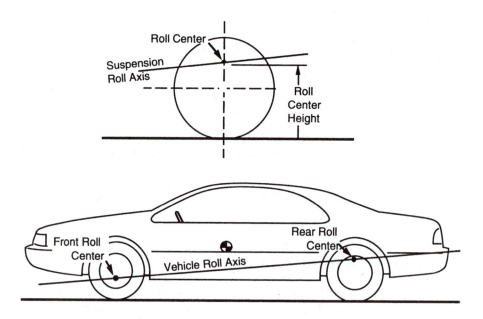

Fig. 7.14 Definitions of suspension roll center and roll axis.

Solid Axle Roll Centers

The suspension roll axis and roll center can be determined from the layouts of the suspension geometry in the plan and elevation views. For the analysis we draw again on the concept of a "virtual reaction point." (The virtual reaction point is analogous to the "instant center" used in kinematic analysis of linkages, but that term is not used here because of implication that it defines a center of motion, when in fact, it does not.) Physically, the virtual reaction point is the intersection of the axes of any pair of suspension control arms. Mechanistically, it is the point where the compression/tension forces in the control arms can be resolved into a single lateral force.

Four-Link Rear Suspension

Consider the case of a four-link suspension with a solid axle, as shown in Figure 7.15. The lateral force acting on the wheel in the top view must react as tension and compression forces in the control arms. The two long arms establish a virtual reaction point ahead of the axle at B, while the two short arms have a virtual reaction point behind the axle at A. In effect, each pair of arms acts like a triangular member pivoting at their respective virtual reaction points with these points establishing the suspension roll axis. Consequently, the lateral force will be distributed between the two points in inverse proportion to the length of the arms in order to achieve moment equilibrium on the axle (i.e., a large force at A and a small force at B).

The two forces at A and B must add up to F_y acting in the transverse vertical plane through the wheel centers. Given that points A and B are at different heights above the ground, their resultant at the axle centerline must be on the line connecting the two. This is the roll center for the axle.

A general procedure for finding roll centers then is as follows:

1) In a plan view of the suspension find the linkages that take the side forces acting on the suspension. Determine the reaction points A and B on the centerline of the vehicle for forces in the links. In the case of paired control arms, this is a virtual reaction point.

2) Locate the points A and B in the side elevation view, thereby identifying the suspension roll axis.

3) The roll center is the point in the side view where the roll axis crosses the vertical centerline of the wheels.

259

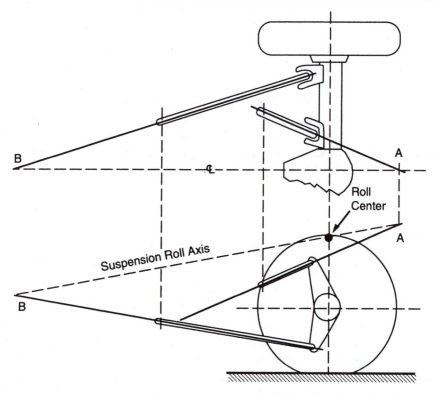

Fig. 7.15 Roll center analysis of a four-link rear suspension.

In the four-link geometry, the change in slope of the roll axis during cornering is often relatively large compared to other live axles. This means considerable change in roll steer and lateral load transfer, which are undesirable effects. Also, the roll center is located relatively high compared to other suspensions, putting excessive roll moment on the rear wheels. On the other hand, the high roll center helps to reduce the tramp and shake of the axle.

Three-Link Rear Suspension

Figure 7.16 shows a three-link suspension consisting of a track bar and two lower control arms. Because the track bar picks up lateral force directly, point A is established at the location where the track bar crosses the centerline of the vehicle. Point B is established as the virtual reaction point for the two lower control arms. Note that the upper link which reacts the axle windup torque does not react lateral forces and is therefore ignored in the analysis.

Due to the location of the track bar, this suspension usually has a roll center that is lower than the four-link geometry. Also, the slope of the roll axis remains relatively unchanged during rolling of the body and with load variations.

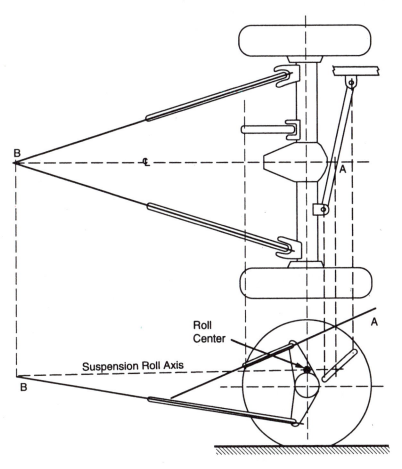

Fig. 7.16 Roll center analysis of a three-link rear suspension.

Four-Link with Parallel Arms

Figure 7.17 shows a four-link rear suspension with lower control arms that are parallel. This geometry is a special case of the four-link suspension discussed previously. In the top view, the virtual reaction point of the upper links is used to find point A in the usual manner. Because the lower arms are parallel, their virtual reaction point is at infinity. Although point B is not defined, we know that this point in the side view must lie on the extended lower arm centerline somewhere at infinity. Therefore, the roll axis of this geometry (obtained by connecting points A and B in the side view) must be a line parallel to the lower links as indicated in the figure. The slope of the lower arm thus is very important in this type of suspension.

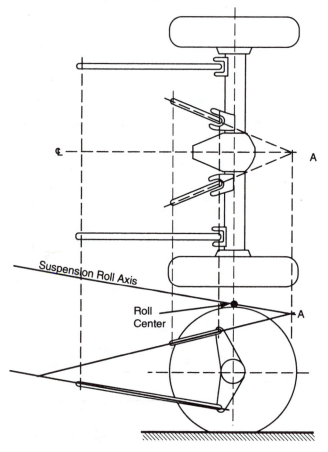

Fig. 7.17 Roll center analysis of a four-link, parallel lower arm rear suspension.

Hotchkiss Suspension

The design of this suspension is quite different from those discussed previously, but the general rules for determining the roll axis and center still apply. Referring to Figure 7.18, it is seen that the leaf springs are the members that react the side thrust. Because they are parallel to the centerline of the vehicle in the top view, the points A and B lie on the centerline of the car, both at infinity.

The lateral forces are applied to the body at the front spring eye and the rear shackle attachment point on the frame. The roll axis of the suspension is established by these points and the roll center is found on the line connecting the points. Although this analysis may seem less obvious than those discussed previously, it should be clear that a side force applied at this point will not roll the body, which is the essential definition of the roll center. Experimental measurements of leaf spring suspensions have generally confirmed the validity of this method in establishing the roll center.

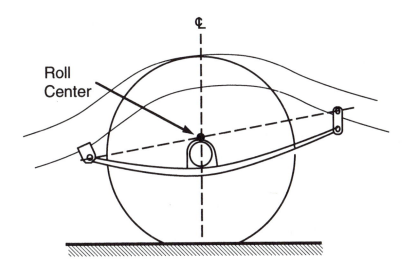

Fig. 7.18 Roll center analysis of a Hotchkiss rear suspension.

Independent Suspension Roll Centers

Determining the roll center of an independent suspension requires a slightly different application of the virtual reaction point concept. Consider the double A-arm suspension shown in Figure 7.19. The virtual reaction point for the A-arms holding the left wheel is located at point A on the right of the vehicle. Mechanistically, the linkage behaves as if the wheel were held by a rigid lateral swing arm pinned to the vehicle body at that point. Understanding the behavior is aided by imagining that the swing arm directly connects the tire contact patch (C) to the pivot point.

A lateral force in the contact patch of the left wheel reacts along the line from the contact patch to the pivot point as illustrated in the left diagram in the figure. Its elevation where it crosses the center plane of the vehicle establishes the roll center, R. Note that a lateral force from the left-side wheel reacting along that line must have an upward (vertical) force component, thus explaining the source of "jacking" forces inherent to independent suspensions. If the right-hand wheel experiences a lateral force of equal magnitude in the same direction, its reaction will involve a downward force component cancelling the lifting effect from the left wheel. In general, both wheels do not generate equal lateral forces in cornering, so some lifting force is usually present on the suspension.

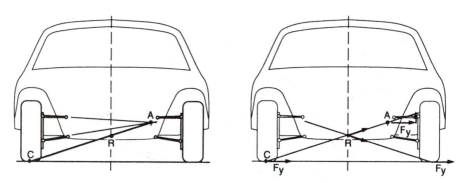

Fig. 7.19 Roll center analysis of an independent suspension.

The procedure for finding the roll center of a symmetric independent suspension is as follows:

1) Find the virtual reaction point of the suspension links (point A).

2) Draw a line from the tire-ground contact patch to the virtual reaction point.

3) The point where this line crosses the centerline of the body is the roll center (R).

Note that this procedure can be used for determining the roll center when the body is rolled; however, the suspensions are no longer symmetrical so both sets must be analyzed.

Positive Swing Arm Geometry

The virtual reaction point of the upper and lower links is first obtained as shown in Figure 7.20. A line is drawn from the tire contact patch to the reaction point. The roll center is established where the line crosses the centerline of the vehicle. This suspension geometry is referred to as the "positive swing arm" because the roll center is located above the ground.

As the vehicle rolls in cornering, the virtual reaction point of the outside wheel moves downward due to jounce of the wheel, while that of the inside wheel moves upward as it goes into rebound. With the loss of symmetry the roll centers for the two wheels no longer coincide. The lateral force from the outside wheel (which usually dominates cornering forces) moves downward on the body, while weaker force from the inside wheel moves upward. As a consequence, the resultant lateral force reaction on the body moves downward, lowering the effective roll center height.

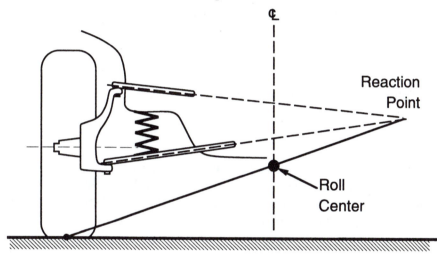

Fig. 7.20 Positive swing arm independent suspension.

Negative Swing Arm Geometry

Negative swing arm geometry is shown in Figure 7.21. The virtual reaction point of the links is first obtained and connected to the tire contact patch as shown. The line is then projected downward to the car centerline below the ground. The roll center is negative; hence, the name "negative swing arm" geometry.

Note that the two independent suspension geometries discussed so far have roll centers either above or below the ground. Consequently, the tread will change during jounce and rebound, and some lateral scrub of the tire contact patch occurs. This scrub introduces friction which (in the past) was considered beneficial for reducing bouncing of the body, but at the expense of tire wear.

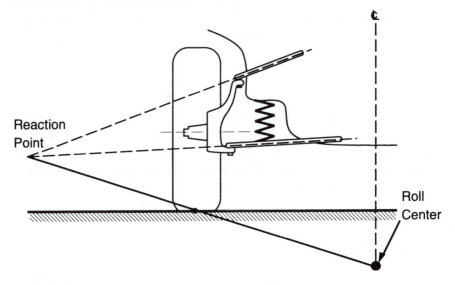

Fig. 7.21 Negative swing arm independent suspension.

Parallel Horizontal Links

A suspension with parallel links that are horizontal (at design load) is shown in Figure 7.22. The virtual reaction point of the two links is therefore at infinity. Drawing a line from the tire contact patch toward infinity places the roll center in the ground plane.

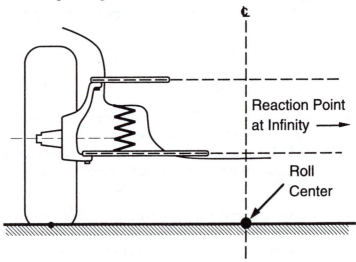

Fig. 7.22 Parallel horizontal link independent suspension.

Inclined Parallel Links

Another possibility is use of parallel links, which are not horizontal at design load as shown in Figure 7.23. The virtual reaction point is at infinity. The line from the tire contact patch to the roll center is inclined at the same angle as the control arms. The roll center is elevated above the ground at the car centerline as shown. In this geometry the roll center moves on the centerline of the car during rolling because the wheels camber with respect to the body. If the links of the suspension are equal there will be no camber change with respect to the body and the roll center will remain stationary.

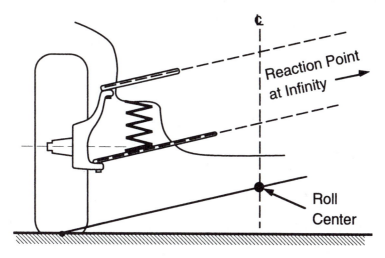

Fig. 7.23 Inclined parallel link independent suspension.

With all of the swing arm geometries the jounce of the outside wheel in cornering lowers the roll center location for that wheel and, consequently, the point at which the lateral force from the wheel is applied to the sprung mass. This reduces the load transfer onto the outside wheel and, with the consequent reduction of cornering force of the wheel, induces an understeer influence on the vehicle.

MacPherson Strut

The MacPherson strut is a combination of a strut with a lower control arm as shown in Figure 7.24. The virtual reaction point must lie at the intersection of the axis of the lower control arm and a line perpendicular to the strut. The roll center is located on the centerline of the vehicle at the intersection with the line from the center of tire contact to the virtual reaction point.

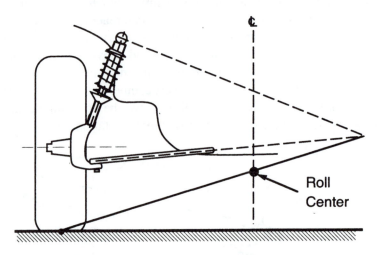

Fig. 7.24 MacPherson strut independent suspension.

Swing Axle

A rear suspension swing axle is generically equivalent to that shown in Figure 7.25. The location of the roll center is easily obtained for this configuration because the virtual reaction point is the actual pivot of the axle. The line from the tire contact passes through the pivot and the roll center is located above the wheel center on the vehicle centerline.

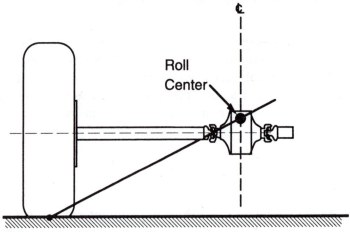

Fig. 7.25 Swing axle independent suspension.

ACTIVE SUSPENSIONS

In the interest of improving the overall performance of automotive vehicles in recent years, suspensions incorporating active components have been developed. The designs may cover a spectrum of performance capabilities [11], but the active components alter only the vertical force reactions of the suspensions, not the kinematics. (Active components that alter kinematic behavior do so to steer the wheels, and would be covered under steering systems.)

Suspension Categories

The various levels of "active" qualities in suspensions may be divided into the categories shown below, listed in order of increasing capabilities.

Passive suspensions consist of conventional components with spring and damping (shock absorber) properties which are time-invariant. Passive elements can only store energy for some portion of a suspension cycle (springs) or dissipate energy (shock absorbers). No external energy is directly supplied to this type of suspension.

Self-leveling suspensions are a variation of the passive suspension in which the primary lift component (usually air springs) can adjust for changes in load. Air suspensions, which are self-leveling, are used on many heavy trucks and on a few luxury passenger cars. A height control valve monitors the suspension deflection, and when its mean position has varied from normal ride height for a designated period of time (typically more than 5 seconds), the air pressure in the spring is adjusted to bring the deflection within the desired range. The most notable feature of an air suspension is that as the pressure changes with load, the spring stiffness changes correspondingly causing the natural frequency of the suspension to remain constant.

Semi-active suspensions contain spring and damping elements, the properties of which can be changed by an external control. A signal or external power is supplied to these systems for purposes of changing the properties. There are several sub-categories of semi-active systems:

- Slow-active—Suspension damping and/or spring rate can be switched between several discrete levels in response to changes in driving conditions. Brake pressure, steering angle or suspension motions are typically used to trigger control changes to higher levels of damping or stiffness. Switching occurs within a fraction of a second giving the system the capability to control pitch, bounce, and roll motions of the

269

sprung mass under more severe road or maneuvering conditions. However, the switch back to softer settings occurs after a time delay. Thus the system does not adjust continuously during individual cycles of vehicle oscillation. Slow-active systems may also be called "adaptive" suspensions.

- <u>Low-bandwidth</u>—Spring rate and/or damping are modulated continuously in response to the low-frequency sprung mass motions (1-3 Hz).

- <u>High-bandwidth</u>—Spring rate and/or damping are modulated continuously in response to both the low-frequency sprung mass motions (1-3 Hz) and the high-frequency axle motions (10-15 Hz).

<u>Full-Active suspensions</u> incorporate actuators to generate the desired forces in the suspension. The actuators are normally hydraulic cylinders. External power is required to operate the system. Full active systems may be classified as low-bandwidth or high-bandwidth according to the definitions given above.

Functions

The interest in active or semi-active suspensions derives from the potential for improvements to vehicle ride performance with no compromise (and perhaps enhancement) in handling. The modes of performance that can be improved by active control are:

<u>Ride Control</u>—Ride improvements can be obtained by several methods. The system may sense and control pitch and bounce motions of the vehicle body directly. Ride improvements are also obtained indirectly when active control is applied to the modes described below. Suspension properties that optimize ride always degrade performance in other modes, thus necessitating a compromise in design. With active suspensions, however, the control can be applied only during the maneuver, and ride performance need not be compromised during other modes of travel. Specifically, the suspension can be trimmed for optimal ride performance during steady, straight-ahead travel, and ride isolation properties superior to that obtained with purely passive elements can be achieved without compromise of handling behavior.

<u>Height Control</u>—Automatic control of vehicle height offers several advantages in performance. By adjusting to keep height constant despite changes in load or aerodynamic forces the suspension can always operate at the design ride height, providing maximum stroke for negotiating bumps, and eliminating changes in handling that would arise from operation at other than the design

ride height. A height control can lower the vehicle for reduced drag at high speeds or alter the pitch attitude to modify aerodynamic lift. Height can be elevated for increased ground clearance and suspension stroke on bad roads. Height elevation can also be convenient for changing tires and to provide clearance for tire chains.

Roll Control—Roll control in cornering is improved by increasing damping or exerting anti-roll forces in the suspension during cornering. Vehicle speed, steer angle, steer rate and/or lateral acceleration may be sensed to determine when roll control is appropriate. With the use of active force-generating components it is possible to eliminate roll in cornering entirely, and thereby eliminate any roll-induced understeer or oversteer effects from the suspensions. In addition, the roll moments may be selectively applied at either the front or rear axles to alter the understeer gradient by action of the change in cornering stiffness due to lateral load transfer.

Dive Control—Control of dive (forward pitch) during braking can be improved by increasing damping or exerting anti-pitch forces in the suspension during braking. Control may be activated by the brake light signal, brake pressure and/or longitudinal acceleration. Dive control in an active suspension relieves the need to design anti-dive geometry in the suspension linkages.

Squat Control—Control of squat (rearward pitch) during acceleration can be improved by increasing damping or exerting anti-pitch forces in the suspension during acceleration. Control may be activated by the throttle position, gear selection and/or longitudinal acceleration. Squat control in an active suspension relieves the need to design anti-squat geometry in the suspension linkages of drive wheels, and can overcome the squat or lift action on the non-driven wheels.

Road Holding—In addition to control of body motions during maneuvers in the modes described above, active suspensions have the potential to improve road holding by reducing the dynamic variations in wheel loads that are caused from road roughness. Generally, cornering performance is improved when dynamic load variations are minimized. The road damage caused by motor vehicles, particularly heavy trucks, is also reduced by minimizing dynamic wheel loads.

Performance

In general, the semi-active and full-active suspension systems have the greatest capability to achieve optimum performance in the modes described above, but at a penalty in weight, cost, complexity and reliability. Thus the challenge to vehicle designers is to achieve the benefits of active control with a minimum of hardware. The table below characterizes the relative performance that can be obtained with each level of sophistication in the design.

Performance Potential of Various Types of Suspension Systems

Suspension Type	Performance Mode					
	Ride	Height	Roll	Dive	Squat	Road-holding
Passive	Performance is a compromise between all modes					
Self-leveling	High	High	NA	NA	NA	NA
Semi-active	Medium	NA	Low	Low	Low	Medium
Full-active	High	High	High	High	High	High

With semi-active systems, even slow-active variable damping allows improvement in roll, dive and squat control along with ride and road-holding. Variable stiffness can provide similar benefits, albeit at greater cost due to the need to use air springs or adjustable mechanical springs.

With low-bandwidth stiffness or damping control a more responsive system can be achieved. High-bandwidth control is effective for maintaining the constant wheel loads beneficial to handling, but there is little additional ride benefit from a high-bandwidth control system. Only with a full-active system can the broadest range of improvements be obtained in all performance modes.

The performance of a full-active system optimized for ride contrasts with that of a passive suspension by much better control of the vertical, pitch and roll motions at the sprung mass resonant frequencies [12]. Figure 7.26 compares the response behavior in these three modes for the two types of systems. Whereas the passive system shows sprung mass resonance near 1 Hz in the vertical, pitch and roll directions, a much reduced response occurs with the active system. In effect, the sprung mass motions in these directions (which are sensed by accelerometers) can be heavily damped by control forces developed in the active suspension system.

With control characteristics optimized for ride, there is no significant change in response at the unsprung mass resonant frequency near 10 Hz. This is rationalized by the fact that for the suspension to exert control forces which will reduce unsprung mass motions, those forces must be reacted against the

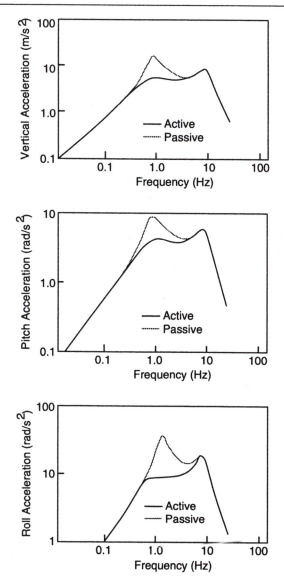

Fig. 7.26 Comparison of responses for active and passive suspension systems [12].

sprung mass, thus increasing the ride vibrations. Handling is affected by system response at the wheel hop frequency because of the associated load variations on the tires. Since the performance of the active and passive systems are identical in this region, little handling benefit is realized. To enhance handling, the control system design should be changed to reduce wheel hop response, although some penalty in ride must be expected.

273

REFERENCES

1. Bastow, D., <u>Car Suspension and Handling</u>, Second Edition, Pentech Press, London, 1990, 300 p.

2. Kami, Y., and Minikawa, M., "Double-Wishbone Suspension for Honda Prelude," SAE Paper No. 841186, 1984, 7 p.

3. Iijima, Y., and Noguchi, H., "The Development of a High-Performance Suspension for the New Nissan 300ZX," SAE Paper No. 841189, 1984, 9 p.

4. Sorsche, J.H., Encke, K., and Bauer, K., "Some Aspects of Suspension and Steering Design for Modern Compact Cars," SAE Paper No. 741039, 1974, 9 p.

5. Goodsell, D., <u>Dictionary of Automotive Engineering</u>, Butterworths, London, 1989, 182 p.

6. Leggat, J.W., "Steering and Handling of the Automobile," paper delivered before the Case Institute of Technology, October 21, 1953, 29 p.

7. Olley, M., "Independent Wheel Suspensions—Its Whys and Wherefores," *SAE Journal,* Vol. 34, No. 3, 1934, pp. 73-81.

8. Olley, M., "Road Manners of the Modern Car," Institution of Automobile Engineers, 1946, pp. 147-182.

9. Nader, R., <u>Unsafe at any Speed: the Designed-in Dangers of the American Automobile</u>, Grossman Publishers, New York, 1965, 365 p.

10. "Vehicle Dynamics Terminology," SAE J670e, Society of Automotive Engineers, Warrendale, PA (see Appendix A).

11. Sharp, R.S., and Crolla, D.A., "Road Vehicle Suspension System Design - A Review," *Vehicle Systems Dynamics,* Vol. 16, No. 3, 1987, pp. 167-192.

12. Chalasani, R.M.,"Ride Performance Potential of Active Suspension Systems — Part II: Comprehensive Analysis Based on a Full-Car Model," Proceedings, <u>Symposium on Simulation and Control of Ground Vehicles and Transportation Systems,</u> AMD-Vol. 80, DSC-Vol 2, American Society of Mechanical Engineers, pp. 205-226.

CHAPTER 8
THE STEERING SYSTEM

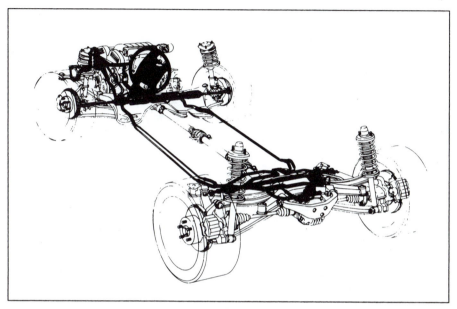

Stealth four-wheel steering system. (Courtesy of Chrysler Corp.)

INTRODUCTION

The design of the steering system has an influence on the directional response behavior of a motor vehicle that is often not fully appreciated. The function of the steering system is to steer the front wheels in response to driver command inputs in order to provide overall directional control of the vehicle. However, the actual steer angles achieved are modified by the geometry of the suspension system, the geometry and reactions within the steering system, and in the case of front-wheel drive (FWD), the geometry and reactions from the drivetrain. These phenomena will be examined in this section first as a general analysis of a steering system and then by considering the influences of front-wheel drive.

THE STEERING LINKAGES

The steering systems used on motor vehicles vary widely in design [1, 2, 3] , but are functionally quite similar. Figure 8.1 illustrates some of these.

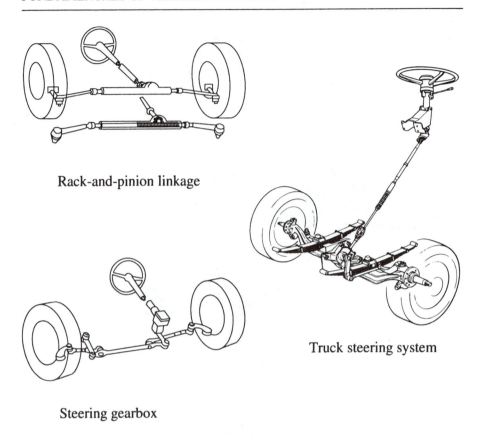

Rack-and-pinion linkage

Truck steering system

Steering gearbox

Fig. 8.1 Illustration of typical steering systems.

The steering wheel connects by shafts, universal joints, and vibration isolators to the steering gearbox whose purpose is to transform the rotary motion of the steering wheel to a translational motion appropriate for steering the wheels. The rack-and-pinion system consists of a linearly moving rack and pinion, mounted on the firewall or a forward crossmember, which steers the left and right wheels directly by a tie-rod connection. The tie-rod linkage connects to steering arms on the wheels, thereby controlling the steer angle. With the tie rod located ahead of the wheel center, as shown in Figure 8.1, it is a forward-steer configuration.

The steering gearbox is an alternative design used on passenger cars and light trucks. It differs from the rack-and-pinion in that a frame-mounted steering gearbox rotates a pitman arm which controls the steer angle of the left and right wheels through a series of relay linkages and tie rods, the specific

276

configuration of which varies from vehicle to vehicle. A rear-steer configuration is shown in the figure, identified by the fact that the tie-rod linkage connects to the steering arm behind the wheel center.

Between these two, the rack-and-pinion system has been growing in popularity for passenger cars because of the obvious advantages of reduced complexity, easier accommodation of front-wheel-drive systems, and adaptability to vehicles without frames. The primary functional difference in the steering systems used on heavy trucks is the fact that the frame-mounted steering gearbox steers the left road wheel through a longitudinal drag line, and the right wheel is steered from the left wheel via a tie-rod linkage [1].

The gearbox is the primary means for numerical reduction between the rotational input from the steering wheel and the rotational output about the steer axis. The steering wheel to road wheel angle ratios normally vary with angle, but have nominal values on the order of 15 to 1 in passenger cars, and up to as much as 36 to 1 with some heavy trucks. Initially all rack-and-pinion gearboxes had a fixed gear ratio, in which case any variation in ratio with steer angle was achieved through the geometry of the linkages. Today, rack-and-pinion systems are available that vary their gear ratio directly with steer angle.

The lateral translation produced by the gearbox is relayed through linkages to steering arms on the left and right wheels. The kinematic geometry of the relay linkages and steering arms is usually not a parallelogram (which would produce equal left and right steer angles), but rather a trapezoid to more closely approximate "Ackerman" geometry which steers the inside wheel to a greater angle than the outside wheel. Ackerman geometry is illustrated in Figure 8.2. From analysis of the triangles it can be readily shown that correct Ackerman geometry requires that:

$$\delta_o = \tan^{-1} \frac{L}{(R + t/2)} \cong \frac{L}{(R + t/2)} \qquad (8\text{-}1)$$

$$\delta_i = \tan^{-1} \frac{L}{(R - t/2)} \cong \frac{L}{(R - t/2)} \qquad (8\text{-}2)$$

For small angles, as are typical of most turning, the arctangent of the angle is very nearly equal to the angle itself (in radians), justifying the approximations shown on the right side of the equations.

Perfect Ackerman is difficult to achieve with practical linkage designs, but is closely approximated by a trapezoidal arrangement as shown in Figure 8.3. When the wheels steer right or left, the asymmetry in the geometry causes the

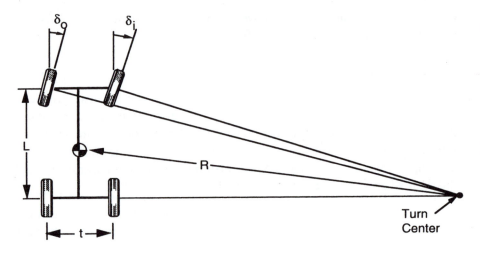

Fig. 8.2 Ackerman turning geometry.

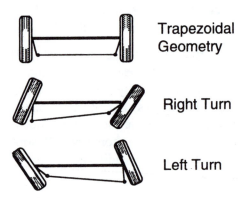

Trapezoidal
Geometry

Right Turn

Left Turn

Fig. 8.3 Differential steer from a trapezoidal tie-rod arrangement.

inside wheel to steer to a greater angle than the outside wheel. When the tie rods are located behind the wheel centers, as shown, the steering arm ball joints are located inboard of the steer axis and provide good wheel clearances. If the steering is designed with the tie rods forward of the wheel centers, the steering arm ball joints must be outboard of the steer rotation axis at the wheels in order to get close to Ackerman geometry. Interference with the wheel usually prevents design for good Ackerman in this case. Proper design of the Ackerman geometry is a function of the vehicle wheelbase and front axle tread. Design methods are straightforward and are available in the literature [1]. The degree to which the Ackerman geometry is achieved on a vehicle has little influence on high-speed directional response behavior, but does have an influence on the self-centering torque during low-speed maneuvers [4]. With Ackerman, the steer resisting torque will grow consistently with steer angle. However, with parallel steer (zero Ackerman), the torque will initially grow with angle, but may then diminish (and even become negative) at sufficiently large angles.

STEERING GEOMETRY ERROR

In the typical steering system the relay linkages transfer the steering action from the gearbox on the body of the vehicle to the steering arms on the wheels. The steering action is achieved by translational displacement of the relay linkage in the presence of arbitrary suspension motions. There is obvious potential for steering actions to arise from suspension motions, which are known as steering geometry errors.

For an ideal steering system the relay linkage is designed such that the arc described by its ball connection to the steering arm exactly follows the arc of the steering arm during suspension deflections. In that case, no steer action results during the normal ride and handling motions of the suspension. In practice, it is not always possible to achieve this ideal because of packaging problems, nonlinearities in the motion of the suspension, and because of geometry changes when the wheels are steered. Consequently, errors will occur that may result in a change in toe angle with suspension deflections, a systematic steer at both wheels, or a combination of both.

The geometry to achieve the "ideal" of no interaction on an independent front suspension is illustrated in Figure 8.4. Regardless of the suspension type used, for the vertical wheel motion around the normal ride point, the motion will be defined by some linkage constraints. For the upper and lower control arm configuration shown, relative to the vehicle body the outboard end of each arm will follow an arc centered at the pivot point on the body. This defines the motion of the upper and lower ends of the steering knuckle and, ultimately, the motion of the steering arm ball at its intermediate location on the knuckle. If the steering arm ball is located in close proximity to one of the control arms, its ideal center will be close to the inboard pivot point of that arm. If located at some intermediate position on the knuckle, its ideal center will be found at a point that is intermediate between the inboard pivots of the two arms. Although the ideal center can be estimated by eye, a geometric study is needed to precisely determine its location. Many of the computer-aided-design (CAD) programs have the capability to identify this point. Alternatively, various geometrical methods (inflection circle, Hartmann's Construction or Bobillier's Construction) may be used to locate the center [5, 6].

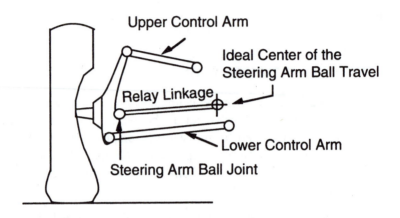

Fig. 8.4 Ideal steering geometry for an independent front suspension.

It should be noted here that the ideal center for the steering arm ball is determined by the kinematic (motion) behavior of the suspension linkages, the analysis for which is subtly different from that used to model anti-dive and anti-squat behavior. In the analysis of anti-dive and anti-squat, the objective is to determine suspension response to force and torque inputs. The conclusion in that case is that the linkages behave like a single arm pivoted at the virtual center of the upper and lower control arms, located at the intersection of the projections of the control arm axes. However, analysis of the motion of the linkages results in a different center because of the angular changes in the linkages as a result of the motion.

Toe Change

The arc that will be followed by the tie-rod end at the wheel is established by the inboard joint of the relay linkage (tie rod)—the joint being the center of the arc. If the linkage joint is either inboard or outboard of this point, the steering geometry error will cause a steer action as the wheel moves into jounce or rebound. Consider the case illustrated in Figure 8.5 which shows the inboard joint of the tie rod located outboard of the ideal center.

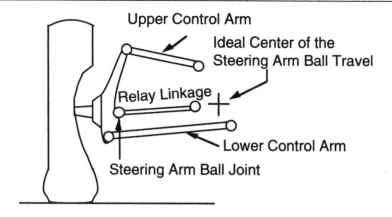

Fig. 8.5 Geometry error causing toe changes.

When the left wheel moves into either jounce or rebound, the end of the relay linkage follows an arc which pulls the steering arm to the right as viewed in Figure 8.5. This produces a left-hand steer when the linkage is located behind the wheel centers. By a similar argument it can be seen that the right wheel will steer to the right in jounce and rebound. Thus, a toe-out error occurs when the wheels are at any position other than the design ride height, and proper toe will be difficult to maintain due to its dependence on front-wheel load condition.

When the relay linkage joint is located too far inboard, the wheels will steer in the opposite direction during jounce and rebound and a toe-in error will occur. Because of the nature of this error, the toe-out or toe-in conditions will also be experienced when the body rolls in cornering. Inasmuch as these are undesirable effects, these may legitimately be considered as steering geometry errors.

Roll Steer

A second type of steering geometry error, which may be used intentionally to alter handling behavior, is to locate the inboard joint of the relay linkage either above or below the ideal center. Figure 8.6 illustrates this case, showing a rear view of a left-hand wheel with the inboard joint located below the ideal center. For the case where the linkage is located aft of the wheel, the arc followed by the relay linkage end will produce a left-hand steer on the wheel as it goes into jounce and a right-hand steer when it goes into rebound. Steer in the opposite directions will be produced on the right-side wheel. Thus, toe-in and toe-out will occur with each cycle of bouncing when the vehicle travels down the road.

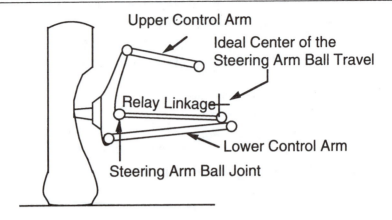

Fig. 8.6 Steering geometry error to add understeer.

Because of the symmetry of this case, both wheels will steer in the same direction when the body rolls. For example, in a turn to the right (positive steer by the SAE convention) the body rolls to the left inducing jounce on the left wheel and rebound on the right. Thus both wheels steer to the left (out of the turn) adding an understeer effect to the vehicle's directional response. Locating the inboard joint of the tie rod above the ideal center produces an oversteer effect.

Figure 8.7 shows the roll steer behavior experimentally measured on a vehicle. Lines sloping upward to the right reflect a roll steer which is understeer in direction (i.e., in a left-hand turn, as the vehicle body rolls to the right, the wheels steer to the right reducing the severity of turn). At any steer angle the slope of the curve is the roll steer coefficient, ε. The understeer gradient is then given by:

$$K_{\text{roll steer}} = \varepsilon \frac{d\phi}{d a_y} \tag{8-3}$$

Before leaving this subject, it should be noted that most suspensions swing in a plane that is skewed with respect the the vehicle longitudinal axis. The analysis, as illustrated above, should actually be made in the swing plane and transferred to the transverse plane.

FRONT WHEEL GEOMETRY

The important elements of a steering system consist not only of the visible linkages just described, but also the geometry associated with the steer rotation axis at the road wheel. This geometry determines the force and moment

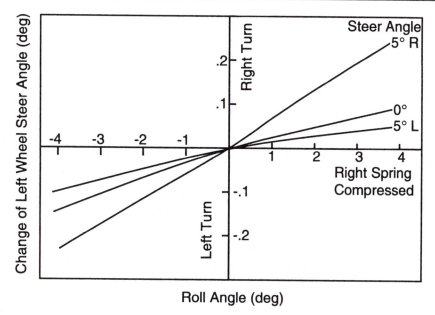

Fig. 8.7 Roll steer behavior experimentally measured on a vehicle.

reactions in the steering system, affecting its overall performance. The important features of the geometry are shown in Figure 8.8.

The steer angle is achieved by rotation of the wheel about a steer rotation axis. Historically, this axis has the name "kingpin" axis, although it may be established by ball joints or the upper mounting bearing on a strut. The axis is normally not vertical, but may be tipped outward at the bottom, producing a lateral inclination angle (kingpin inclination angle) in the range of 0-5 degrees for trucks and 10-15 degrees on passenger cars.

It is common for the wheel to be offset laterally from the point where the steer rotation axis intersects the ground. The lateral distance from the ground intercept to the wheel centerline is the offset at the ground (sometimes called "scrub") and is considered positive when the wheel is outboard of the ground intercept. Offset may be necessary to obtain packaging space for brakes, suspension, and steering components. At the same time, it adds "feel of the road" and reduces static steering efforts by allowing the tire to roll around an arc when it is turned [7].

Caster angle results when the steer rotation axis is inclined in the longitudinal plane. Positive caster places the ground intercept of the steer axis ahead of the center of tire contact. A similar effect is created by including a

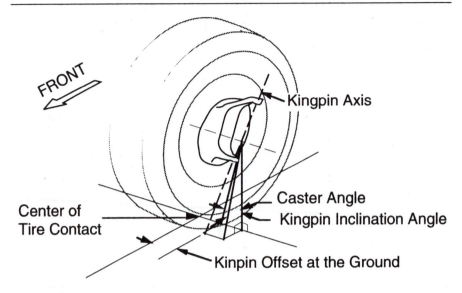

Fig 8.8 Steer rotation geometry at the road wheel.

longitudinal offset between the steer axis and the spin axis of the wheel (spindle), although this is only infrequently used. Caster angle normally ranges from 0 to 5 degrees and may vary with suspension deflection.

Wheel camber angles and toe-in normally have only secondary effects on steering behavior and high-speed directional response. The typical fractional angles specified for camber are selected to achieve near-zero camber angle for the most common load conditions of the vehicle. The small static toe angles are normally selected to achieve zero angle when driving forces and/or rolling resistance forces are present on the road. The selection of these angles is normally dominated by considerations of front tire wear rather than handling [8, 9].

STEERING SYSTEM FORCES AND MOMENTS

The forces and moments imposed on the steering system emanate from those generated at the tire-road interface. The SAE has selected a convention by which to describe the forces on a tire, as shown in Figure 8.9. The forces are measured at the center of the contact with the ground and provide a convenient basis by which to analyze steering reactions.

The ground reactions on the tire are described by three forces and moments, as follows:

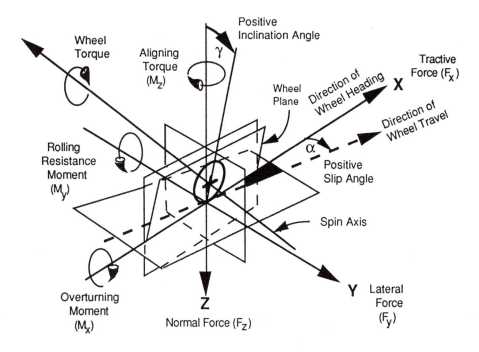

Fig. 8.9 SAE tire force and moment axis system.

Normal force	Aligning torque
Tractive force	Rolling resistance moment
Lateral force	Overturning moment

On front-wheel-drive cars, an additional moment is imposed by the drive torque. This will be discussed separately, along with other factors unique to front-wheel-drive cars that affect handling behavior.

The reaction in the steering system is described by the moment produced on the steer axis, which must be resisted to control the wheel steer angle. Ultimately, the sum of moments from the left and right wheels acting through the steering linkages with their associated ratios and efficiencies account for the steering-wheel torque feedback to the driver.

Figure 8.10 shows the three forces and moments acting on a right-hand road wheel. Each will be examined separately to illustrate its effect on the steering system.

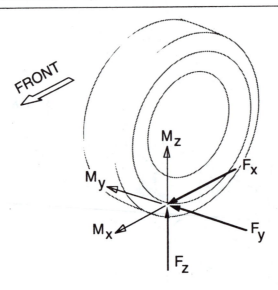

Fig. 8.10 Forces and moments acting on a right-hand road wheel.

Vertical Force

The vertical load, F_z, acts vertically upward on the wheel and by SAE convention is considered a positive force. Because the steering axis is inclined, F_z has a component acting to produce a moment attempting to steer the wheel. The moment arises from both the caster and lateral inclination angles. Assuming small angles and neglecting camber of the wheel as it steers, the total moment from the two can be approximated by:

$$M_V = - (F_{zl}+F_{zr}) \, d \sin \lambda \sin \delta + (F_{zl} - F_{zr}) \, d \sin v \cos \delta \qquad (8\text{-}4)$$

where:

M_V = Total moment from left and right wheels
F_{zl}, F_{zr} = Vertical load on left and right wheels
d = Lateral offset at the ground
λ = Lateral inclination angle
δ = Steer angle
v = Caster angle

The first expression on the right side of the above equation arises from lateral inclination angle, and the last from caster angle. The source of each of these moments is most easily visualized by considering the effects of lateral inclination angle and caster angle separately.

286

The vertical force acting on lateral inclination angle, illustrated in Figure 8.11, results in a sine angle force component, $F_{zr} \sin \lambda$, which nominally acts laterally on the moment arm "d sin δ" when the wheel is steered. The moment is zero at zero steer angle. With a steer angle, the moments on both the left and right wheels act together producing a centering moment, as shown in Figure 8.12. The net moment is proportional to the load but independent of left and right load imbalance. When steering, both sides of the vehicle lift, an effect which is often described as the source of the centering moment.

The caster angle results in a sine angle force component, $F_{zr} \sin v$, which nominally acts forward on the moment arm "d cos δ" as shown in Figure 8.13. The moments on the left and right wheels are opposite in direction, as shown in Figure 8.14, and tend to balance through the relay linkages. The balance depends on equal right and left wheel loads. Hence, load and caster angle may affect wheel toe-in, and imbalances due to load or geometric asymmetry may result in steering pull. With steer angle, one side of the axle lifts and the other drops, so that the net moment produced depends also on the roll stiffness of the front suspension as it influences the left and right wheel loads.

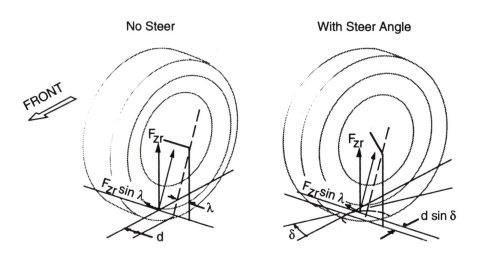

Fig. 8.11 Moment produced by vertical force acting on lateral inclination angle.

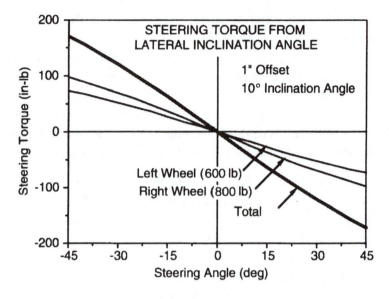

- Torque = $- (F_{zl} + F_{zr})\, d \sin \lambda \sin \delta$
- Axle <u>lifts</u> when steered
- Unaffected by left-right load differences
- Torque gradient depends on:
  — wheel offset at the ground
  — inclination angle
  — axle load

Fig. 8.12 Steering torques arising from lateral inclination angle.

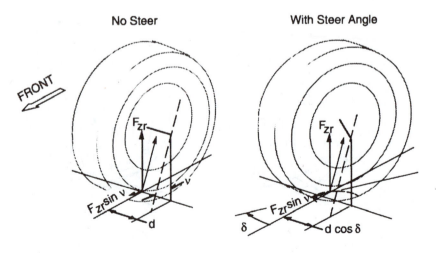

Fig. 8.13 Moment produced by vertical force acting on caster angle.

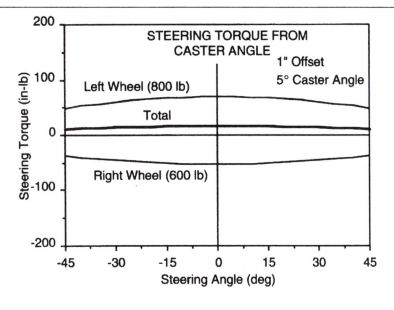

• Torque = $(F_{zl} - F_{zr})$ d sin ν cos δ
• Axle rolls when steered
• Sensitive to left-right load imbalance (load or spring asymmetry)
• Torque gradient depends on
    — wheel offset at the ground
    — caster angle
    — left-right load difference in cornering
        ·front and rear suspensions roll stiffnesses
        ·suspension roll center height
        ·center of gravity height
        ·lateral acceleration level

Fig. 8.14 Steering torques due to caster angle.

Lateral Force

The lateral force, F_y, acting at the tire center produces a moment through the longitudinal offset resulting from caster angle, as shown in Figure 8.15. The net moment produced is:

$$M_L = - (F_{yl} + F_{yr}) \, r \tan ν \qquad\qquad (8\text{-}5)$$

where:

    F_{yl}, F_{yr} = Lateral forces at left and right wheels (positive to the right)
    r   = Tire radius

The lateral force is generally dependent on the steer angle and cornering condition, and with positive caster produces a moment attempting to steer the vehicle out of the turn. Hence, it is a major contributor to understeer.

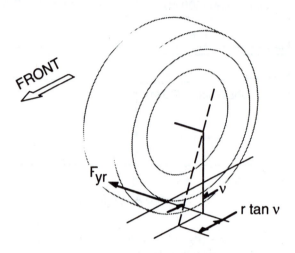

Fig. 8.15 Steering moment produced by lateral force.

Tractive Force

The tractive force, F_x, acts on the kingpin offset to produce a moment as shown in Figure 8.16.

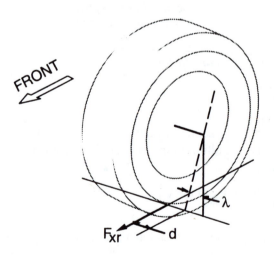

Fig. 8.16 Steering moment produced by tractive force.

Then net moment is:

$$M_T = (F_{xl} - F_{xr})\, d \qquad\qquad (8\text{-}6)$$

where:

F_{xl}, F_{xr} = Tractive forces on left and right wheels (positive forward)

The left and right moments are opposite in direction and tend to balance through the relay linkage. Imbalances, such as may occur with a tire blowout, brake malfunction, or split coefficient surfaces, will tend to produce a steering moment which is dependent on the lateral offset dimension.

Aligning Torque

The aligning torque, M_z, acts vertically and may be resolved into a component acting parallel to the steering axis. Since moments may be translated without a change in magnitude, the equation for the net moment is:

$$M_{AT} = (M_{zl} + M_{zr}) \cos \sqrt{\lambda^2 + v^2} \qquad\qquad (8\text{-}7)$$

where:

M_{zl}, M_{zr} = Aligning torques on the left and right wheels

Under normal driving conditions, the aligning torques always act to resist any turning motion, thus their effect is understeer. Only under high braking conditions do they act in a contrary fashion.

Rolling Resistance and Overturning Moments

These moments at most only have a sine angle component acting about the steer axis. They are second-order effects and are usually neglected in analysis of steering system torques.

STEERING SYSTEM MODELS

Equations (8-4) through (8-7) from the preceding discussion describe the moments input to the steer axis of each road wheel coming from the forces and moments acting on the tires. The reactions can be summed directly to determine the torque feedback to the steering wheel if desired. To quantify the influence on open-loop directional response, however, a model of the steering compliances is required [2]. Figure 8.17 shows the simplest model that is suitable for describing low-frequency behavior. The significant properties of the linkages are the stiffnesses shown here as the composite values between the gearbox and the road wheels. For various reasons, the front suspensions will also exhibit compliance in the lateral direction that adds to the effective compliance interacting with the steering displacements. These effects can be taken into account by appropriately increasing the "lumped" compliance values of the linkages.

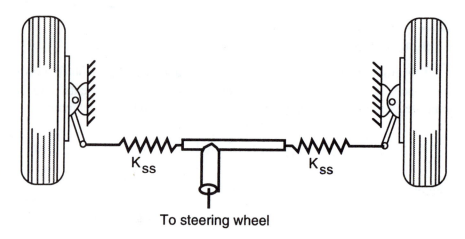

K_{ss} K_{ss}

To steering wheel

Fig. 8.17 Steering linkages model.

Depending on the purpose of the analysis, the steering gearbox might be represented by the effective steering ratio and by the appropriate input/output torque relationships and efficiencies. Similarly, the model may be expanded by the addition of Ackerman steer angle relationships between the road wheels with nonlinearities when desired. The properties of integral power-steering systems might also be introduced at this point.

The modeling equations are not developed here, but are simple to formu-late from the geometric relationships [10] and the application of Hooke's Law to relate forces and displacements across the compliances. Models of steering systems have proved most useful when combined with existing vehicle simulation models, such as the ADAMS models. The directional response simulations compute the vehicle dynamic motions and the associated force and moment conditions imposed on each tire, needed as input to the steering system model. These are used as input to the model then to determine the incremental steer angles produced, which in turn alter the turning behavior of the vehicle. This makes it possible to examine the precise influences of steering system properties on overall handling behavior.

EXAMPLES OF STEERING SYSTEM EFFECTS

The specific design of a steering system geometry has a well-recognized influence on steering performance measures such as center feel, returnability, and steering efforts as normally evaluated by vehicle manufacturers. Additionally, in the systematic study of directional response, other phenomena are observed, ranging from simple influences on steering ratio to cornering and even braking.

Steering Ratio

The steering ratio is defined as the ratio of steering wheel rotation angle to steer angle at the road wheels. Normally these range from 15 or 20 to 1 on passenger cars, and 20-36 to 1 on trucks. Because of the compliance and steer torque gradients with increasing steer angles, the actual steering ratio may be as much as twice the designed ratio. Figure 8.18 shows experimental measure-

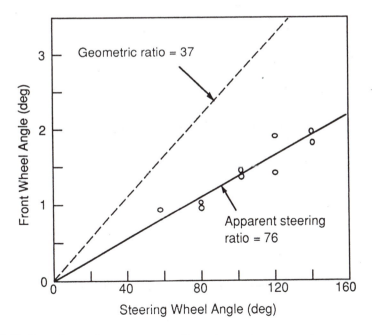

Fig. 8.18 Experimental measurement of steering ratio on a truck.

293

ments on a truck which illustrate the phenomenon. While the compliance property is constant on a vehicle, the torque gradient will vary with load on the front tires, tire type, pressure, coefficient of friction, etc. Hence, the actual steering ratio may vary (always exceeding the design value) and influencing the low-speed maneuverability of the vehicle.

Understeer

The steady-state cornering performance of a vehicle is frequently characterized by the understeer gradient measured at the steering wheel. Because compliance in the steering system allows the road wheels to deviate from the steering wheel input, the results obtained are influenced by the steering system properties. Figure 8.19 shows the steer angle gradient measured at the steering wheel and the left road wheel of a loaded truck with manual steering [11].

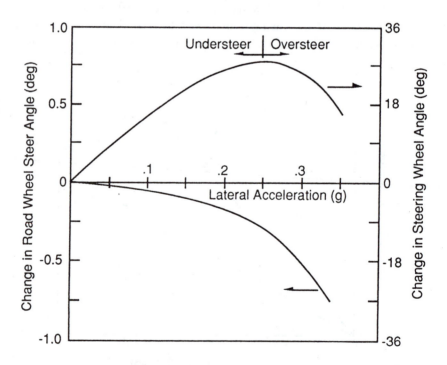

Fig. 8.19 Understeer gradient measured at the steering wheel and road wheel of a truck.

The vehicle has a very high understeer gradient at the steering wheel equal to approximately 150 degrees/g (the initial slope of the steering wheel gradient on the plot). Corrected for the ratio of the steering system (36 to 1) it is equivalent to an apparent gradient of 4 degrees/g at the road wheel. However, independent measurements of the road wheel angle indicated an initial slope that is nearly horizontal—equivalent to neutral steer at the road wheel. The difference arises from deflections in the steering linkage as the reactions on the road wheels act against the steering compliances.

The magnitude of the steering system contribution is dependent on the front-wheel load and caster angle. From a simple analysis for the understeer influences in which the lateral forces and aligning torques are dominant (neglecting vertical force effects), it can be shown that the understeer gradient is:

$$K_{strg} = \frac{W_f(r\,v + p)}{K_{ss}} \qquad (8\text{-}8)$$

where:

K_{strg} = Understeer increment (deg/g) due to steering system
W_f = Front wheel load (lb)
r = Wheel radius (in)
p = Pneumatic trail associated with aligning torque (in)
v = Caster angle (rad)
K_{ss} = Steering stiffness (in-lb/deg) between road wheel and steering wheel

As seen here, caster angle and aligning torque effects add to the understeer in the presence of a compliant steering system. For typical values of the above parameters, Eq. (8-8) would account for understeer increments on the order of 4-6 deg/g.

Braking Stability

Braking is a special case in which steering system design plays an important role in directional response [2]. Specifically, the design has a direct influence on stability and resistance to brake imbalance effects. It was shown that caster angle influences stability by way of its action to resist steering deviations caused by front brake imbalance. Yet, the benefits of caster angle are particularly vulnerable during braking conditions. Vehicle pitch and front suspension windup may overcome the few degrees of caster angle designed into the system at normal trim conditions. Further, the tire aligning torques which effectively act like 4-8 degrees of caster angle under free-rolling conditions can also reverse in direction during braking. This tire effect is illustrated in Figure 8.20 by measurements from a truck tire. The tire caster effect in this plot is obtained by dividing the aligning torque by the lateral force to determine the pneumatic trail at each data point. The pneumatic trail normalized by the tire radius then yields the effective angle at which the lateral force acts under the tire, which is the tire caster effect. At low braking coefficients the aligning torque acts in the direction to steer the tire in its direction of travel, which on the steered wheels attempts to steer the vehicle out of the turn (an understeer influence). But at high braking coefficient, the aligning torque reverses direction and may reach elevated negative levels, which will attempt to steer the tire into the direction of turn (an oversteer influence). As a result, the normal stabilizing effects of positive caster and tire aligning torque may be substantially reduced or eliminated during high-level braking.

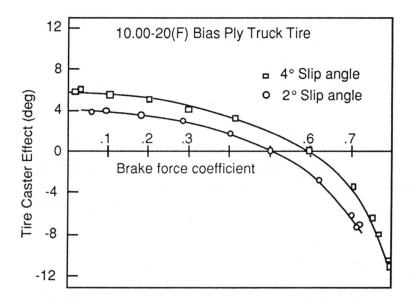

Fig. 8.20 Change of tire aligning torques with braking coefficient.

Brake force imbalance (due to brake malfunction or a split-coefficient surface) will also act on the compliant steering system, attempting to steer the vehicle. Using the split coefficient surface as an example, the higher brake forces on the high-coefficient surface will attempt to rotate the vehicle onto that surface by virtue of the moment produced on the vehicle. With a positive lateral offset, the dominant front wheel brake force on the high-coefficient surface will also attempt to steer the vehicle onto that surface. The brake force "steering" effect may be as much as 2 to 3 times greater than the direct moment on the vehicle in causing the vehicle to veer onto the high-coefficient surface. Negative offsets have been used on certain cars to counteract this mechanism on split coefficient braking, and when diagonal-split brake failure modes are employed.

INFLUENCE OF FRONT-WHEEL DRIVE

It is generally recognized that with front-wheel-drive (FWD) vehicles, the turning behavior varies with the application of engine power. In most cases, throttle-on produces understeer, and throttle-off produces oversteer. The turning equation developed in Chapter 6 for FWD vehicles would suggest just the opposite behavior. Obviously, other mechanisms must be at work. Three have been identified and will be discussed here. The discussion focuses on handling influences unique to front-wheel drive. All the other influences considered in earlier sections will still be present and acting on the vehicle.

Driveline Torque About the Steer Axis

Even in straight-ahead driving, the torque in the driveline produces a moment about the steer axis. This comes out of the model shown in Figure 8.21. In it a constant-velocity joint connects the halfshaft to the wheel spindle. The front wheel is in the straight-ahead position.

Neglecting the rolling resistance moment and the moments deriving from the normal force between the tire and road, the net moment about the steer axis of one wheel is:

$$M_{SA} = F_x \, d \cos v \cos \lambda + T_d \sin (\lambda+\zeta) \qquad (8-9)$$

Since

$$T_d = F_x \, r \qquad (8-10)$$

equation (8-9) can be rewritten:

$$M_{SA} = F_x \, [d \cos v \cos \lambda + r \sin (\lambda+\zeta)] \qquad (8-11)$$

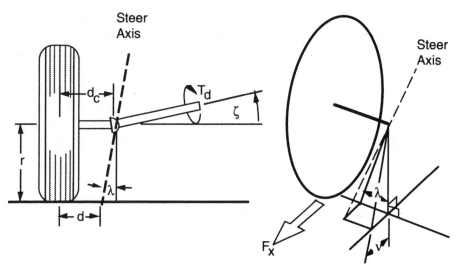

Fig. 8.21 Drive forces and moments acting on a front wheel.

Normally, the lateral inclination and caster angles (λ and v) are small enough that the cosine function can be assumed unity, in which case:

$$M_{SA} = F_x \, [d + r \sin (\lambda + \zeta)] \tag{8-12}$$

In effect, the arm about which the drive force acts to create a moment on the steering axis is $[d + r \sin (\lambda + \zeta)]$. Now d is the lateral offset at the ground. The term "$r \sin (\lambda + \zeta)$" is the additional distance out to the perpendicular from the halfshaft at the constant-velocity joint. That is, envision a plane through the constant-velocity joint which is perpendicular to the halfshaft. The offset determining the moment arm extends from the tire contact patch to that plane.

When the halfshaft is horizontal (most often the case in straight-ahead driving), ζ is zero. Then the moment arm is $d + r \sin \lambda$ which is the same as the offset at the wheel center, d_c. Hence, the expression "the drive force acts at the wheel center." (Note, because the brake torque acts through the suspension, it can be shown that the brake force moment arm is simply the lateral offset at the ground, d. When inboard brakes are used, the moment arm is again $d + r \sin \lambda$.)

When a vehicle goes into a turn, body roll causes the halfshaft on the outside wheel to reduce its inclination angle, ζ (going negative if it were already zero), while the angle on the inside wheel increases. Thus the moment arm about which the drive force acts gets smaller on the outside wheel and larger on the inside wheel. With a drive force (in the forward direction) this imbalance introduces a moment in the steering system which opposes the steer angle, trying to steer the vehicle out of the turn (understeer). The magnitude of the moment is dependent on: the degree of body roll, and how much difference in halfshaft angles is created; the difference between kingpin inclination angles on the left and right side during body roll; caster angles; and any geometric differences between the left- and right-hand sides (tire radius, etc.).

The understeer influence is proportional to the magnitude of the moment divided by the stiffness of the steering system. Thus minimizing body roll and stiffening the steering system minimizes the effect. Although the influence is specific to each vehicle, the understeer change from throttle-on to throttle-off is estimated to be on the order of 1 degree/g for a typical vehicle [12].

Influence of Tractive Force on Tire Cornering Stiffness

It is well known that a tire loses cornering force when a tractive force is present as well. Figure 8.22 shows typical behavior of lateral force as a function of tractive force. The effect is most pronounced with bias-ply tires, and less so with radial tires. The application of throttle (a demand for drive force) causes the front tires to lose cornering force, and the tires must seek a higher slip angle. This, of course, produces understeer.

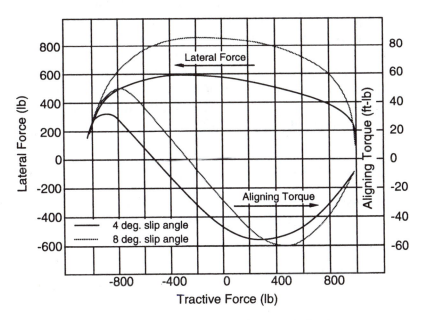

Fig. 8.22 Effect of tractive force on tire lateral force and aligning moment.

299

The magnitude will be proportional to the tractive force level. It is estimated [13] that the understeer change will be in the range of 0 to 2 degrees/g for the throttle change going from 0.2 g acceleration to 0.05 g deceleration. The influence is lowest with radial-ply tires.

Influence of Tractive Force on Aligning Moment

As seen in Figure 8.22, tractive force demand tends to increase the aligning moment produced by a tire (again less significantly with radial-ply tires). The additional aligning moment tends to steer the vehicle out of the turn, and is thus understeer. The magnitude of the understeer depends on the change in aligning moment divided by the stiffness of the steering system. It has been estimated that this mechanism contributes on the order of 0.5 to 1 degree/g of understeer [12].

Fore/Aft Load Transfer

Understeer is normally characterized only in steady speed turns. Yet it has been necessary to consider the difference between accelerating and decelerating conditions for purposes of describing the influence of FWD on understeer. When the vehicle accelerates, load is transferred to the rear wheels dynamically. This causes the rear wheels to achieve a higher cornering stiffness, while the front wheels lose cornering stiffness (causing understeer). It is worthwhile to estimate this effect for comparison purposes (to see to what degree it accounts for the understeer change that we ascribe to the FWD car because it would also be present on RWD cars, if they were tested in the same fashion). For a typical car, the understeer influence for the throttle changes described above is on the order of 1 degree/g [12].

Summary of FWD Understeer Influences

In summary, the primary mechanisms responsible for throttle on/off changes in understeer of a FWD vehicle are:

1) The lateral component of drive thrust—While this mechanism is relatively weak (<0.5 deg/g), it is oversteer in direction.

2) Drive torque acting about the steer axis—Highly dependent on driveline geometry and the degree of body roll in cornering, this mechanism is understeer in direction (about 1 deg/g).

3)   Loss of lateral force—A tire property which causes understeer (about 1-1.5 deg/g).

4)   Increase in aligning moment—A tire property which causes understeer (about 0.5-1 deg/g).

5)   Fore/aft load transfer—Although present on FWD and RWD vehicles, it is always understeer in direction (about 1 deg/g).

The total understeer due to above effects is approximately 4-5 deg/g. The mechanisms present in items 2-4 generate torques that feed back into the steering system and are the primary sources of "torque steer" often noted with FWD vehicles. Finally, it should be noted that friction in a differential can be significant (10 to 15%) when a driveline is under load. Although not treated here, under certain circumstances it could be an additional mechanism contributing to throttle-on understeer in FWD vehicles.

FOUR-WHEEL STEER

Vehicle performance in turning can be enhanced by actively steering the rear wheels as well as the front wheels (4WS). Active steering is accomplished by steering action applied directly to the rear wheels, in contrast to passive steering in which compliances are purposely designed into the suspension to provide incremental steer deviations that improve cornering [14]. Four-wheel steering may be used to improve low-speed maneuverability and/or high-speed cornering.

Low-Speed Turning

Low-speed turning performance is improved by steering the rear wheels out-of-phase with the front wheels to reduce the turn radius, thus improving maneuverability as shown in Figure 8.23. Rear-wheel steer is accomplished by mechanical, hydraulic or electronic means [14-18]. Normally, the rear-wheel steer angles are a fraction of that at the front (typically limited to about 5 degrees of steer), and may only be applied at low speeds [17] or at high steer angles typical of low-speed turns [15]. Analysis of the turning performance is simplified by assuming average angles for the front and rear wheels, analogous to the bicycle model approximation.

With the rear-wheel steer angle proportional to the front-wheel angle, the turning equations are as follows:

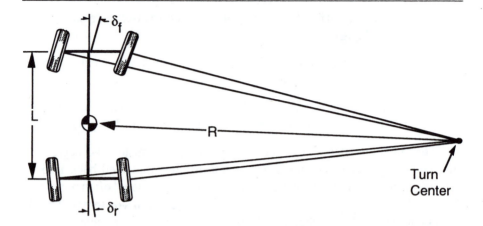

Fig. 8.23 Turning geometry of a four-wheel-steer vehicle.

$$\delta_r = \xi\,\delta_f \tag{8-13}$$

$$\delta_f + \delta_r = \delta_f + \xi\,\delta_f = \delta_f\,(1 + \xi) = L/R \tag{8-14}$$

Then the turn radius is:

$$R = \frac{L}{\delta_f(1 + \xi)} \tag{8-15}$$

Equation (8-15) gives the explicit expression for the way in which the turn radius is reduced by the use of rear steer. At 50 percent rear steer, a one-third reduction in turn radius (1/1.5) is achieved. At 100 percent rear steer (steering the rear wheels to the same magnitude as the front wheels), a 50 percent reduction in turn radius (1/2) occurs.

The expression for off-tracking with four-wheel steer is somewhat more complicated than that for two-wheel steer. Recognizing that the front and rear turn radii, R_f and R_r, respectively, are related by the expression:

$$R_f \cos \delta_f = R_r \cos \delta_r = R_r \cos (\xi\,\delta_f) \tag{8-16}$$

it is possible to obtain an equation in the following form as an approximation of the off-tracking distance:

$$\Delta \cong \frac{L^2}{2\,R}\,\frac{(1 - \xi^2)}{(1 + \xi^2)} \tag{8-17}$$

302

With no rear steer ($\xi = 0$) the off-tracking is the same as developed earlier in Eq. (6-4). With the rear wheels steered to the same angle as the front (100 percent), the off-tracking distance becomes zero.

High-Speed Cornering

The out-of-phase rear steer used for low-speed maneuverability would be inappropriate for high-speed turning because the outward movement of the rear wheels would constitute an oversteer influence. Thus an in-phase rear steer is used at high speed (e.g., 20 mph and above), although limited to a few degrees of steer. The transition between out-of-phase and in-phase steering is accomplished by sensing vehicle speed and changing the steering control algorithm in electronically controlled systems [17], or in mechanical systems by a mechanism that produces in-phase steer at small front wheel angles (0 to 250 degrees at the steering wheel) typical of high-speed driving [15].

The primary advantages of four-wheel steer are derived from the better control of transient behavior in cornering [19-24]. In general, 4WS systems yield a quicker response with better damping of the yaw oscillation that occurs with initiation of a turn. This can be seen in the lateral acceleration response to a step steer as shown in Figure 8.24 when the behavior of a "proportional" 4WS is compared to two-wheel steer. Other schemes for improving performance such as adding advances or delays into the steering action at front or rear wheels can provide additional options for tailoring the performance on 4WS vehicles.

Another picture of the advantages of 4WS can be obtained by examining the response in sideslip angle as a turn is initiated. Figure 8.25 compares the behavior of a two-wheel-steer system to various implementations of four-wheel steer. Depending on the amount of rear-wheel-steer action, the body sideslip angle can be arbitrarily reduced in cornering. The reduced amount of oscillation in sideslip angle with 4WS, as seen in Figure 8.25, adds the general feeling of better stability during transient maneuvers.

Overall, a properly implemented four-wheel steer can result in a vehicle which is more maneuverable at low speeds, and more responsive and stable in high-speed transient maneuvers. In other high-speed driving, however, its presence is imperceptible.

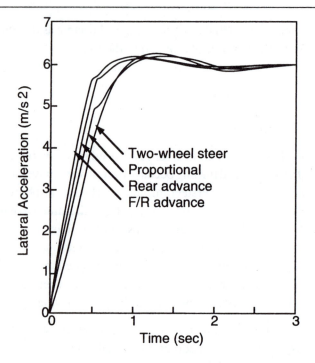

Fig. 8.24 Lateral acceleration response with different 4WS systems [20].

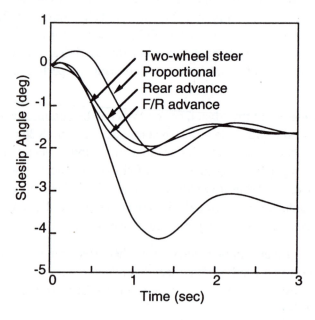

Fig. 8.25 Body sideslip angle with different 4WS systems [20].

REFERENCES

1. Durstine, J.W., The Truck Steering System from Hand Wheel to Road Wheel, SAE SP-374, January 1974, 76 p.

2. Gillespie, T.D., "Front Brake Interactions with Heavy Vehicle Steering and Handling during Braking," SAE Paper No. 760025, 1976, 16 p.

3. Dwiggins, B.H., Automotive Steering Systems, Delmar Publisher, Albany, NY, 1968, 248 p.

4. Pitts, S., and Wildig, A.W., "Effect of Steering Geometry on Self-Centering Torque and 'Feel' During Low-Speed Maneuvers," *Automotive Engineer,* Inst. of Mech. Engr., July-July 1978.

5. Hall, A.S., Jr., Kinematics and Linkage Design, Prentice Hall, Inc., Englewood Cliffs, NJ, 1961.

6. Dijksman, E.A., Motion Geometry of Mechanisms, Cambridge University Press, Cambridge, England, 1976.

7. Taborek, J.J., Mechanics of Vehicles, Towmotor Corporation, Cleveland, OH, 1957, 93 p.

8. Lugner, P., and Springer, H., "Uber den Enfluss der Lenkgeometrie auf die stationare Kurvenfahrt eines LKW," (Influence of Steering Geometry on the Stationary Cornering of a Truck), *Automobil-Industrie*, April 1974, 5 p.

9. Wheel Alignment—Modern Setting for Modern Vehicles, SAE SP-249, December 1963, 13 p.

10. MacAdam, C.C., *et al.*, "A Computerized Model for Simulating the Braking and Steering Dynamics of Trucks, Tractor-Semitrailers, Doubles, and Triples Combinations—User's Manual, Phase 4," Highway Safety Research Institute, University of Michigan, Report No. UM-HSRI-80-58, September 1980, 355 p.

11. Gillespie, T.D., "Validation of the MVMA/HSRI Phase II Straight Truck Directional Response Simulation," Highway Safety Research Institute, University of Michigan, Report No. UM-HSRI-78-46, October 1978, 58 p.

12. Gillespie, T.D., and Segel, L., "Influence of Front-Wheel Drive on Vehicle Handling at Low Levels of Lateral Acceleration," Road

Vehicle Handling, Mechanical Engineering Publications Ltd., London, 1983, pp. 61-68.

13. Braess, H.H., "Contributions to the Driving Behavior of Motor Vehicles with Front Wheel Drive Throttle Change During Cornering," Institute of Internal Combustion Machines and Motor Vehicles, Munich (Germany), 1970, 15 p.

14. Sharp, R.S., and Crolla, D.A., "Controlled Rear Steering for Cars - A Review," Proceedings of the Institution of Mechanical Engineers, International Conference on Advanced Suspensions, 1988, pp. 149-163.

15. Sano, S., et al., "Operational and Design Features of the Steer Angle Dependent Four Wheel Steering System," 11th International Conference on Experimental Safety Vehicles, Washington, D.C., 1988, 5 p.

16. Nakaya, H., and Oguchi, Y., "Characteristics of the Four-Wheel Steering Vehicle and Its Future Prospects," Vehicle System Dynamics, Vol. 8, No. 3, 1987, p. 314-325.

17. Takiguchi, T., et al., "Improvement of Vehicle Dynamics by Vehicle-Speed-Sensing Four-Wheel Steering System," SAE Paper No. 860624, 1986, 12 p.

18. Eguchi, T., et al., "'Super HICAS' - A New Rear Wheel Steering System with Phasereversal Control," SAE Paper No. 891978, 1989, 10 p.

19. Fukui, K., et al., "Analysis of Driver and a 'Four Wheel Steering Vehicle' System Using a Driving Simulator," SAE Paper No. 880641, 1988, 13 p.

20. Nalecz, A.G., and Bindemann, A.C., "Analysis of the Dynamic Response of Four Wheel Steering Vehicles at High Speed," International Journal of Vehicle Design, Vol. 9, No. 2, 1988, pp. 179-202.

21. Nalecz, A.G., and Bindemann, A.C., "Investigation into the Stability of Four Wheel Steering Vehicles," International Journal of Vehicle Design, Vol. 9, No. 2, 1988, pp. 159-179.

22. Ohnuma, A., and Metz, L.D., "Controllability and Stability Aspects of Actively Controlled 4WS Vehicles," SAE Paper No. 891977, 1989, 14 p.

23. Whitehead, J.C., "Four Wheel Steering: Maneuverability and High Speed Stabilization," SAE Paper No. 880642, 1988, 14 p.

24. Whitehead, J.C., "Rear Wheel Steering Dynamics Compared to Front Steering," *Journal of Dynamic Systems, Measurement and Control,* Vol. 112, No. 1, March 1990, pp. 88-93.

CHAPTER 9
ROLLOVER

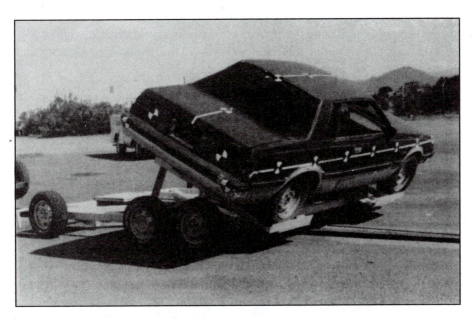

Dolly rollover test. (SAE Paper No. 900366.)

Among the dynamic maneuvers a motor vehicle can experience, rollover is one of the most serious and threatening to the vehicle occupants. Rollover may be defined as any maneuver in which the vehicle rotates 90 degrees or more about its longitudinal axis such that the body makes contact with the ground. Rollover may be precipitated from one or a combination of factors. It may occur on flat and level surfaces when the lateral accelerations on a vehicle reach a level beyond that which can be compensated by lateral weight shift on the tires. Cross-slope of the road (or off-road) surface may contribute along with disturbances to the lateral forces arising from curb impacts, soft ground, or other obstructions that may "trip" the vehicle.

The rollover process is one that involves a complex interaction of forces acting on and within the vehicle, as influenced by the maneuver and roadway. The process has been investigated analytically and empirically using models that cover a range of complexities. The process is most easily understood by starting with the fundamental mechanics involved in a quasi-static case (neglecting the inertial terms and accelerations in the roll plane), and progressing to the more complex models.

QUASI-STATIC ROLLOVER OF A RIGID VEHICLE

The most rudimentary mechanics involved in rollover of a motor vehicle can be seen by considering the balance of forces on a rigid vehicle in cornering. By rigid vehicle it is meant that the deflections of the suspensions and tires will be neglected in the analysis.

In a cornering maneuver the lateral forces act in the ground plane to counterbalance the lateral acceleration acting at the CG of the vehicle, as shown in Figure 9.1. The difference in the position at which these forces act creates a moment on the vehicle, which attempts to roll it toward the outside of the turn.

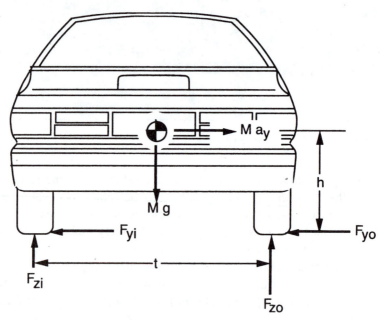

Fig. 9.1 Forces acting to roll over a vehicle.

For the purpose of analyzing the behavior, assume the vehicle is in a steady turn so that there is no roll acceleration, and let the tire forces shown in the figure represent the total for both the front and rear wheels. In many highway situations it is also appropriate to consider a transverse slope, known as cross-slope or superelevation. For the analysis, the angle will be denoted by the symbol "φ," with a slope downward to the left representing a positive angle. A cross-slope in this direction helps to counterbalance the lateral acceleration. Cross-slope angles are normally quite small, justifying the use of small angle approximations ($\sin \varphi = \varphi$, $\cos \varphi = 1$) in the analysis that follows. Taking moments about the center of contact for the outside tires yields:

$$M \, a_y \, h - M \, \varphi \, h + F_{zi} \, t - M \, g \, t/2 = 0 \qquad (9\text{-}1)$$

from which we can solve for a_y to get:

$$\frac{a_y}{g} = \frac{t/2 + \varphi \, h - \dfrac{F_{zi}}{M \, g} \, t}{h} \qquad (9\text{-}2)$$

On a level road ($\varphi = 0$) with no lateral acceleration, this equation is satisfied when the load on the inside wheels, F_{zi}, is one-half of the weight of the vehicle ($M \, g$). Further, F_{zi} can be maintained at half the weight of the vehicle in the presence of lateral acceleration by judicious choice of the cross-slope angle. That is, by choosing:

$$\varphi = \frac{a_y}{g} \qquad (9\text{-}3)$$

In highway design, cross-slope is used in curves exactly for this purpose. Given the radius of turn and an intended travel (design) speed, the cross-slope will be chosen to produce a lateral acceleration in the range of zero to 0.1 g's. The speed at which zero lateral acceleration is experienced on a superelevated curve is called the "neutral speed."

Returning again to Eq. (9-2), as the lateral acceleration builds up, the load on the inside wheels must diminish. It is through this process that the vehicle acts to resist, or counterbalance, the roll moment in cornering. The limit cornering condition will occur when the load on the inside wheels reaches zero (all the load has been transferred to the outside wheels). At that point, rollover will begin because the vehicle can no longer maintain equilibrium in the roll plane. The lateral acceleration at which rollover begins is the "rollover threshold" and is given by:

$$\frac{a_y}{g} = \frac{t/2 + \varphi \, h}{h} \qquad (9\text{-}4)$$

With no cross-slope the lateral acceleration that constitutes the "rollover threshold" is simply "t over 2h." This simple measure of rollover threshold is often used for a first-order estimate of a vehicle's resistance to rollover. It is especially attractive because it requires knowledge of only two vehicle parameters—the tread and the CG height. However, the estimates are very conservative (predicting a threshold that is greater than the actual) and are more useful for comparing vehicles rather than predicting absolute levels of performance. (Some dynamicists use the inverse form of this threshold, "h over t/2" as a measure of rollover propensity, in which case a higher value corresponds to a lower rollover threshold.)

The rollover threshold differs distinctively among the various types of vehicles on the road. As examples, typical values fall in the following ranges [1]:

Vehicle Type	CG Height	Tread	Rollover Threshold
Sports car	18-20 inches	50-60 inches	1.2-1.7 g
Compact car	20-23	50-60	1.1-1.5
Luxury car	20-24	60-65	1.2-1.6
Pickup truck	30-35	65-70	0.9-1.1
Passenger van	30-40	65-70	0.8-1.1
Medium truck	45-55	65-75	0.6-0.8
Heavy truck	60-85	70-72	0.4-0.6

The rigid-vehicle model suggests that the lateral acceleration necessary to reach the rollover of passenger cars and light trucks exceeds the cornering capabilities arising from the friction limits of the tires (typical peak coefficients of friction are on the order of 0.8). That being the case, it is possible for the car to spin out on a flat surface without rolling over. From that one might conclude that rollover with these types of vehicles should be rare; however, the accident statistics [2] prove otherwise and motivate the more in-depth analysis of rollover phenomena that will be addressed later in this chapter. In the case of heavy trucks it is equally obvious that it is possible to reach the rollover threshold within the friction limits of the tires [3]. As a consequence, a heavy vehicle is at risk of rollover if the driver allows the vehicle to spin out on a dry road surface.

Rigid-body rollover can be illustrated more fully by way of a plot of the lateral acceleration as a function of roll angle, ϕ, for equilibrium of the vehicle, as shown in Figure 9.2. Because of the rigid vehicle assumption, while at zero

312

roll angle the lateral acceleration can be any value up to the rollover threshold. Once this threshold is reached, the inside wheels lift. The vehicle begins to roll and the equilibrium lateral acceleration decreases with angle because the center of gravity is lifting and shifting toward the outside wheels.

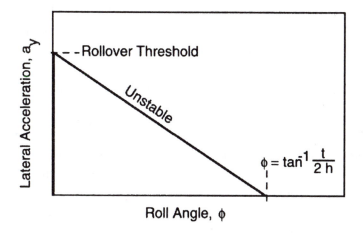

Fig. 9.2 Equilibrium lateral acceleration in rollover of a rigid vehicle.

This region is an inherently unstable roll condition. Consider a vehicle tipped up on two wheels in a turn. In order to be in equilibrium, the vehicle roll angle must be at the precise value on the above curve where the equilibrium lateral acceleration matches the actual. Any slight disturbance that increases the roll angle reduces the equilibrium lateral acceleration, and the excess lateral acceleration produces a roll acceleration that further increases the angle driving away from the equilibrium point. If left to continue, the vehicle roll attitude accelerates rapidly to complete the rollover in a matter of a second or two.

This brings up the issue of defining when rollover begins. Because of the inherent instability of the vehicle when the inside wheels leave the ground, it is appropriate to consider wheel lift-off as the <u>beginning</u> of rollover. Nevertheless, it is possible for a driver to halt the action by quickly steering out of the turn, thereby reducing the lateral acceleration to a level that will return the vehicle to an upright position. Quick response (within a fraction of a second) is necessary because of the speed with which rollover proceeds. Theoretically, rollover becomes irrecoverable only when the roll angle becomes so large that the center of gravity of the vehicle passes outboard of the line of contact of the outside wheels. This limit corresponds to the point in the figure where the equilibrium lateral acceleration reaches zero ($\phi = \tan^{-1}(t/2h)$).

It is well recognized that "stunt" drivers can take a vehicle up to this point and drive on two wheels for extended distances despite the instability. Yet, it is a rare event for a typical motorist to avoid rollover if the vehicle should inadvertently roll to this extreme position. Taking a conservative viewpoint, the automotive engineer should assume that the great majority of drivers will not have the reflexes or skills to deal with the instability once the wheels on one side of the vehicle leave the ground, and should focus on optimizing behavior of the vehicle up to that point.

QUASI-STATIC ROLLOVER OF A SUSPENDED VEHICLE

Neglecting the compliances in the tires and suspensions, as was done in the previous analysis, overestimates the rollover threshold of a vehicle [4]. In cornering, the lateral load transfer unloads the inside wheels of the vehicle and increases load on the outside wheels. Concurrently the body rolls with a lateral shift of the center of gravity toward the outside of the turn. The offset of the center of gravity reduces the moment arm on which the gravity force acts to resist the rollover.

Figure 9.3 illustrates these mechanisms on a vehicle with a suspension system. The body is represented by its mass, M_s, connected to the axle at an imaginary point known as the roll center. The roll center is the pivot around which body roll occurs, and is also the point at which lateral forces are transferred from the axle to the sprung mass.

A simple analytical solution for the rollover threshold is possible if the mass and roll of the axles are neglected [4, 5]. Taking moments about the point where the right wheel contacts the ground, and assuming the left wheel load has gone to zero gives:

$$\Sigma \, M_o = 0 = M_s \, a_y \, h - M_s \, g[t/2 - \phi \, (h - h_r)] \qquad (9\text{-}5)$$

Now the roll angle of the sprung mass, ϕ, is simply the roll rate, R_ϕ, times the lateral acceleration, a_y. The roll rate is the rate of change of roll angle with lateral acceleration expressed in units of radians per g. Substituting to eliminate the roll angle and solving for lateral acceleration yields:

$$\frac{a_y}{g} = \frac{t}{2 \, h} \frac{1}{[1 + R_\phi(1 - h_r/h)]} \qquad (9\text{-}6)$$

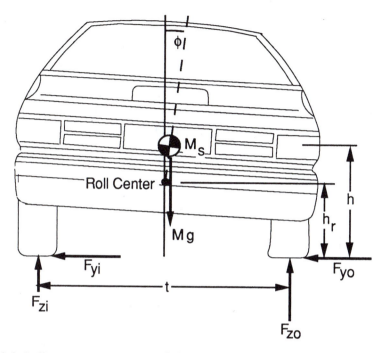

Fig. 9.3 Roll reactions on a suspended vehicle.

where:

> h = Height of the center of gravity above the ground
> h_r = Height of the roll center above the ground at the longitudinal CG
>   location
> t = Tread
> R_ϕ = Roll rate (radians/g)

   Thus taking into account the lateral shift of the CG, the rollover threshold
is "t over 2h" reduced by the second term on the right-hand side of the above
equation. For a passenger car with $h_r/h = 0.5$ and a roll rate of 6 degrees per
g (0.1 radians/g), the second term evaluates to approximately 0.95. That is, the
rollover threshold is reduced approximately 5 percent due to this mechanism.
Sports vehicles with a low roll rate and low center of gravity experience less
of these effects, whereas luxury cars with a higher roll rate and higher center
of gravity will experience more. Solid axles (which tend to have a high roll
center) also reduce the effect of lateral shift compared to independent suspensions (which have low roll centers) due to the reduced distance from the CG to
the roll center.

A similar mechanism arises from lateral deflections of the outside tires, which allows the load center under the tires to move inboard during cornering, effectively reducing the tread. For typical passenger cars the lateral shift of the tire contact point may contribute another 5 percent reduction to the threshold.

A more precise analysis of the lateral shift and the effect on the rollover threshold requires detailed modeling of the tire and suspension systems. Among the mechanisms that must be considered are:

- Lateral shift of the sprung mass center of gravity caused directly by roll about the suspension roll center.

- Lateral shift of the suspension roll center with respect to the tread, due to roll of a solid axle or camber of independently sprung wheels.

- Lateral movement of the action point of the tire vertical force due to cornering forces and deflections (these factors being reflected in changes to the overturning moment under combined cornering and camber).

- Differences in behavior of the front and rear suspensions and wheels.

Taking into account all of these effects is less amenable to analytical solution. Particularly, if the front and rear suspensions are much different in load or roll stiffness, it is necessary to model simultaneous behavior of both front and rear suspensions. Computer programs [6] are the normal approach to calculating quasi-static rollover threshold when these effects are to be included.

When these mechanisms are precisely modeled, the quasi-static roll response of a motor vehicle will take the form shown in Figure 9.4. At low levels of lateral acceleration the vehicle roll response increases linearly with a slope equal to the roll rate. This proceeds until one of the inside wheels lifts off. (Both the front and rear wheels will not necessarily leave the ground at precisely the same instant on an actual vehicle due to differences between the front and rear suspensions and their loads. In the case of multi-axle trucks, the slope will change with the lift-off of each inside wheel, resulting in a curve with three or four line segments in this area.) At this point the response changes to a lower slope because the roll rate is reduced to that provided by the one suspension that remains in contact with the ground. When the second inside wheel lifts, the rollover threshold has been reached. Thereafter, the roll response follows the downward sloping line, closely equivalent to that discussed in the case of the rigid vehicle.

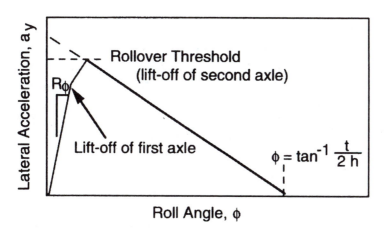

Fig. 9.4 Equilibrium lateral acceleration in rollover of a suspended vehicle.

The plot indicates that for a vehicle with a given tread and CG height, the highest rollover threshold will be achieved by maintaining the sprung mass roll rate at the highest possible level (using suspensions with high roll stiffness), and by designing the front and rear suspensions so as to have the inside wheels lift off at the same roll angle condition.

Experimental methods have been developed to measure the quasi-static rollover threshold by use of a "tilt-table." As suggested by the name, the table tilts the vehicle in the lateral, or roll, plane and, from measurement of the angle at which rollover occurs, the threshold is determined. The method is reasonably accurate for heavy trucks which have a high center of gravity and experience rollover at small angles (on the order of 20 to 25 degrees).

For passenger cars, however, the rollover threshold may well be on the order of 45 degrees. At high angles the component of the vehicle weight acting downward in the vehicle plane is reduced substantially (30 percent at 45 degrees). The reduced loading on the suspensions and tires raises the body above its normal ride position causing premature rollover and invalidating the test. In order to avoid these errors, test procedures must be devised which impose a lateral force at the center of gravity location (the "cable pull" test [5]) or by applying a pure moment to the body of the vehicle.

TRANSIENT ROLLOVER

Heretofore, the analyses have been quasi-static, and model rollover only when the vehicle is in a steady turn. (The quasi-static assumption is reasonable only when the lateral acceleration is changing much more slowly than the vehicle responds in roll.) In order to examine vehicle response to rapidly changing lateral acceleration conditions, a transient model is necessary. A transient response model attempts to represent the way the vehicle roll varies with time. At the most elementary level, a simple roll model may be used to examine response to analytically simple examples of time-varying lateral accelerations. Alternately, more comprehensive models combining motions in the yaw and roll planes have been developed to examine roll response associated with specific maneuvering conditions.

Simple Roll Models

The first and simplest approach for investigating transient roll response is with a model similar to the suspended vehicle discussed previously, to which is added a roll moment of inertia for the sprung mass as shown in Figure 9.5. The body is represented by its mass, M_s, and roll moment of inertia, I_{xxs}. Not shown is the suspension stiffness and damping on the left and right side of the vehicle. Again, the properties of the front and rear tires and suspensions may be combined to simplify the analysis.

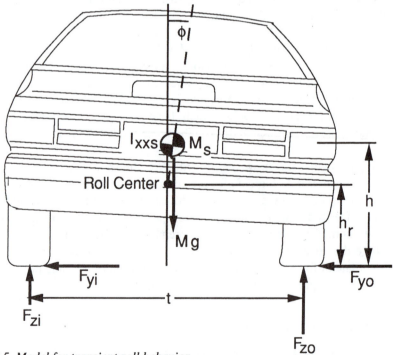

Fig. 9.5 Model for transient roll behavior.

318

This model can be useful for examining vehicle response to suddenly applied lateral accelerations in the nature of a step input. It is also representative of the transient that occurs when a vehicle goes into a slide with the brakes locked up and then experiences a sudden return of the cornering forces when the brakes are released. Or, it may simulate the effect of sliding from a low-friction surface onto one of a high-friction level.

The differential equations for motion in the roll plane can be written and solved analytically for the case of a step input [4]. The response of the system will be similar to that of a damped single-degree-of-freedom system exposed to a step input as shown in Figure 9.6.

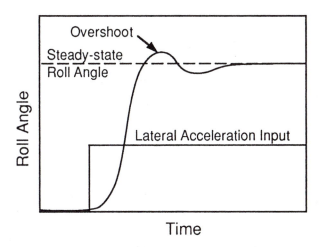

Fig. 9.6 Roll response to a step input.

With the sudden acceleration input the roll angle responds like a second-order system. With less than critical damping the angle rises toward the equilibrium point, but because it has roll velocity when it reaches equilibrium, it overshoots. Thereafter it reverses and may oscillate before settling to a steady-state angle at the equilibrium position.

The fact that the roll angle can overshoot means that wheel lift-off may occur at lower levels of lateral acceleration input in transient maneuvers than for the quasi-static case. A step steer maneuver that produces a lateral acceleration level just below the quasi-static threshold can result in rollover in the transient case because of the overshoot. Thus the rollover threshold is lower in transient maneuvers.

The extent to which overshoot occurs is dependent on roll damping. Figure 9.7 shows the calculated rollover threshold as a function of the damping ratio for a passenger car, utility vehicle,[1] and a heavy truck. The lowest rollover threshold occurs when there is no damping. It rises with the damping ratio but at a diminishing rate. Even so, the benefits of roll damping are evident. The rollover threshold of the automobile increases by nearly one-third in going from zero to 50 percent of critical damping. For the automobile and the utility vehicle, the transient in a step steer will reduce the rollover threshold by about 30 percent from the "t over 2h" value, compared to only about 10 percent for the quasi-static suspended vehicle. For the heavy truck the reduction is nearly 50 percent [4].

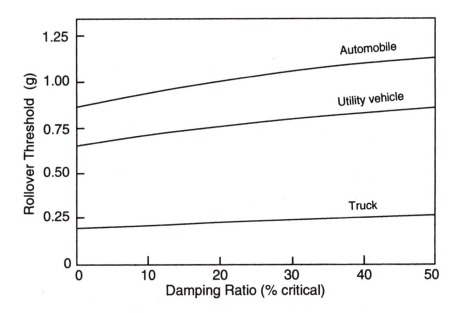

Fig. 9.7 Effect of damping ratio on rollover threshold in a step steer [4].

Exercising this model with a sinusoidal acceleration input illustrates the effect of roll resonance on the rollover threshold. A sinusoidal acceleration is similar to the input that would be experienced in a slalom course.

---

[1] Utility vehicles are defined as multipurpose passenger vehicles (other than passenger cars) which have a wheelbase of 110 inches or less and special features for occasional off-road operation.

Under a sinusoidal lateral acceleration the response of the vehicle will be dependent on the frequency of the input. Figure 9.8 shows the frequency dependence of the lateral acceleration threshold at which rollover (wheel lift-off) occurs for an automobile, utility vehicle and a heavy truck. At zero frequency the thresholds approach the steady-state values that would be obtained from the quasi-static model of the suspended vehicle. With increasing frequency the thresholds drop, going through a minimum which corresponds to the roll resonant frequency.

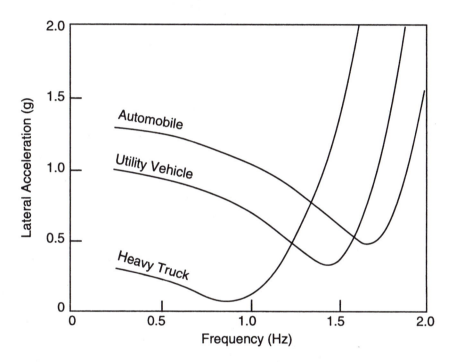

Fig. 9.8 Rollover threshold as a function of frequency in a sinusoidal steer [4].

The roll resonant frequency for a heavy truck, which is less than one cycle per second because of its high center of gravity, makes it especially vulnerable to these dynamics. Experience has shown that "lane-change type" maneuvers executed over two seconds (one-half Hz) are well capable of exciting roll dynamics that can precipitate rollover of heavy trucks [7]. The two-second timing is easily accomplished by the driver [8] and corresponds to the steering frequency necessary to move 8 to 10 feet laterally to avoid a road obstacle at normal highway speeds. As a result, lane change maneuvers have been identified as a common cause of heavy-truck rollover accidents [7].

The utility vehicle and automobile, which by comparison have a lower CG height to tread width ratio, have roll resonant frequencies of 1.5 Hz and greater. In order to tune into roll resonance, a very rapid steer oscillation is necessary. Studies of driver behavior show that steering inputs at these frequencies are normally of low amplitude [8]. Further, they produce only minor deviations in lateral position because of the attenuation of yaw response at these frequencies. (Even a relatively high-amplitude steer oscillation at 2 Hz will only cause the vehicle to move about one foot laterally.) Thus a logical conclusion is that simple roll resonance is of less significance to rollover with passenger cars and utility vehicles. In order to perform lane change maneuvers or negotiate slalom courses the timing of side-to-side oscillations is much slower (on the order of 4 seconds). Exciting frequencies below 1 Hz elicits vehicle roll response that is close to the quasi-static behavior. Therefore, from the rollover perspective, the step steer actually represents a more challenging maneuver to these vehicles than the sinusoidal steer.

Yaw-Roll Models

To develop the most complete and accurate picture of vehicle roll behavior, it is necessary to rely on more comprehensive vehicle models which simulate both yaw and roll response. Yaw motions produce the lateral accelerations causing roll motions, and roll motion in turn alters yaw response through the modification of tire cornering forces arising from lateral load transfer and suspension action. A number of computer models have been developed by the vehicle dynamics community to investigate this behavior [9, 10, 11].

Using a more comprehensive model to examine sinusoidal steer reveals an additional phenomenon of importance to vehicle roll response—the phasing of front and rear tire forces. On vehicles steered by the front wheels only, a steering action causes the front wheels to develop lateral force rather immediately (delayed only by the relaxation length of the tires), but the rear wheels do not develop a force until a sideslip angle builds up. As a result the rear wheel tire forces exhibit a phase lag in a sinusoidal steer. The phenomenon is illustrated for a passenger car in Figure 9.9.

In the one-cycle-per-second sinusoidal steer shown, the lateral forces on the rear wheels lag the fronts by approximately 0.2 seconds, corresponding to about a 70-degree phase lag. The lateral acceleration, which depends on the sum of the forces, is diminished by the phase lag. If the lateral force from both front and rear tires peaked simultaneously, the lateral acceleration would reach

0.8 g rather than 0.5 g in this maneuver. At higher frequencies the attenuation is even greater.

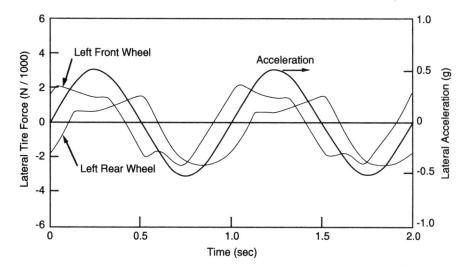

Fig. 9.9 Phasing of tire forces and lateral acceleration in a sinusoidal steer [4].

The effect of the phase lag is to allow the vehicle to yaw and change direction while moderating the level of lateral acceleration by spreading the acceleration over a longer time period. With passenger cars this effect contributes to a perception of lack of responsiveness (or sluggishness) in transient cornering. Since the time lag increases with the wheelbase of the vehicle, large cars do not feel as responsive in these maneuvers as small cars. Four-wheel-steer cars invariably steer the rear wheels in the same direction as the front wheels (albeit at a lesser steer angle) to eliminate the phase lag, thereby improving responsiveness in transient cornering. It could be argued that four-wheel steer—like any other feature that enhances cornering response—may therefore contribute to behavior that increases the potential for rollover. Keeping in mind that the roll resonance frequency of passenger cars is in the range of 1.5 to 2 Hz, the absence of the phase lag with four-wheel steer makes it easier for a motorist to inadvertently excite roll resonance in an evasive maneuver.

On very long vehicles like school buses, trucks and tractor-trailers, the phase lag may be very pronounced. Figure 9.10 shows the lateral accelerations experienced on the tractor and full trailer of a doubles combination. (A "doubles" is a tractor-semitrailer pulling a full trailer.)

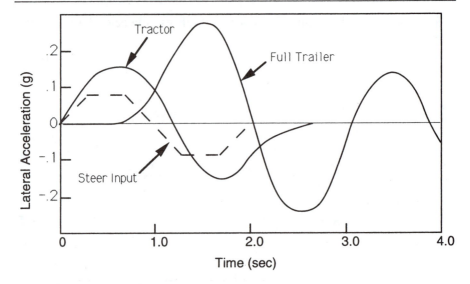

Fig. 9.10 Lateral accelerations on a tractor and full trailer [3].

A sinusoidal steer two seconds in duration excites both a rearward amplification of the yaw response and roll resonance of the full trailer, such that the full trailer experiences a much larger lateral acceleration than the tractor. Because of the vehicle length, the lateral acceleration on the full trailer is almost exactly 180 degrees out of phase with the tractor. The rearward amplification, which amounts to "cracking the whip," is recognized as very detrimental to safety performance of doubles combinations because low-level maneuvers on the tractor are amplified and can cause the full trailer to roll over. One way to prevent this is to use a hitch arrangement between the tractor-semitrailer and the full trailer which provides roll coupling. With roll coupling the out-of-phase lateral acceleration allows the full trailer to help the tractor-semitrailer resist rollover during the beginning of the maneuver, and the tractor-semitrailer helps the full trailer during the end of the maneuver. This feature is being used in the new generation of hitches for doubles combinations that are being developed at this time.

Tripping

A final class of rollover accidents that requires special modeling is the case where a vehicle skids laterally impacting an object, such as a curb or soft ground, which trips the vehicle into rollover. Engineering models for this phenomenon have been developed [12], although the understanding of the phenomenon is in the embryonic stage. A nonlinear eight-degree-of-freedom simulation model was developed which utilizes simple linear sub-systems to model tires, suspension and impact forces. The vehicle is represented by a sprung mass and an unsprung mass (combining the front and rear suspensions), as shown in Figure 9.11. The masses have degrees of freedom in roll, lateral and vertical translation, while vehicle yaw and pitch are analyzed using a single lumped mass. The impact force for lateral wheel/curb impact is modeled using both plastic and elastic deformations. Damping effects are included through energy dissipative forces in the tires, the lateral bushing between the sprung and unsprung masses, the shock absorbers in the suspension, and the wheel/curb impact force. This model was developed for the National Highway Traffic Safety Administration with public funds, and therefore should be available to any user by request to the Administration.

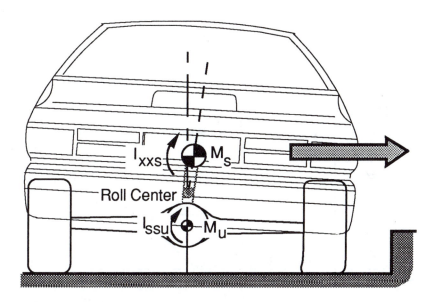

Fig. 9.11 Vehicle approaching a tripped rollover.

Models of this type have been used to investigate the conditions under which vehicles can experience tripped rollover, focusing on whether sufficient energy is developed in the curb impact to raise the CG of the vehicle to the rollover point [13]. At impact with the curb, rotation of the vehicle produces kinetic energy equal to one-half times the moments of inertia of the sprung and unsprung masses about their rotation points, times their respective rotational velocities squared. Concurrently, the lifting of the vehicle CG adds potential energy equal to the mass times the increase in CG height. If the total of these two exceeds the potential energy necessary to lift the CG over the outside wheels, rollover is predicted.

From an engineering viewpoint this energy approach has many weaknesses because of the assumption that all of the kinetic energy is transformed to potential energy, raising the CG to the rollover point. It neglects additional energy input or dissipation from wheel contact with the ground during the process, and energy storage or dissipation in the tires and suspensions.

Figure 9.12 shows typical examples of the results from an energy analysis of the curb impact process. The vertical axis plots the net rollover energy, which is the instantaneous total of the rotational kinetic energy plus the potential energy of the elevated CG. The rollover threshold is the potential energy level associated with the CG passing over the outside wheels. If the rollover energy exceeds the threshold, rollover will occur.

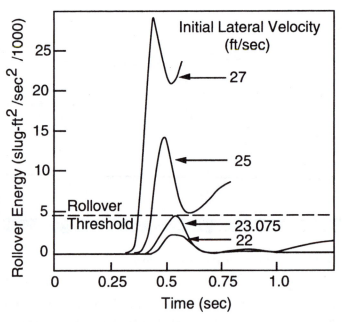

Fig. 9.12 Rotational energy during curb impact.

In the analysis, the simulated vehicle is given an initial lateral velocity while still 7.5 feet from the curb. With a 22 ft/sec initial velocity, the impact causes a brief rise in the rollover energy level due to the kinetic energy of rotation and potential energy from elevation of the CG. However, the net energy always remains well below the threshold, so rollover does not occur. With time the energy is damped out by the suspension system.

The velocity of 23.075 ft/sec represents the case just sufficient to take the vehicle to rollover. The rollover energy rises to the rollover point, where the kinetic energy component nearly reaches zero. Thereafter, the energy drops as the vehicle completes the rollover. At the higher initial speeds of 25 and 27 ft/sec, rollover occurs.

This methodology was used to examine the effect of vehicle parameters on the propensity for rollover. Not surprisingly, the geometric parameters of track width and CG height are found to be most influential. The second most important variable is the deformation characteristics of the vehicle at impact. By spreading the impact deformation over a large distance, the energy dissipated in crush reduces the amount of energy which can contribute to the vehicle's rolling motion. The weight of the vehicle appears to have little effect except as it affects the ride height—greater weight lowering the CG height. Likewise, the suspension stiffness and damping properties were found to be of little influence.

ACCIDENT EXPERIENCE

The primary motivation for giving attention to the mechanics of rollover in vehicle design is to reduce or prevent rollover accidents. In recent years, analysts have examined the accident records in an effort to identify those characteristics of vehicles that appear to be most closely correlated with rollover experience—the presumption being that the frequency of rollover accidents can be reduced by altering those correlated properties of the vehicles.

It is common practice in these studies to stratify the analyses both in the types of accidents and the types of vehicle. In the simplest treatment, the frequency of rollover in all accidents of a given make of vehicle might be considered on the assumption that all vehicles are exposed to the same general spectrum of accident types. Therefore, any atypical characteristics of those vehicles are potential causes of rollover and good design practice would argue that they should be eliminated. A flaw in this approach, however, becomes evident when it is recognized that utility vehicles experience more off-road rollover accidents than passenger cars, in part because they operate more frequently in this environment. Improving their rollover experience by making them lower and wider can only be achieved with a penalty in off-road mobility.

In the interest of normalizing accident statistics, it is then necessary to distinguish between on-road and off-road accidents, rollovers as the first or only event of an accident, rollovers as a subsequent event, and the use or exposure factors for the class of vehicle. With regard to vehicle types, they are often classified as passenger cars, utility vehicles (high CG four-wheel-drive vehicles used for personal transport), light trucks (used for personal transport and light hauling) and heavy trucks.

Systems Technology, Inc., in the work they have done for the NHTSA [14], examined the rollover accident experience of small cars as a function of the rollover potential. A replot of some of their data is shown in Figure 9.13. The rollover rate (fatal accidents per 100,000 new car years) is plotted against rollover threshold for accidents where rollover was the first event or a subsequent event in the accident. The data show a trend of decreasing rollover involvement as the threshold increases. However, the degree of scatter in the plot suggests that more than just the rollover threshold is needed to explain the accident experience. For example, the Mercury Capri has three times as many rollover accidents as the Vega, even though they both have the same rollover threshold. Because of this broad disparity in behavior, there is no guarantee to the automotive designer that good rollover experience is assured by increasing the rollover threshold.

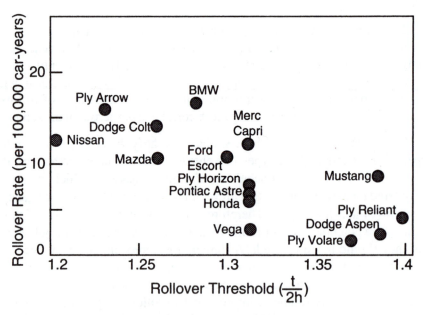

Fig. 9.13 Rollover accident rates of small cars [14].

Observations of this nature are common when examining rollover accident rates and prompt analysts to hypothesize explanations for the differences in behavior between vehicle makes. A methodical analysis of rollover accident experience for passenger cars and utility vehicles was recently conducted by Robertson and Kelley [15] in which some of the potential explanatory factors were examined. In their work, a broader range of vehicles was considered. Figure 9.14 shows their data for number of accidents per 100,000 vehicle-years in which rollover was the "first harmful event."

As plotted here the data would appear to show a much more direct relationship between rollover threshold and accident rates. This impression comes about because of the inclusion of the utility vehicles (CJ-5, CJ-7, Blazer, and Bronco) which have much higher accident rates. Among the automobiles, which have thresholds ranging from 1.25 to 1.6, there is no trend evident. The high involvement of the utility vehicles has prompted proposals for Federal Motor Vehicle Safety Standards (FMVSS) requiring new vehicles to have a minimum rollover threshold of 1.2. Over-involvement of utility vehicles is not unique to this study, but has been found in other studies [16, 17] as well.

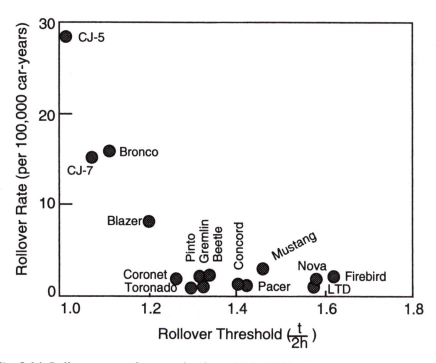

Fig. 9.14 Rollover rates of cars and utility vehicles [15].

The authors used the accident records to examine other factors which might be peculiar to the use of the different kinds of vehicles, to determine if they might be correlated to the accident experience, thereby providing other potential explanations for the high frequency of rollover accidents with utility vehicles. For example, it might be argued that the type of driver most attracted to utility vehicles is one willing to take more risks and therefore is more likely to have accidents. However, when the non-rollover fatal accident rates are compared, the utility vehicle rates are seen to be no higher than the passenger cars. When driver characteristics—suspended licenses, history of traffic violations, DWI convictions, or blood alcohol level at the time of the accident—were examined, no explanations were found.

Similarly, the road environment—urban vs. rural, interstate vs. other roads, straight vs. curved, dry vs. wet, etc.—had no correlation. The only significant environmental factor was whether the vehicle crashed on the road or after leaving it. The ratio of rollover fatal crashes on the road relative to those that left the road was substantially higher among the utility vehicles. Finally, the possibility that utility vehicles accumulate higher mileage as a potential explanation of higher rollover accident rates was examined, but with the conclusion that the usage rates necessary to account for the higher accident rates were unreasonable.

In general, other reviews of accident experience confirm the conclusions reached above, even though the Robertson study is easily criticized. Numerous studies show conflicting results and interpretations [16, 17, 18] because of the uncertainty in identifying rollover threshold as the most significant variable related to rollover frequency for these classes of vehicles. In particular, the concern that other vehicle factors play a significant role appears justified. Vehicle control and handling stability are identified as important associated variables. Likewise, vehicle wheelbase can be shown to be correlated with rollover experience, suggesting that it is a combination of factors that must be controlled before effective action to reduce rollovers can be taken. Until those factors and interactions are known, it is considered premature to impose an arbitrary limit on rollover threshold on the industry.

Another class of vehicle that has received special attention for rollover accident experience is the heavy truck. For tractor-semitrailer vehicles, the frequency of rollover in single-vehicle accidents has been related to rollover threshold as shown in Figure 9.15. This curve is derived from accident data for three-axle tractors pulling two-axle van-type semitrailers reported to the Bureau of Motor Carrier Safety (BMCS) of the U.S. Department of Transportation. The data were resolved into the illustrated format assuming a rollover threshold for each vehicle combination based on the gross vehicle weight reported to BMCS with each accident. Knowing the gross vehicle weight, the analysis assumed that payload was placed in a fashion representing medium-density freight. Typical values for tire, spring, and geometric properties were then employed to calculate rollover threshold for each increment of gross weight in the accident file. The relationship shown here has proved very useful in heavy truck design for assessing the potential benefits of design alternatives which can reduce rollover experience.

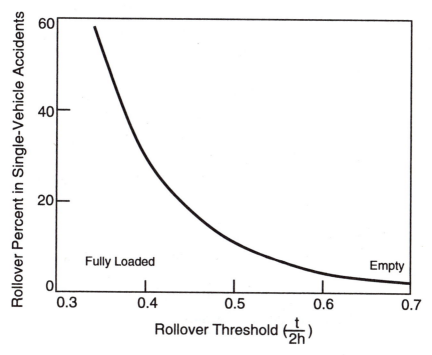

Fig. 9.15 Rollover frequency of tractor-semitrailers in single-vehicle accidents [3].

REFERENCES

1.   Gillespie, T.D., and Ervin, R.D., "Comparative Study of Vehicle Roll Stability," The University of Michigan Transportation Research Institute, Report No. UMTRI-83-25, May 1983, 42 p.

2.   "NCSS for Passenger Cars, January 1977-March 1979," Leda Ricci, Editor, The University of Michigan, Report No. UM-HSRI-80-36, 1980, 102 p.

3. Ervin, R.D., "The Influence of Size and Weight Variables on the Roll Stability of Heavy Duty Trucks," SAE Paper No. 831163, 1983, 26 p.

4. Bernard, J., Shannan, J., and Vanderploeg, M., "Vehicle Rollover on Smooth Surfaces," SAE Paper No. 891991, 1989, 10 p.

5. Bickerstaff, D., "The Handling Properties of Light Trucks," SAE Paper No. 760710, 1976, 29 p.

6. Gillespie, T.D., *et al.*, "Roll Dynamics of Commercial Vehicles," *Vehicle Systems Dynamics,* Vol. 9, 1980, pp. 1-17.

7. Ervin, R.D., *et al.*, "Ad Hoc Study of Certain Safety-Related Aspects of Double Bottom Tankers," The University of Michigan Transportation Research Institute, Report No. UM-HSRI-78-18, 1978, 78 p.

8. McLean, J.R., and Hoffman, E.R., "The Effects of Restricted Preview on Driver Steering Control and Performance," *Human Factors,* Vol. 15, No. 4, 1973, pp. 421-30.

9. Gillespie, T.D., and MacAdam, C.C., "Constant Velocity Yaw/Roll Program: User's Manual," The University of Michigan Transportation Research Institute, Report No. UM-HSRI-82-139, October 1982, 119 p.

10. McHenry, R.R., "Research in Automobile Dynamics—Computer Simulation of General Three-Dimensional Motions," SAE Paper No. 710361, 1971, 20 p.

11. Orlandea, N.V., "ADAMS Theory and Applications," Proceedings of Advanced Vehicle Systems Dynamics Seminar, Swets and Zeitlanger, 1987, pp. 121-166.

12. Rosenthal, R.J., *et al.,* "User's Guide and Program Description for a Tripped Roll Over Vehicle Simulation," Report No. DOT HS 807140, Systems Technology Inc., 1987, 76 p.

13. Nalecz, A. G., Bindemann, A.C., and Bare, C., "Sensitivity Analysis of Vehicle Tripped Rollover Model," Final Report, Contract No. DTNH22-8-Z-07621, University of Missouri-Columbia, July 1988, 100 p.

14. Wade Allen, R., *et al.,* "Validation of Tire Side Force Coefficient and Dynamic Response Analysis Procedures: Field Test and Analysis Comparison of a Front Wheel vs. a Rear Wheel Drive Subcompact,"

Systems Technology Inc., Working Paper No. 1216-6, Contract No. DTNH22-84-D-17080 - Task 4, 17 February 1986, 50 p.

15. Robertson, L.S., and Kelley, A.B., "Static Stability as a Predictor of Overturn in Fatal Motor Vehicle Crashes," *Journal of Trauma,* Vol. 29, No. 3, 1988, pp. 313-319.

16. Terhune, K.W., "A Comparison of Light Truck and Passenger Car Occupant Protection in Single Vehicle Crashes (MVMA)," Calspan Corporation, Report No. 7438-1, September 1986, 58 p.

17. Ajluni, K.K., "Rollover Potential of Vehicles on Embankments, Sideslopes, and Other Roadside Features," *Public Roads,* Vol. 52, No. 4, March 1989, pp. 107-113

18. Malliaris, A.C., Discerning the State of Crash Avoidance in the Accident Experience, Proceedings, Tenth International Technical Conference on Experimental Safety Vehicles, Technical Session No. 2, Crash Avoidance, July 1985, pp. 199-220.

CHAPTER 10
TIRES

Eagle CS-C. (Photo courtesy of Goodyear Tire & Rubber Co.)

In modern highway vehicles all the primary control and disturbance forces which are applied to the vehicle, with the exception of aerodynamic forces, are generated in the tire-road contact patch. Thus it has been said that "the critical control forces that determine how a vehicle turns, brakes and accelerates are developed in four contact patches no bigger than a man's hand." A thorough understanding of the relationship between tires, their operating conditions, and the resulting forces and moments developed at the contact patch is an essential aspect of the dynamics of the total vehicle.

The tire serves essentially three basic functions:

1) It supports the vertical load, while cushioning against road shocks.
2) It develops longitudinal forces for acceleration and braking.
3) It develops lateral forces for cornering.

While the tire is a simple visco-elastic toroid, with modern refinement and optimization of its properties it is a very complex nonlinear system which is

335

difficult and complex to quantify [1]. Numerous simplified tire models have been developed in the past to approximate various performance properties [2, 3], but for the purposes of understanding their role in vehicle dynamics it is sufficient to look to empirical data to quantify essential properties.

As a mechanical structure, the elastic torus of a tire is composed of a flexible carcass of high-tensile-strength cords fastened to steel-cable beads that firmly anchor the assembly to the rim. The internal (inflation) pressure stresses the structure in such a way that any external force causing deformation in the carcass results in a tire reaction force. The behavioral characteristics of the tire depend not only on the operating conditions, but on the type of construction as well.

TIRE CONSTRUCTION

Two basic types of tire construction are broadly used—radial- and bias-ply tires. The two types are illustrated in Figure 10.1. Bias-ply tires were the standard in the early years of the American automotive industry until about the 1960s when the advantages of radial tires (developed in Europe) became recognized. Over several decades radial tires gradually displaced bias-ply tires on passenger cars, such that they are the standard today. The acceptance on trucks has lagged that of passenger cars, such that radial and bias tires see about equal use today. Bias-belted tires had a brief life as a cross between radial and bias tires during the transition period, but are seen very little today.

Radial construction is characterized by parallel plies (rubberized fabric reinforced by cords of nylon, rayon, polyester or fiberglass) running directly across the tire from one bead to the other at a nominal 90-degree angle to the circumference. These plies are referred to as the "carcass." This type of construction makes for an extremely flexible sidewall and a soft ride but provides little or no directional stability. Directional stability is supplied by a stiff belt of fabric or steel wire that runs around the circumference of the tire between the carcass and the tread. The cord angle in the belts is normally within about 20 degrees of the tread. In cornering, the belts help to "stabilize" the tread, keeping it flat on the road despite lateral deflection of the tire. Most radial passenger-car tires have a two-ply fabric sidewall, and either one or two steel belts, or two to six fabric belts.

In bias tire construction the carcass is made up of two or more plies extending from bead to bead with the cords at high angles (35 to 40 degrees to the circumference) and alternating in direction from ply to ply. High angles

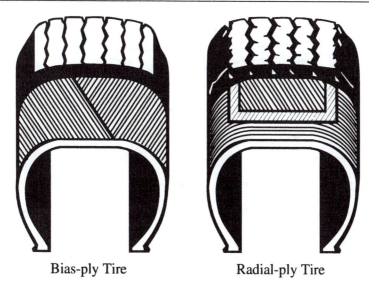

Bias-ply Tire   Radial-ply Tire

Fig. 10.1 Illustrations of bias and radial tire construction. (Courtesy of Goodyear Tire & Rubber Co.)

result in tires which are soft for ride comfort, but low angles are best for directional stability. Although a bias-ply carcass is laterally much stiffer than a radial-ply tire, in cornering, the bias plies allow the tread to roll under, putting more load on the outer ribs. Bias construction also causes more distortion in the contact patch as the toroid deforms into a flat shape, causing the tread to squirm in the contact patch [4] when rolling, as seen in Figure 10.2.

SIZE AND LOAD RATING

Tire size is denoted by one of several methods, variously including the section height (distance from the bead seat to the outer extreme of the tread), section width (maximum width across the shoulder), aspect ratio (ratio of height to width) and rim diameter [5]. The Tire & Rim Association [6] defines the method of designating tire sizes and the load range for which they should be designed by the manufacturers. Bias tire size (e.g., 6.95-14) denotes the section width with the first number and the rim size with the second, both in inch units. Radial tire size (e.g., 175R14) denotes the section width in millimeters and the rim diameter in inches. The more recent P-metric designation method (e.g., P175/70R14) denotes passenger car (P), 175-mm section width, 70 aspect ratio, radial (R-radial, B-belted, D-bias), and 14-inch rim.

337

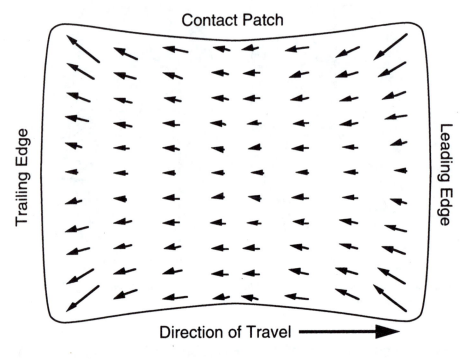

Fig. 10.2 Squirm in the contact patch of a bias-ply tire.

TERMINOLOGY AND AXIS SYSTEM

To facilitate precise description of the operating conditions, forces, and moments experienced by a tire, the SAE [5] has defined the axis system shown in Figure 10.3. The X-axis is the intersection of the wheel plane and the road plane with the positive direction forward. The Z-axis is perpendicular to the road plane with a positive direction downward. The Y-axis is in the road plane, its direction being chosen to make the axis system orthogonal and right-hand.

The following definitions are of importance in describing the tire and its axis system.

- Wheel Plane—central plane of the tire normal to the axis of rotation.

- Wheel Center—intersection of the spin axis and the wheel plane.

- Center of Tire Contact—intersection of the wheel plane and projection of the spin axis onto the road plane.

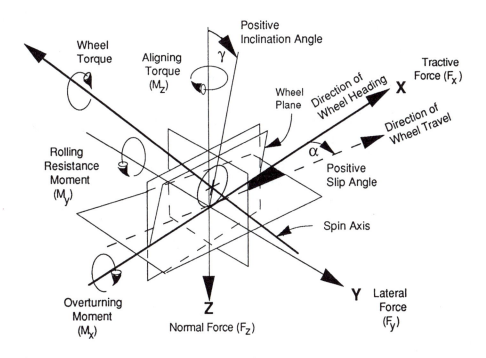

Fig. 10.3 SAE tire axis system.

- Loaded Radius—distance from center of tire contact to the wheel center in the wheel plane.

- Longitudinal Force (F_x)—component of the force acting on the tire by the road in the plane of the road and parallel to the intersection of the wheel plane with the road plane. The force component in the direction of wheel travel (sine component of the lateral force plus cosine component of the longitudinal force) is called tractive force.

- Lateral Force (F_y)—component of the force acting on the tire by the road in the plane of the road and normal to the intersection of the wheel plane with the road plane.

- Normal Force (F_z)—component of the force acting on the tire by the road which is normal to the plane of the road. The normal force is negative in magnitude. The term vertical load is defined as the negative of the normal force, and is thus positive in magnitude.

- Overturning Moment (M_x)—moment acting on the tire by the road in the plane of the road and parallel to the intersection of the wheel plane with the road plane.

- Rolling Resistance Moment (M_y)—moment acting on the tire by the road in the plane of the road and normal to the intersection of the wheel plane with the road plane.

- Aligning Moment (M_z)—moment acting on the tire by the road which is normal to the plane of the road.

- Slip Angle (α)—angle between the direction of wheel heading and the direction of travel. Positive slip angle corresponds to a tire moving to the right as it advances in the forward direction.

- Camber Angle (γ)—angle between the wheel plane and the vertical. Positive camber corresponds to the top of the tire leaned outward from the vehicle.

MECHANICS OF FORCE GENERATION

The forces on a tire are not applied at a point, but are the resultant from normal and shear stresses distributed in the contact patch. The pressure distribution under a tire is not uniform but will vary in the X and Y directions. When rolling, it is generally not symmetrical about the Y-axis but tends to be higher in the forward region of the contact patch. Both of these phenomena are shown in Figure 10.4.

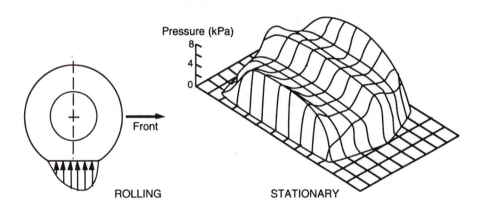

Fig. 10.4 Pressure distributions under a rolling and non-rolling tire.

340

Because of the tire's visco-elasticity, deformation in the leading portion of the contact patch causes the vertical pressure to be shifted forward. The centroid of the vertical force does not pass through the spin axis and therefore generates rolling resistance. With a tire rolling on a road, both tractive and lateral forces are developed by a shear mechanism. Each element of the tire tread passing through the tire contact patch exerts a shear stress which, if integrated over the contact area, is equal to the tractive and/or lateral forces developed by the tire.

There are two primary mechanisms responsible for the friction coupling between the tire and the road [4] as illustrated in Figure 10.5.

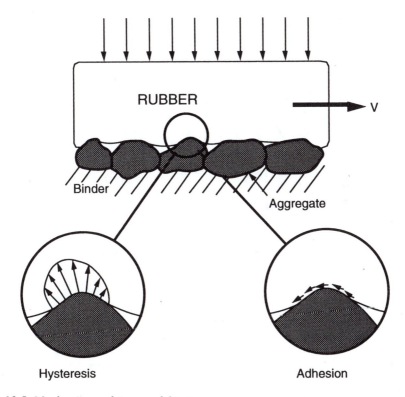

Fig. 10.5 Mechanisms of tire-road friction.

Surface adhesion arises from the intermolecular bonds between the rubber and the aggregate in the road surface. The adhesion component is the larger of the two mechanisms on dry roads, but is reduced substantially when the road surface is contaminated with water; hence, the loss of friction on wet roads.

The hysteresis mechanism represents energy loss in the rubber as it deforms when sliding over the aggregate in the road. Hysteresis friction is not so affected by water on the road surface, thus better wet traction is achieved with tires that have high-hysteresis rubber in the tread. Both adhesion and hysteresis friction depend on some small amount of slip occurring at the tire-road interface.

TRACTIVE PROPERTIES

Under acceleration and braking, additional slip is observed as a result of the deformation of the rubber elements in the tire tread as they deflect to develop and sustain the friction force. Figure 10.6 illustrates the deformation mechanism in the contact patch under braking conditions.

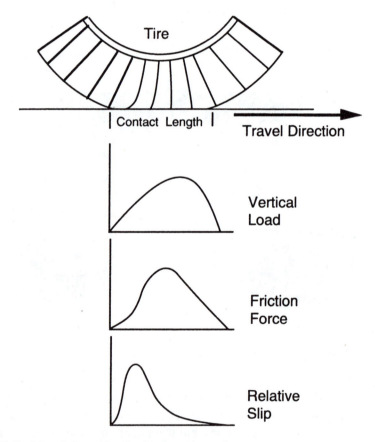

Fig. 10.6 Braking deformation in the contact patch.

As the tread elements first enter the contact patch they cannot develop a friction force because of their compliance—they must bend to sustain a force. This can happen only if the tire is moving faster than the circumference of the tread. As the tread element proceeds back through the contact patch its deflection builds up concurrently with vertical load and it develops even more friction force. However, approaching the rear of the contact patch the load diminishes and there comes a point where the tread element begins to slip noticeably on the surface such that the friction force drops off, reaching zero as it leaves the road.

Thus acceleration and braking forces are generated by producing a differential between the tire rolling speed and its speed of travel. The consequence is production of slip in the contact patch. Slip is defined non-dimensionally, as a percentage of the forward speed, as:

$$\text{Slip } (\%) = (1 - \frac{r \, \omega}{V}) \times 100 \qquad (10\text{-}1)$$

where:

r = Tire effective rolling radius
ω = Wheel angular velocity
V = Forward velocity

Under typical braking conditions the longitudinal force produced by a tire will vary with slip as shown in Figure 10.7. As slip is applied (e.g., by brake application) the friction force rises with slip along an initial slope that defines a longitudinal stiffness property of the tire. In general, this property is not directly critical to braking performance, except at the detailed level of the design of anti-lock systems where the cycling efficiency may be affected by this property. Longitudinal stiffness tends to be low when the tire is new and has full tread depth, increasing as the tire wears. For the same reasons, rib-type treads produce a higher stiffness than lug (traction) tires.

On a dry road, when the slip approaches approximately 15-20 percent, the friction force will reach a maximum (typically in the range of 70 to 90 percent of the load) as the majority of tread elements are worked most effectively without significant slip. Beyond this point friction force begins to drop off as the slip region in the rear of the contact patch extends further forward. The force continues to diminish as the tire goes to lockup (100% slip).

Performance on slippery roads is qualitatively similar to dry roads, differing primarily in the peak level of friction force that can be achieved. Since

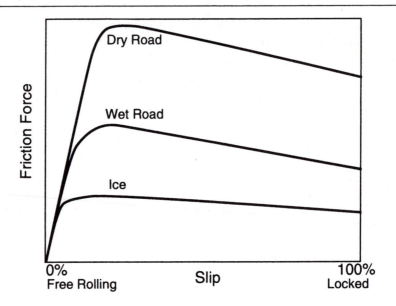

Fig. 10.7 Brake force versus slip.

the initial rate at which friction force builds up with slip is dependent on tire stiffness properties, the initial slope is the same. On wet roads the peak friction force will typically be in the range of 25 to 50 percent of the vertical load. On ice-covered roads the peak friction will be only 10 to 15 percent of the vertical load and will be reached at only a few percent slip. A part of the treachery of driving on ice is that not only is the friction level low, but the tire is very quick to brake through its maximum friction level.

For purposes of characterizing the traction properties of tires it is common to refer to the coefficient of friction (traction force divided by load) at the peak and slide conditions. These are referred to as μ_p and μ_s. The peak and slide coefficients will be dependent on a number of variables.

Vertical Load

Increasing vertical load is known categorically to reduce friction coefficients under both wet and dry conditions [7]. That is, as load increases, the peak and slide friction forces do not increase proportionately. Typically, in the vicinity of a tire's rated load, both coefficients will decrease on the order of 0.01 for a 10% increase in load. The general range of the coefficients of friction for passenger-car tires on dry roads as a function of load is shown in Figure 10.8.

344

Truck tires generally exhibit lower coefficient values because of their higher unit loading in the contact patch and different tread rubber compounds.

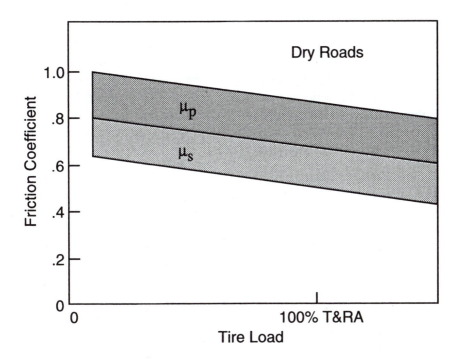

Fig. 10.8 Typical variation of friction coefficient with tire load.

Inflation Pressure

On dry roads, peak and slide coefficients are only mildly affected by inflation pressure. On wet surfaces, inflation pressure increases are known to significantly improve both coefficients.

Surface Friction

The road surface and its condition have a direct effect on the friction coefficient that can be achieved. Strictly speaking, a tire alone does not have a coefficient of friction; it is only the tire-road friction pair that has friction properties. But in the interest of characterizing the relative friction qualities of road surfaces, the highway community has developed a test method in which a "standard" tire is dragged at lockup over the surface.

This method has been standardized by the American Society for Testing and Materials as ASTM Standard Method E-274 using an ASTM Standard E-501 test tire [8]. Equipment for making these measurements is generically referred to as a "skid tester." The testers are typically configured as a trailer pulled behind a light truck with the capability to brake one wheel of the trailer while measuring the friction force and load. The ratio of friction force to load is a coefficient of friction, which is typically less than 1. For convenience, the coefficient is multiplied by 100 and given the name Skid Number. Thus the Skid Number of 81, which is specified for dry surfaces in government braking regulations, means that the surface should exhibit a 0.81 coefficient of friction when measured with the ASTM E-274 test method. Most clean, dry roads have a Skid Number close to this value.

Skid testers also have the capability to distribute water in front of the locked wheel in a controlled fashion, such that a wet road skid number can be measured. The 30 Skid Number specification for wet road friction in government braking regulations then means that the surface should exhibit a 0.3 coefficient of friction when measured with the ASTM E-274 method under wet conditions. Bituminous asphalt roads with smooth, polished aggregate generally fall in this range when lightly wetted. Portland Cement concrete road surfaces with good texture will have a higher wet Skid Number, in the range of 45 to 50. Road surfaces coated with a bitumen material like driveway sealer (e.g., Jennite) will have Skid Numbers in the range of 20 to 25.

Speed

On dry roads, both peak and slide coefficients decrease with velocity as illustrated in Figure 10.9. Under wet conditions, even greater speed sensitivity prevails because of the difficulty of displacing water in the contact patch at high speeds. When the speed and water film thickness are sufficient, the tire tread will lift from the road creating a condition known as hydroplaning.

Relevance to Vehicle Performance

Longitudinal traction properties are the properties of the tire/vehicle system that determine braking performance and stopping distance. The peak coefficient, μ_p, determines the limit for braking when the wheels do not lock up. Because of the weight transfer during deceleration, all wheels cannot be brought to the peak traction condition except by careful design of the braking system so as to proportion the front and rear braking forces in accordance with the prevailing loads under these dynamic conditions. In situations where one or more wheels lock up, the sliding coefficient of friction, μ_s, determines the braking contributions from those wheels. Since it is practically impossible to design a conventional braking system that can achieve exact proportioning under all conditions of load, center of gravity location, and road condition, it is inevitable that the driver will experience occasions of lockup. Therefore, the sliding coefficient of friction is an important tire performance property. With the use of anti-lock braking systems (ABS) the brake system maintains the wheels near the peak of the traction curve and does not allow lockup. Thus with ABS the dominant tire performance parameter is the peak coefficient.

Longitudinal traction properties may also determine limiting acceleration or hill-climbing performance of a vehicle. Again, the peak coefficient is of primary importance and can be effectively utilized by the skilled driver or a traction control system. The sliding coefficient is only related in an uncontrolled attempt at acceleration. Even then the sliding coefficient of friction (defined for the wheels-locked condition) is only an approximate indicator of the traction level that can be achieved when uncontrolled wheel spin occurs.

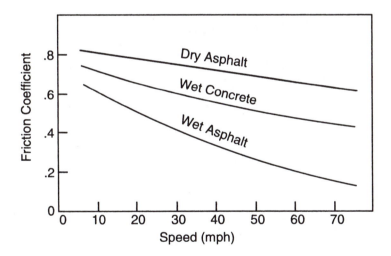

Fig. 10.9 Sliding coefficient as a function of speed on various surfaces.

CORNERING PROPERTIES

One of the very important functions of a tire is to develop the lateral forces necessary to control the direction of the vehicle, generate lateral acceleration in corners or for lane changes, and resist external forces such as wind gusts and

347

road cross-slope. These forces are generated either by lateral slip of the tire (slip angle), by lateral inclination (camber angle), or a combination of the two.

Slip Angle

When a rolling pneumatic tire is subjected to a lateral force, the tire will drift to the side. An angle will be created between the direction of tire heading and the direction of travel. This angle is known as slip angle. The mechanisms responsible can be appreciated by considering the simplified illustration of the tire's behavior as shown in Figure 10.10.

As the tire advances and tread elements come into contact with the road, they are undeflected from their normal position and therefore can sustain no lateral force. But as the tire advances further at the angle of its direction of travel, the tread elements remain in the position of their original contact with the road and are therefore deflected sideways with respect to the tire. By this process the lateral force builds up as the element moves rearward in the contact patch up to the point where the lateral force acting on the element overcomes the friction available and slip occurs. Thus the profile of the lateral force developed throughout the contact patch takes the form shown in Figure 10.10.

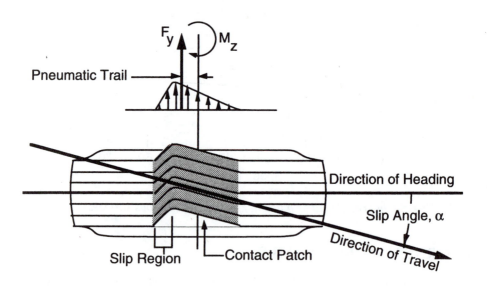

Fig. 10.10 Rolling tire deformation under a lateral force.

The integration of the forces over the contact patch yields the net lateral force with a point of action at the centroid. The asymmetry of the force buildup in the contact patch causes the force resultant to be positioned toward the rear of the contact patch by a distance known as the pneumatic trail. By SAE convention the lateral force is taken to act at the center of tire contact. At this position the net resultant is a lateral force, F_y, and an aligning moment, M_z. The magnitude of the aligning moment is equal to the lateral force times the pneumatic trail.

The mechanism is not an instantaneous phenomenon, but lags the actual development of slip angle because of the necessity of deflecting the tire sidewalls in the lateral direction [9]. The lag is closely related to the rotation of the tire, typically taking between one-half and one full revolution of the tire to effectively reach the steady-state force condition. The phenomenon is seen under low-speed test conditions when the tire is given a step change in steer angle. The lateral force response is then similar to that shown in Figure 10.11. With the change in steer angle the tire must roll through a half-turn or more for the lateral deflection and force to build up. This distance is often referred to as the "relaxation length." The time lag in development of lateral force necessarily depends on the speed of rotation of the tire. At a highway speed corresponding to 10 revolutions per second of the tire, the time lag will be only about 0.05 (1/20) second, which is imperceptible to many motorists. The effect, however, may be perceptible to expert drivers as a lag or sluggishness in turning response.

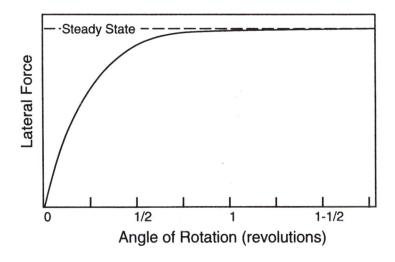

Fig. 10.11 Lateral force response to a step steer.

The relaxation effect is instrumental in the loss of cornering force when a tire operates on a rough road surface and experiences variations in its vertical load. When the load diminishes, slip occurs over the entire length of the contact patch and the tire sidewalls straighten out. The tire must then roll through its relaxation length in order to again build up a lateral force. As a consequence the tire is observed to have lower lateral force capability on rough roads. To achieve the best road-holding performance, the suspension should be designed to minimize tire load variations under rough road conditions.

Most commonly, the lateral force behavior of rolling tires is characterized only in the steady state (constant load and slip angle). Experimental measurements invariably exhibit the characteristic relationship to slip angle like that shown in Figure 10.12. When the slip angle is zero (the tire is pointed in its direction of travel) the lateral force is zero. With the first 5 to 10 degrees of slip angle the lateral force builds rapidly and linearly as the mechanisms shown in the previous figures take effect. In the range of 15 to 20 degrees the lateral force reaches a maximum (nominally equal to $\mu_p \cdot F_z$) and begins to diminish as the slip region grows in the contact area. At large angles it approaches the behavior of a locked wheel, which has a lateral force equal to the sine angle resultant of the sliding coefficient of friction, μ_s, times the vertical load, F_z.

A property of primary importance to the turning and stability behavior of a motor vehicle is the initial slope of the lateral force curve. The slope of the curve evaluated at zero slip angle is known as the "cornering stiffness," usually denoted by the symbol C_α.

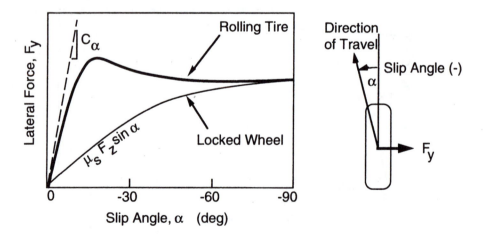

Fig. 10.12 Tire lateral force properties.

$$C_\alpha = -\left.\frac{\partial F_y}{\partial \alpha}\right|_{\alpha = 0} \qquad (10\text{-}2)$$

It should be noted that by the SAE convention, a positive slip angle produces a negative force (to the left) <u>on</u> the tire, implying that C_α must be negative. For that reason the slip angles in the figure are labeled with negative angles. In order to get around this problem, the SAE defines cornering stiffness as the negative of the slope, such that C_α takes on a positive value. A positive convention for C_α is used throughout this text.

Tire cornering properties as a function of load and slip angle are often shown in the form of a carpet plot as seen in Figure 10.13.

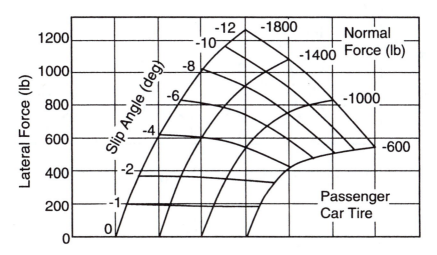

Fig. 10.13 Carpet plot of lateral force due to slip angle for a bias-ply tire.

The vertical axis is the scale of lateral force. The horizontal axis is a scale for both the slip angle and normal force. Note that both the slip angle and normal force are shown as negative numbers—negative slip angle produces a positive lateral force, and negative normal force is a positive vertical load. The load can be shown as positive if it is labeled "vertical load." The carpet plot provides a convenient format for mapping the properties of a tire.

The cornering stiffness is dependent on many variables. Tire size and type (radial versus bias construction), number of plies, cord angles, wheel width, and tread design are significant variables. For a given tire, the load and inflation pressure are the main variables.

351

Tire Type

On average, radial tires have a higher cornering stiffness than bias-ply tires. This is illustrated in Figure 10.14, which shows the cornering coefficient for the population of radial, bias, and bias-belted tires used on passenger cars.

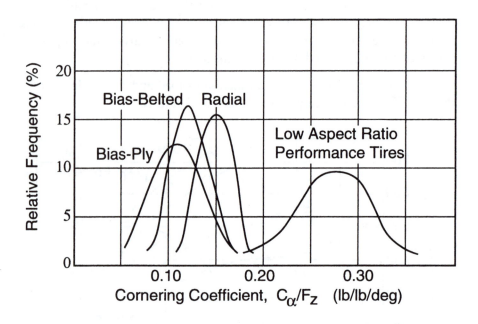

Fig. 10.14 Frequency distribution of cornering coefficient for passenger-car tires.

The cornering coefficient is the cornering stiffness normalized by the vertical load. For bias-ply tires the cornering coefficient is just about 0.12 lb/lb/deg. This means that at 1 degree of slip angle the average bias tire will produce a lateral force that is 10 percent of the vertical load. Radial tires of similar aspect ratio on average are stiffer, producing a lateral force at the same condition that is approximately 15 percent of the vertical load. As seen in Figure 10.14, the properties of bias-belted tires fall between those of radial and bias-ply tires. Although it is a general rule that radial tires have greater cornering stiffness than bias tires, as the distribution illustrates, the opposite may be true among some tires in the population.

By far one of the greatest influences on tire cornering properties is the aspect ratio—the ratio of section height to section width. For years the more common passenger-car tires have had aspect ratios from 0.78 down to 0.70, yielding the cornering coefficient properties just described. The trends toward lower-aspect-ratio performance tires (0.60 and lower) on passenger cars have resulted in tires with much higher cornering coefficients. These tires with cornering coefficients in the range of 0.25 to 0.30 and above have given passenger cars much quicker and precise cornering behavior.

Load

Although cornering force at a given slip angle rises with vertical load on the tire, it does not rise proportionately with load. By and large, the maximum cornering force per unit load occurs at the lightest loads. The effect of load can be seen in Figure 10.15 as it affects both the cornering stiffness and the cornering coefficient. Characteristically, the stiffness versus load curve is always concave down, a property that has some significance to the understeer gradient.

Load also decreases the peak coefficient of friction that can be achieved in cornering. Over the range of 50 percent to 125 percent of rated load, the peak friction level may decrease by more than 20 percent.

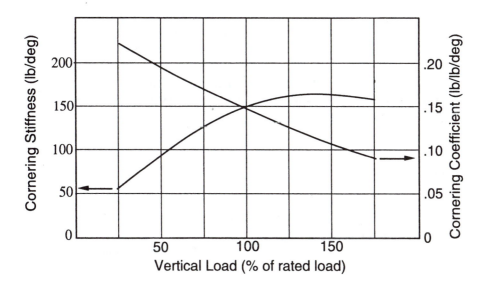

Fig. 10.15 Effect of load on cornering performance.

Inflation Pressure

Since inflation pressure increases carcass stiffness but reduces contact length, the net influence on cornering stiffness cannot be generalized across all types of tires [7]. It is generally accepted that increasing inflation pressure results in increasing cornering stiffness for passenger-car tires. Because of the monotonic and rather strong relationship between cornering stiffness and inflation pressure, it was common practice in the past to control the low-g directional behavior of passenger cars through the specification of different inflation pressures for front and rear tires. This practice is no longer common. In the case of truck tires, the influence of inflation pressure on cornering stiffness is varied and dependent on obscure sensitivities to details of the tire carcass design.

Inflation pressure also has a strong influence on the peak traction level that can be achieved under slip angle conditions. The pressure most influences lateral force production at high loads, and tires at reduced inflation pressures arrive at lateral force saturation at substantially higher values of slip angle.

Size and Width

For a given load condition, larger or wider tires exhibit a greater cornering stiffness. This effect is attributable to the contribution of carcass stiffness to cornering stiffness. Larger tires generally have a higher load capacity. Thus at the same load as a smaller tire, the larger tire will be operating at a lower percentage of its rated load and will effect greater cornering stiffness. For tires of the same nominal size but greater width (lower aspect ratio), that with the greater width will generally have a higher carcass stiffness resulting in greater cornering stiffness. The same effect can be accomplished by increasing rim width for a given tire.

Tread Design

The lateral compliance of the tread rubber acts as a series spring in the generation of lateral force response to slip angle; therefore, tread design has a potential influence on cornering stiffness. Among bias-ply tires it is generally recognized that snow-type tread designs produce lower cornering stiffness levels than do rib-tread designs. It is not clear that this trend holds true for radial tires. At best one can only be confident that all other things being held fixed, changes in tread design producing a more open pattern with deeper grooves and less support from one tread block or rib element to another will effect greater lateral compliance in the tread, and thus a reduction in cornering stiffness.

It is not known that sensitivities of cornering stiffness to tread compound have been demonstrated. Presumably, increases in durometer produce higher stiffness. Accordingly, one could expect a stiffness increase with rubber durometer, although the sensitivity would be expected to be small.

Other Factors

Velocity does not significantly affect cornering stiffness of tires in the normal range of highway speeds. This basic insensitivity is due to the kinematic nature of the mechanisms determining cornering stiffness. Surface properties also have little effect on cornering stiffness, as long as the surface itself is sufficiently rigid to react the shear forces without appreciable shear deflection of its own. This is also true of wet surfaces as well. The surface effects have their strongest influence on the peak traction that can be achieved in cornering under wet conditions. Harsh, gritty textures which can penetrate the water film provide much higher friction levels than smooth, polished surfaces.

Relevance to Vehicle Performance

Cornering stiffness is one of the primary variables affecting steady-state and transient cornering properties of vehicles in the normal driving regime. Understeer gradient, the characteristic commonly used to quantify turning behavior, is directly influenced by the balance of cornering stiffness on front and rear tires, as normalized by their loads. A higher relative cornering stiffness on the rear wheels is necessary to achieve understeer. Higher stiffness on the front wheels will produce oversteer, unless compensated by other factors, and results in a vehicle that has a critical speed above which it becomes unstable.

CAMBER THRUST

A second means of lateral force generation in a tire derives from rolling at a non-vertical orientation, the inclination angle being known as camber angle. With camber, a lateral force known as "camber thrust" is produced. The inclination angle is defined with respect to the perpendicular from the ground plane, positive corresponding to an orientation with the top of the wheel tipped to the right when looking forward along its direction of travel. (Note that on a vehicle, positive camber occurs when the wheel leans outward at the top. Thus the direction of positive camber differs from side-to-side on a vehicle, and the relationship to a positive lateral force is complicated by that fact. For discussion it is sufficient to recognize that the force is always oriented in the direction the tire is inclined.)

As with slip angle the lateral force due to camber angle is characterized by the initial slope of the curve, termed the camber stiffness, $C\gamma$, and is defined as in the equation:

$$C_\gamma = \left. \frac{\partial F_y}{\partial \gamma} \right|_{\gamma = 0} \tag{10-3}$$

In absolute value, the camber stiffness of a tire is typically in the range of 10 to 20 percent of the cornering stiffness. Figure 10.16 provides a carpet plot of lateral force as a function of camber and load for a typical passenger-car tire.

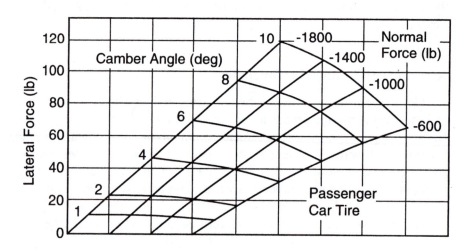

Fig. 10.16 Carpet plot of lateral force due to camber angle for a bias-ply tire.

Tire Type

Large changes in camber stiffness are known to accompany differences in tire construction. Figure 10.17 shows the camber stiffness distributions for a population of radial, bias, and bias-belted tires. The camber stiffness for low-aspect-ratio performance tires is in the same range as other radial tires shown. The stiffness for radial passenger-car tires is generally only about 40 to 50 percent of that for bias-ply tires.

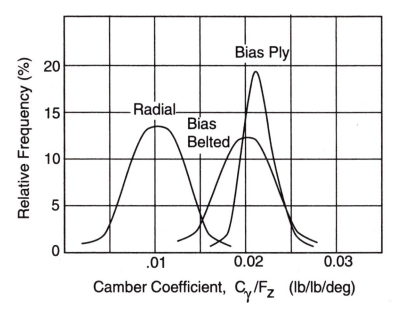

Fig. 10.17 Frequency distribution of camber coefficient for passenger-car tires.

The camber coefficient is a particularly important tire property with regard to how a tire responds to surface discontinuities oriented along the direction of travel, such as ruts in the wheelpath, faulting between lanes, shoulder dropoffs, railroad tracks, etc. When a vertically oriented tire operates on a surface with a cross-slope (such as the side of a rut in the wheelpath), the horizontal component of its load acts to push the wheel toward the lowest part of the rut as shown in Figure 10.18.

The lateral force per unit load is:

$$\frac{F_y}{W} = \sin \gamma' \cong \gamma' \qquad (10\text{-}4)$$

where:

W = Weight on the tire
γ' = Inclination angle of the road surface

357

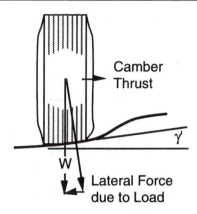

Fig. 10.18 Forces acting on a tire on a cross-slope surface.

Thus at 1 degree of surface cross-slope angle, a lateral force of 1/57.3 = 0.0175 lb/lb is produced in the "downhill" direction by the gravitational component. On the other hand, the camber thrust from the tire acts to push the tire "up" the slope in proportion to its camber coefficient. If the camber coefficient is greater than 0.0175 lb/lb/deg, the tire will try to climb out of the rut. If it is less it will tend to run down in the bottom of the rut and track in that position.

As seen from Figure 10.17, radial and bias-ply tires fall on opposite sides of that boundary, such that radial tires will tend to run in a rut, but bias-ply tires will tend to climb out of the rut. With the high proportion of radial tires being used on modern trucks, it has been postulated that this tracking tendency may be one of the primary factors responsible for the dual wheel ruts that frequently develop on asphalt roads.

Load

As indicated by the typical bias-ply car tire in Figure 10.16, camber stiffness is only slightly affected by vertical load in the vicinity of the design load. Over the range of 600 to 1800 lb load, the stiffness increases only about 20 percent.

Inflation Pressure

There is no general rule concerning the sensitivity of camber stiffness to inflation pressure. Limited data available tend to show some increase in camber stiffness with inflation pressure for bias-ply tires (0.25 lb/deg per psi), whereas radial tires are relatively insensitive to changes in pressure [7].

Tread Design

Camber stiffness is sensitive to the gross compliance properties of the tire tread, increasing substantially with tread stiffness. Thus more compliant treads as is often true of open pattern snow traction treads will be lower in camber stiffness by factors of 20 to 40 percent. For the same reasons, camber stiffness will go up as the tread wears.

Other Factors

Surface texture has no effect on camber stiffness except as it affects the limiting level of the frictional coupling. Likewise, the presence of water on the surface should have no significant effect. Because of the mechanisms involved, speed should have negligible effect on camber stiffness, except at very high values where centrifugal loading may act to stiffen the tire.

Relevance to Vehicle Performance

Camber thrust is the primary cornering force by which motorcycles and other two-wheeled vehicles are controlled. On passenger cars and trucks, camber thrust contributes to understeer behavior, but normally as a secondary source. On vehicles with independent suspensions where significant camber angles may be achieved, this mechanism may contribute up to about 25 percent of the understeer gradient. On vehicles with solid axles, little camber can occur, such that its contribution to turning performance is even less.

ALIGNING MOMENT

Because the shear forces in the contact patch of a tire operating at a slip angle develop with their centroid aft of the tire centerline, an aligning moment or torque is generated about the vertical axis. Although this moment is only a small contribution to the total yaw moments on a vehicle, it contributes to reactions in the steering system which may have a more substantial effect. It should be noted that positive aligning moment always attempts to steer the tire in the direction it is traveling, thus it has a stabilizing influence on a vehicle.

Slip Angle

A typical carpet plot showing the influence of both slip angle and vertical load on aligning moment is given in Figure 10.19. Both radial and bias-ply tires possess aligning moment characteristics similar to those shown, although on average the moment is larger with radial tires. Typically, bias-ply tires will have an aligning moment coefficient (aligning moment per pound load) of approximately 0.033 ft-lb/lb/deg, whereas radial tires have a coefficient on the order of 0.043 ft-lb/lb/deg.

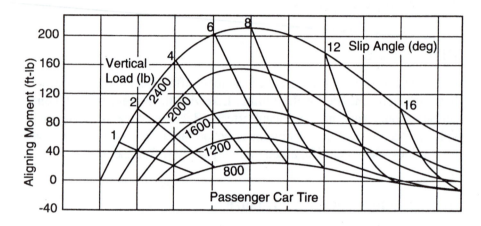

Fig. 10.19 Carpet plot of aligning moment versus slip angle and load.

The aligning moment is very sensitive to the size of the contact patch and the growth of the slip region. The shear stress and the torque arm responsible for the moment are both proportional to the distance from the tire center. Thus the major contributors to aligning moment are the tread elements at the extremes of the contact patch. The moment rises with the increasing shear forces up to about 8 degrees of slip angle. At greater angles, however, the growing slip region erodes the extremities and causes a decrease in the aligning moment. At very high slip angles the slip region advances forward such a distance that the aligning moment can actually become negative.

A high sensitivity to vertical load is seen in the plots. This sensitivity arises from the influence of contact area on the moment. Although doubling the load nominally doubles the contact area, all the increase in area occurs at the extreme regions of the contact patch—the regions that most affect the aligning moment. As a result, aligning moment increases in an accelerated fashion with load.

360

An aligning moment is also produced when a tire rolls at a non-zero camber angle. As shown in Figure 10.20, a bias-ply tire produces aligning moments due to camber which are on the order of 10 percent of the magnitude produced in response to slip angle. For radial tires, aligning moments due to camber angle are substantially lower than those measured for bias tires. The aligning moment coefficient (aligning moment per pound load) due to camber will be approximately 0.003 ft-lb/lb/deg for bias-ply tires, versus 0.001 ft-lb/lb/deg for radial tires. With regard to sensitivity to operating variables, as a general rule, influences which increase side force level also tend to increase aligning moment due to camber.

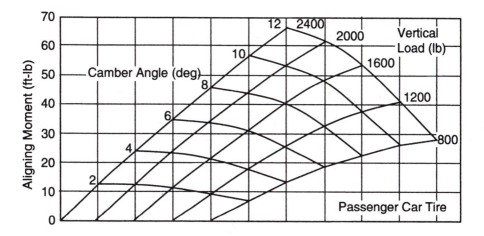

Fig. 10.20 Carpet plot of aligning moment versus camber angle and load.

Path Curvature

Whereas the aligning moments discussed thus far arise from a tire generating lateral force by traveling in a straight line with either slip or camber angle, aligning moments are also created when a tire is forced to roll on a curved path, even though not generating a lateral force. In effect, a torque must be applied to the tire about the vertical axis to force it to roll on a curve when under load. Such conditions are seen when a vehicle is turned at low speed, or when the wheels are steered with the vehicle stationary such that they are forced to roll on the small-radius path of the scrub radius. Little experimental data is available to quantify behavior under this condition, but Figure 10.21 shows the aligning moment generated by a truck tire on a dry road surface [10]. At zero radius, large torques are required to rotate the tire about the steer axis. The

361

aligning moment goes up disproportionately with load because the increase in contact area that occurs with increase in load must always occur at the perimeter of the contact patch, and it is the shear in the perimeter region that contributes the most to the moment.

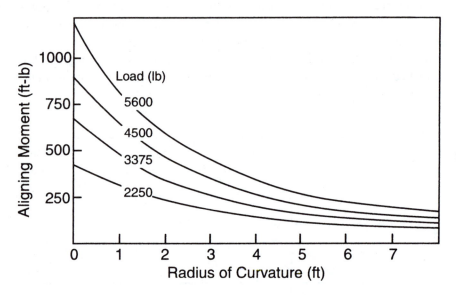

Fig. 10.21 Aligning moment as a function of radius of turn.

Relevance to Vehicle Performance

Aligning moment as a torque acting directly on the vehicle contributes a small component to the understeer of a vehicle. The fact that positive aligning moments attempt to steer the vehicle out of the turn means that they are understeer in direction. Overall, the direct action of the moments contributes only a few percent to the understeer gradient of a vehicle.

The aligning moment has a more direct influence on understeer by its action on the steered wheels. The moment is normally in the direction to turn the steered wheels out of the turn. Even though the steer deflection angles in response to aligning moments may be small (fractions of a degree in normal driving), this is normally an important contribution to understeer gradient.

Aligning moment is also important to steering feel for a moving vehicle. Its contribution is equal to or greater than caster angle in producing returnability torques—the torques acting to return the steering to the straight-ahead position when cornering.

Aligning moment arising from path curvature is primarily important for static steer and very low-speed maneuvering. The moment is the dominant source of steering torque and may be quite large. Because this represents a condition that places highest torque demand on a steering system it must be considered in sizing power-steering hardware and in durability testing. By offsetting the tire outside of the steering axis (positive scrub) the tire can be allowed to roll on a radius that will decrease the magnitude of the aligning moment in static steer situations. In operating situations, simply moving the vehicle at low speed while increasing the steering angle greatly increases the radius of curvature and thereby reduces the steering torque required.

COMBINED BRAKING AND CORNERING

When a tire is operated under conditions of simultaneous longitudinal and lateral slip, the respective forces depart markedly from those values derived under independent conditions. The application of longitudinal slip generally tends to reduce the lateral force at a given slip angle condition, and conversely, application of slip angle reduces the longitudinal force developed under a given braking condition. This behavior is shown in Figure 10.22.

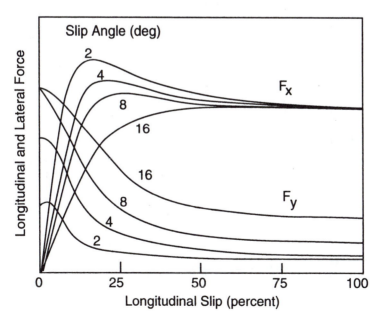

Fig. 10.22 Brake and lateral forces as a function of longitudinal slip.

Friction Circle

The general effect on lateral force when braking is applied is illustrated in the traction field of Figure 10.23. The individual curves represent the lateral force at a given slip angle. As the brake force is applied, the lateral force gradually diminishes due to the additional slip induced in the contact area from the braking demand.

This type of display of a tire traction field is the basis for the "friction circle" (or friction ellipse) concept [11]. Recognizing that the friction limit for a tire, regardless of direction, will be determined by the coefficient of friction times the load, it is clear that the friction can be used for lateral force, or brake force, or a combination of the two, in either the positive or negative directions. But, in no case can the vector total of the two exceed the friction limit. The limit is therefore a circle in the plane of the lateral and longitudinal forces. The portion of the circle in the figure is the friction circle for the positive quadrant of the traction field. The limit is characterized as a friction circle for tires which

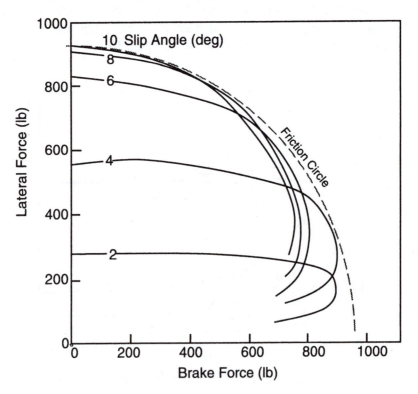

Fig. 10.23 Lateral force versus longitudinal force at constant slip angles.

364

have effectively the same limits for lateral and braking forces. Certain specialized tires, however, may be optimized for lateral traction or braking traction, in which case the limit is not a circle but an ellipse.

The friction circle concept has been used in recent years as a means to evaluate race car drivers by making continuous recordings of the lateral and longitudinal accelerations maintained on a track. For maximum efficiency in getting around a closed course, the tires should be working continuously at either the cornering or braking/acceleration limits. Therefore, the combined lateral and longitudinal accelerations measured on the car should always be pressing the friction limit, and the most effective driver is the one who can most closely maintain this optimum. By plotting the record of the two accelerations on a polar plot similar to Figure 10.23, one gets a visual indication of the performance of the driver by observing the percentage of time spent at the friction limit.

Figure 10.23 illustrates another observation that is frequently made under conditions of combined traction. Note that at intermediate slip angle conditions near 4 degrees, the application of moderate levels of brake force actually increases the lateral force developed at that slip angle. This phenomenon is shown more precisely in the plot of Figure 10.24, which shows the lateral force and aligning moment under tractive forces in both the braking and driving directions [12].

Using the free-rolling (zero tractive force) values of lateral force as a reference point, it is seen that when braking (negative) force is applied, the lateral force increases slightly while the aligning moment decreases. In effect, the presence of the braking force acts to stiffen the tire structure (sidewalls and/ or tread) with respect to the mechanism that generates lateral force. The reduction of aligning moment implies a significant redistribution of the shear forces in the contact patch. As the braking force increases toward its maximum value the lateral force diminishes because the friction limits are being approached. Concurrently, the aligning moment decreases to the point where it may actually go negative near the braking limit. A negative aligning moment attempts to steer the wheel to a greater slip angle, and may adversely affect stability in braking, particularly through its effects on the steering system [13].

Under a moderate driving (positive) traction force the opposite effects are observed. Lateral force decreases slightly, although aligning moment increases markedly. At levels near the friction limit both lateral force and aligning moment decrease. Unlike braking, however, the aligning moment never goes negative near the limit of driving force.

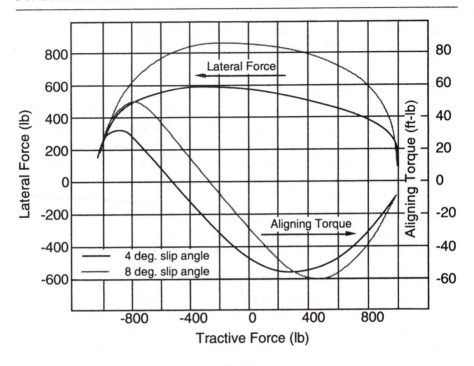

Fig. 10.24 Lateral force and aligning torque versus tractive force.

Variables

Although tire type (radial versus bias-ply) and inflation pressure have significant influences on cornering stiffness, behavior under combined slip is qualitatively similar to that shown above. The behavior is insensitive to velocity, and is only affected by surface conditions through the influence on friction limits.

Relevance to Vehicle Performance

The combined slip behavior of tires is only meaningful in the context of braking-in-a-turn maneuvers. When brakes are applied to a vehicle in a steady turn, the increasing level of tire longitudinal slip produces a loss in tire side force which characteristically serves to disturb the path and/or yaw orientation of the vehicle. Alternatively, if a large steering input is applied while the vehicle is braking, both the braking performance and the cornering performance stand to be degraded in comparison to the performances expected with independent inputs of steering or braking.

366

Minimal degradation of braking performance occurs with concurrent cornering up to levels of about 0.3 g lateral acceleration. However, as limit braking is approached, the directional or yaw response can be degraded to the point of total loss of control. The nature of the control loss will depend on the order in which front and rear tires approach the wheel-lock condition. Front-wheel lockup will render the vehicle unsteerable, whereas rear-wheel lockup precipitates spinout.

CONICITY AND PLY STEER

The behavior of tires in the near-zero lateral slip region has grown more important in recent years with the refinement of high-speed automobiles. The importance derives from the emphasis that is put on on-center feel of the steering system.

For an ideal tire, zero lateral force coincides with zero slip angle, but for actual tires this is not true. For actual tires the behavior of lateral force at small slip angles will be similar to that shown in Figure 10.25. In this plot the lateral force is plotted as the tire is rolled in both directions, arbitrarily labeled as forward and reverse in the figure. The important observation in the figure is that the lateral force behavior differs with the direction of rotation (forward versus reverse) and may be offset from the origin of the graph.

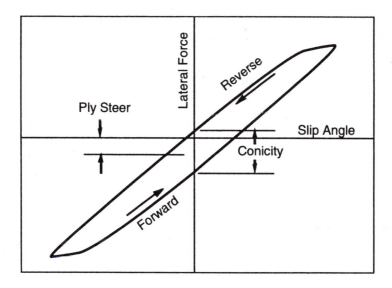

Fig. 10.25 Lateral force behavior around zero slip angle.

One mechanism in the tire accounting for this behavior is conicity in the construction. Conicity derives from small side-to-side differences in the tire such as an asymmetrical offset in the positioning of the belt. As the name implies, these variations are manifest in a tire as a bias toward a conical shape as illustrated in Figure 10.26. Because of this shape, a freely rolling tire will want to follow an arc centered about the apex of the cone, shown at the right of Figure 10.26. Forced to follow a straight line this tire will experience a lateral force toward the right in the figure, regardless of which direction it may roll. By the SAE convention, when the tire rolls upward in the top view the lateral force is to the right and is positive in direction. Rolling downward, the force will again be to the right, but since the longitudinal axis of the tire is now pointed downward it is a negative lateral force. Thus conicity is manifest as a difference in the lateral force at zero slip angle when the tire is rolled in opposite directions. Conicity has the character of being random in direction and is dependent on quality control in tire construction. Turning a tire on the rim will change the direction of the lateral force caused by conicity.

Top view

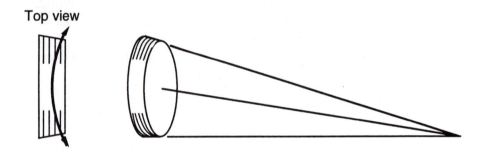

Fig. 10.26 Conicity in a pneumatic tire.

The other mechanism that may be present in a tire is ply steer, which arises from the angle of the cords in the belt layers. To avoid this bias in the tire's preferred rolling direction, belts are constructed with alternating belt layers at opposite angles, but a perfect balance is impossible to achieve. Thus a free-rolling tire will exhibit a tendency to drift from its direction of heading. Instead of following an arc as illustrated in Figure 10.26, it will follow a line that is skewed with respect to its center plane. If, when rolling in one direction, it creates a lateral force to the right in the SAE tire axis system, when rolled in the opposite direction, it will again exhibit a force to the right in the SAE tire axis system. Thus when tire lateral force properties are measured in the vicinity of zero-degree slip angle, ply steer is manifest as a non-zero offset in the lateral force averaged from both directions of travel. Ply steer is dependent on tire design; hence, it will be nearly equal in magnitude and direction for all tires of a common design. Turning the tire on the rim does not change the direction of the lateral force caused by ply steer.

Both conicity and ply steer force magnitudes are dependent on the vertical load carried by a tire. Conicity is more sensitive to inflation pressure and may be reduced by making adjustments to the pressure.

Relevance to Vehicle Performance

The effects of conicity and ply steer are to create a "pull" in the steering system or a "drift" in the tracking of the car. Pull refers to a condition where the driver must apply a continuous torque to the steering wheel holding it off-center to maintain the vehicle on a straight path; or with the wheel free and in the center position, the vehicle will follow a curved path.

Excess conicity on the front wheels may cause the steering to pull to such a degree that it is fatiguing to the driver and becomes a source of customer dissatisfaction. Conicity on the rear wheels will cause the vehicle to track with the rear wheels offset from the front. It may also affect the centering of the steering wheel in the straight-ahead position.

Since ply steer is likely to act on all wheels in the same direction, a vehicle may exhibit a slight drift due to ply steer forces. This may require some steering wheel offset to compensate and keep the vehicle traveling straight; no steering pull is likely.

DURABILITY FORCES

Although tires act as a cushion between the vehicle and the road, the bumps present in most roads transmit forces that are perceptible to the motorist and may contribute to the cyclic loading and fatigue of suspension components. Road bump features that have a characteristic length on the same order as the length of the tire contact patch (e.g., tar strips, faults in concrete surfaces, potholes) generate forces that are dependent on the tire's ability to envelop these features. Because these forces, particularly those attributable to encounters with potholes, may be large in magnitude and thus significant to fatigue and durability of a vehicle, they are often referred to as durability forces.

369

A smooth road surface is seen as a flat plane by a free-rolling tire and generates predominantly vertical force inputs to the tire. Even at high speed the vertical inputs are slow to change in magnitude relative to the time it takes the tire to advance along the length of its contact patch. However, when a tire encounters an abrupt discontinuity in a road surface, such as the edge of a pothole, the forces may change markedly as the feature passes through the contact patch. Dynamic vertical and longitudinal forces are generated in the process. The shape and magnitude of the force inputs depend on the properties and mechanics of the tire.

Tire performance when enveloping road discontinuities has been studied by examining performance when tires negotiate small step changes in road elevation [14]. (In theory, any shape of bump can be approximated by a judiciously selected combination of small steps.) Mechanistically the tire can be thought of as a series of radial springs (sometimes modeled with dampers in parallel with each spring) in contact with the surface, to understand its behavior. When a tire encounters an upward step in the road surface, the vertical and longitudinal forces on the tire will change abruptly as shown in Figure 10.27.

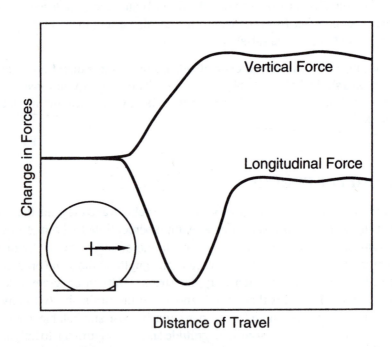

Fig. 10.27 Vertical and longitudinal tire forces produced by a step in the road.

As the tire contacts the leading edge of the bump, the vertical force begins to rise due to depression of the tread. The force rises more or less continuously until the full contact patch has advanced onto the bump. Performed at high speed, the axle on which the tire is mounted is not able to respond and move upward in the time it takes the entire contact patch to move onto the bump. Thus an increase in vertical load occurs that is approximately equal to the bump height times the vertical stiffness of the tire.

The encounter with the bump also creates a longitudinal force as a result of several mechanisms. For the tire to rise onto the bump, a longitudinal force is required. This force must be provided by the axle, thus a negative (opposite to the direction of travel) force is experienced when the tire first encounters the edge of the bump. Once on the bump the increased vertical load causes an increase in the rolling resistance of the tire, so the longitudinal force does not immediately return to its original value, but must wait for the axle to adjust to a new height representing the balance of vertical forces.

At high speed, a second mechanism is at work as well. With the change in effective radius of the tire on the bump, it must assume a new rotation rate in correspondence with its forward speed. With a smaller radius, the tire must increase its rotational speed. This is accomplished by generating a shear force in the contact patch which is ultimately balanced out by an opposite force reaction at the axle. To speed up the rotation rate of the tire, a second component of negative force is imposed on the axle. The rate at which the tire mounts the bump, the rate at which it must speed up, and the magnitude of the longitudinal force created depends on the forward speed of the wheel. Thus this component of longitudinal force depends on the speed of travel.

When a tire encounters a downward step in the road surface, similar forces are created differing only in their direction of action. That is, a downward step causes a decrease in vertical force and creates an impulsive force in the forward direction on the axle.

TIRE VIBRATIONS

Thus far the tire has been treated as a mechanism for generating forces by which a vehicle may be controlled in braking and turning. With regard to ride dynamics it is seen to behave primarily as a spring which absorbs the roughness features in the road and interacts with the vertical motions of the body and unsprung masses. The tire, however, is also a dynamic system with resonances which affect the transmission of vibrations to the vehicle and may interact with vehicle resonances [15].

A relatively large portion of the tire mass is concentrated in the tread which is connected to the wheel by the compliant sidewalls. This combination of mass and compliance permits the tread to resonate when excited by road inputs. Figure 10.28 shows examples of the first three modal resonances of the tire in the vertical plane.

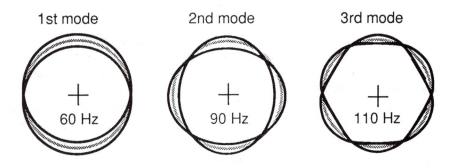

Fig. 10.28 First, second and third modal resonances of a tire.

The first mode, which will occur somewhere near 60 Hz for a passenger-car tire, involves a simple vertical motion of the entire tread band without distortion. The mode is easily excited by vertical input at the proper frequency in the contact patch. Since the entire tread band moves up and down in unison, the force associated with the resonance is transmitted to the wheel and axle.

The second mode contrasts with the first in that the tread band is oscillating in an elliptical fashion always remaining symmetrical about the vertical and horizontal axes. The top and bottom of the tread are always moving out of phase so that no net vertical force is imposed on the wheel. (Likewise, there is no net fore-aft force.) Although the resonance can be excited by vertical inputs at the contact patch, the tire is very effective at absorbing the inputs without transmitting forces to the axle. In a similar fashion the third- and higher-order resonances of the tire are very effective in absorbing road inputs without transmitting them to the wheel and axle.

In between these modal resonances the tire has anti-resonant modes characterized by very asymmetrical tread distortion and little mobility at the contact patch. The asymmetry of the motion results in unbalanced forces being imposed around the circumference of the wheel, such that the resultant force is transmitted to the wheel. The fact that the contact patch is stationary implies that the tire appears as a very stiff, rather than compliant, element with regard to road inputs at this frequency.

From this simple picture of a tire as a resonant system, it is possible to begin building an understanding of the dynamic behavior of the tire in transmitting road vibrations in the chassis of a motor vehicle. The system can be characterized by examining several relevant properties. Figure 10.29 shows experimental measurements on a radial tire mounted on a passenger car exposed to vertical excitation at the contact patch [15].

The transmissibility in this figure is defined as the ratio of acceleration on the axle per unit of road displacement at the contact patch. The first peak just below 20 Hz is axle hop resonance in which the tire acts as the primary stiffness constraining the unsprung mass. More of interest are the several peaks at higher frequencies. Note that they occur at frequencies in between the tire resonances (i.e., the anti-resonant points), corresponding to the peaks in transmissibility and peaks as well in the footprint stiffness of the tire (tire force per unit of road displacement).

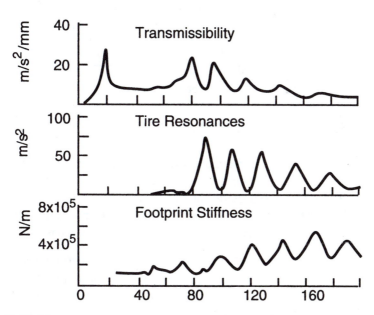

Fig. 10.29 Tire resonance properties measured on a vehicle.

This somewhat simplified picture of tire dynamics hints at the forces that will be input at the wheel of a motor vehicle. Figure 10.30 shows the spectra of forces measured when a passenger-car tire encounters a small obstacle at a speed of 30 feet per second [16]. Data are shown for both a radial and bias-ply tire and for forces in both the vertical and longitudinal directions.

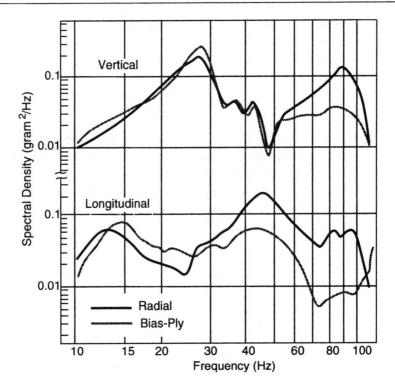

Fig. 10.30 Spectra of forces measured when a tire encounters an obstacle.

In the vertical direction the radial tire is distinguished by the increased amplitude of force in the frequency range of 50 to 100 Hz. From the previous discussion it would be expected that this behavior is a result of the high transmissibility of the anti-resonant modes in this frequency range for radial-ply tires. Obviously, bias-ply tires are much better in this range.

Perhaps the more important distinction between the two types of tires is seen in the spectra of longitudinal response. Except for a narrow band around 15 to 20 Hz, radial tires are more responsive in the longitudinal direction than bias-ply tires. At the higher frequencies the greater transmissibility indicates a higher effective stiffness in the longitudinal direction. The higher transmissibility of the radial tire near 10 Hz is one of the key differences that required "ride tuning" when radials were first introduced to the American automotive market. On vehicles historically developed on bias-ply tires, the application of radials produced more suspension fore/aft vibration requiring the addition of more longitudinal compliance in suspensions to prevent these vibrations from being transmitted to the body.

REFERENCES

1. Mechanics of Pneumatic Tires, Samuel K. Clark, editor, U.S. Department of Transportation National Highway Traffic Administration, U.S. Government Printing Office, Washington, D.C., 1981, 931 pp.

2. Dugoff, H., Fancher, P.S., and Segel, L., "Tire Performance Characteristics Affecting Vehicle Response to Steering and Braking Control Inputs," The University of Michigan Highway Safety Research Institute, August 1969, 105 p.

3. Sharp, R.S., and El-Nashar, M.A., "A Generally Applicable Digital Computer Based Mathematical Model for the Generation of Shear Forces by Pneumatic Tires," Vehicle Systems Dynamics, Vol. 15, No. 4, 1986, pp. 187-209.

4. Meyer, W.E., and Kummer, H.W., "Mechanisms of Force Transmission Between Tire and Road," SAE Paper No. 620407 (490A), 1962, 18 p.

5. "Vehicle Dynamics Terminology," SAE J670e, Society of Automotive Engineers, Warrendale, PA (see Appendix A).

6. 1991 Yearbook, The Tire & Rim Association Inc., Copley, Ohio, 1991.

7. Ervin, R.D., "The State of Knowledge Relating Tire Design to Those Traction Properties which May Influence Vehicle Safety," The University of Michigan Transportation Research Institute, Report No. UM-HSRI-78-31, July 1978, 128 p.

8. Annual Book of ASTM Standards, American Society for Testing and Materials, Philadelphia, PA.

9. Loeb, J.S., Guenther, D.A., Chen, H.H., and Ellis, J.R., "Lateral Stiffness, Cornering Stiffness and Relaxation Length of the Pneumatic Tire," SAE Paper No. 900129, 1990, 11 p.

10. Pacejka, H.B., "Tire Characteristics and Vehicle Dynamics," course notes, University Consortium for Continuing Education, November 2-4, 1988, Washington, D.C., 199 p.

11. Radt, H.S., and Milliken, W.F., "Motions of Skidding Automobiles," SAE Paper No. 600133 (205A), 1960, 21 p.

12. Nordeen, D.L., and Cortese, A.D., "Force and Moment Characteristics of Rolling Tires," SAE Paper No. 640028 (713A), 1964, 13 p.

13. Gillespie, T.D., "Front Brake Interactions with Heavy Vehicle Steering and Handling during Braking," SAE Paper No. 760025, 1976, 16 p.

14. Lippman, S.A., *et al.*, "A Quantitative Analysis of the Enveloping Forces of Passenger Tires," SAE Paper No. 670174, 1967, 10 p.

15. Potts, G.R., *et al.*, "Tire Vibrations," *Tire Science and Technology,* Vol. 5, No. 4, 1977, p. 202-225.

16. Barson, C.W., and Dodd, A.M., "Vibrational Characteristics of Tyres," Institution of Mechanical Engineers, Paper C94/71, 1971, 12 p.

Appendix A
SAE J670e
Vehicle Dynamics Terminology

SAE Recommended Practice
Issued by the Vehicle Dynamics Committee July 1952
Last revised July 1976

NOTE: Italicized words and phrases appearing in a definition are themselves defined elsewhere in this Terminology.

1. Mechanical Vibration-Qualitative Terminology

1.1 Vibration (Oscillation), General — Vibration is the variation with time of the displacement of a body with respect to a specified reference dimension when the displacement is alternately greater and smaller than the reference. (Adapted from ANS Z24.1-1951, item 1.040.)

1.2 Free Vibration — Free vibration of a system is the *vibration* during which no variable force is externally applied to the system. (Adapted from ANS Z24.1-1951, item 2.135.)

1.3 Forced Vibration — Forced vibration of a system is *vibration* during which variable forces outside the system determine the period of the vibration. (Adapted from ANS Z24.1-1951, item 2.130.)

1.3.1 RESONANCE — A *forced vibration* phenomenon which exists if any small change in *frequency* of the applied force causes a decrease in the *amplitude* of the vibrating system. (Adapted from ANS Z24.1-1951, item 2.105.)

1.4 Self-Excited Vibration — *Vibrations* are termed self-excited if the vibratory motion produces cyclic forces which sustain the *vibration*.

1.5 Simple Harmonic Vibration — *Vibration* at a point in a system is simple harmonic when the displacement with respect to time is described by a simple sine function.

1.6 Steady-State Vibration — Steady-state vibration exists in a system if the displacement at each point recurs for equal increments of time. (Adapted from ANS Z24.1-1951, items 11.005 and 1.045.)

1.7 Periodic Vibration — Periodic vibration exists in a system when recurring *cycles* take place in equal time intervals.

1.8 Random Vibration — Random vibration exists in a system when the *oscillation* is sustained but irregular both as to *period* and *amplitude.*

1.9 Transient Vibration — Transient vibration exists in a system when one or more component *oscillations* are discontinuous.

2. Mechanical Vibration-Quantitative Terminology

2.1 Period — Period of an *oscillation* is the smallest increment of time in which one complete sequence of variation in displacement occurs. (Adapted from ANS Z24.1-1951, item 1.050.)

2.2 Cycle — Cycle of *oscillation* is the complete sequence of variations in displacement which occur during a *period.* (Adapted from ANS Z24.1-1951, item 1.055.)

2.3 Frequency — Frequency of *vibration* is the number of *periods* occurring in unit time. (Adapted from ANS Z24.1-1951, item 1.060.)

2.3.1   NATURAL FREQUENCY — Natural frequency of a body or system is a frequency of free vibration. (Same as ANS Z24.1-1951, item 2.140.)

2.3.2   EXCITING FREQUENCY — Exciting frequency is the *frequency* of variation of the exciting force.

2.3.3   FREQUENCY RATIO — The ratio of *exciting frequency* to the *natural frequency.*

2.3.4   RESONANT FREQUENCY — *Frequency* at which *resonance* exists. (Same as ANS Z24.1-1951, item 2.110.)

2.4 Amplitude — Amplitude of displacement at a point in a *vibrating system* is the largest value of displacement that the point attains with reference to its equilibrium position. (Adapted from ANS Z24.1-1951, item 1.070.)

2.4.1   PEAK-TO-PEAK AMPLITUDE (DOUBLE AMPLITUDE) — Peak-to-

Peak amplitude of displacement at a point in a *vibrating system* is the sum of the extreme values of displacement in both directions from the equilibrium position. (Adapted from ANS Z24.1-1951, item 1.075.)

2.4.2   STATIC AMPLITUDE—Static amplitude in *forced vibration* at a point in a system is that displacement of the point from its specified equilibrium position which would be produced by a static force equal to the maximum value of exciting force.

2.4.3   AMPLITUDE RATIO (RELATIVE MAGNIFICATION FACTOR)—The ratio of a *forced vibration* amplitude to the *static amplitude*.

2.5 Velocity — Velocity of a point in a *vibrating system* is the time rate of change of its displacement. (Adapted from ANS Z24.1-1951, item 1.345.)

In *simple harmonic vibration*, the maximum velocity,

$$v_m = \omega x$$

where:

ω $= 2\pi f$
f $=$ frequency
x $=$ amplitude

2.6 Acceleration — Acceleration of a point is the time rate of change of the *velocity* of the point. (Same as ANS Z24.1-1951, item 1.355.)

In *simple harmonic vibration*, the maximum acceleration,

$$a_m = \omega^2 x$$

2.7 Jerk — "Jerk" is a concise term used to denote the time rate of change of *acceleration* of a point.

In *simple harmonic motion*, the maximum jerk,

$$j_m = \omega^3 x$$

2.8 Transmissibility — Transmissibility in *forced vibration* is the ratio of the transmitted force to the applied force.

3. Vibrating Systems

3.1 Degree of Freedom — The number of degrees of freedom of a *vibrating* system is the sum total of all ways in which the masses of the system can be independently displaced from their respective equilibrium positions.

EXAMPLES — A single rigid body constrained to move only vertically on supporting springs is a system of one degree of freedom. If the same mass is also permitted angular displacement in one vertical plane, it has two degrees of freedom; one being vertical displacement of the center of gravity; the other, angular displacement about the center of gravity.

3.2 Linear — Linear *vibrating systems* are those in which all the variable forces are directly proportional to the displacement, or to the derivatives of the displacement, with respect to time.

3.3 Nonlinear — Nonlinear *vibrating systems* are those in which any of the variable forces are not directly proportional to the displacement, or to its derivatives, with respect to time.

EXAMPLE — A system having a variable *spring rate*.

3.4 Undamped — Undamped systems are those in which there are no forces opposing the vibratory motion to dissipate energy.

3.5 Damped — Damped systems are those in which energy is dissipated by forces opposing the vibratory motion.

Any means associated with a *vibrating system* to balance or modulate exciting forces will reduce the vibratory motion, but are not considered to be in the same category as damping. The latter term is applied to an inherent characteristic of the system without reference to the nature of the excitation.

3.5.1   VISCOUS DAMPING — Damping in which the force opposing the motion is proportional and opposite in direction to the velocity.

3.5.2   CRITICAL DAMPING — The minimum amount of *viscous damping* required in a *linear system* to prevent the displacement of the system from passing the equilibrium position upon returning from an initial displacement.

3.5.3   DAMPING RATIO — The ratio of the amount of *viscous damping* present in a system to that required for *critical damping*.

3.5.4   COULOMB DAMPING — Damping in which a constant force opposes the vibratory motion.

3.5.5   COMPLEX DAMPING — Damping in which the force opposing the vibratory motion is variable, but not proportional, to the *velocity*.

In the field of aircraft flutter and vibration, complex damping is also used to denote a specific type of damping in which the damping force is assumed to be *harmonic* and in phase with the *velocity* but to have an *amplitude* proportional to the *amplitude* of displacement.

4. Components and Characteristics of Suspension Systems

4.1 Vibrating Mass and Weight

4.1.1   SPRUNG WEIGHT — All weight which is supported by the suspension, including portions of the weight of the suspension members.

In the case of most vehicles, the sprung weight is commonly defined as the total weight less the weight of *unsprung* parts.

4.1.2   SPRUNG MASS — Considered to be a rigid body having equal mass, the same center of gravity, and the same moments of inertia about identical axes as the total *sprung weight*.

4.1.3   DYNAMIC INDEX — (k^2/ab ratio) is the square of the radius of gyration (k) of the *sprung mass* about a transverse axis through the center of gravity, divided by the product of the two longitudinal distances (a and b) from the center of gravity to the front and rear *wheel centers*.

4.1.4   UNSPRUNG WEIGHT — All weight which is not carried by the suspension system, but is supported directly by the tire or wheel, and considered to move with it.

4.1.5   UNSPRUNG MASS — The unsprung masses are the equivalent masses which reproduce the inertia forces produced by the motions of the corresponding unsprung parts.

4.2 Spring Rate — The change of load of a spring per unit deflection, taken as a mean between loading and unloading at a specified load.

4.2.1   STATIC RATE — Static rate of an elastic member is the rate measured between successive stationary positions at which the member has settled to substantially equilibrium condition.

4.2.2   DYNAMIC RATE — Dynamic rate of an elastic member is the rate measured during rapid deflection where the member is not allowed to reach static equilibrium.

4.3 Resultant Spring Rate

4.3.1   SUSPENSION RATE (WHEEL RATE) — The change of wheel load, at the *center of tire contact,* per unit vertical displacement of the *sprung mass* relative to the wheel at a specified load.

If the *wheel camber* varies, the displacement should be measured relative to the lowest point on the rim centerline.

4.3.2   TIRE RATE (STATIC) — The *static rate* measured by the change of wheel load per unit vertical displacement of the wheel relative to the ground at a specified load and inflation pressure.

4.3.3   RIDE RATE — The change of wheel load, at the *center of tire contact,* per unit vertical displacement of the *sprung mass* relative to the ground at a specified load.

4.4 Static Deflection

4.4.1   TOTAL STATIC DEFLECTION — Total static deflection of a loaded suspension system is the overall deflection under the static load from the position at which all elastic elements are free of load.

4.4.2   EFFECTIVE STATIC DEFLECTION — Effective static deflection of a loaded suspension system equals the static load divided by the *spring rate* of the system at that load.

Total static deflection and effective static deflection are equal when the *spring rate* is constant.

4.4.3   SPRING CENTER — The vertical line along which a vertical load applied to the sprung mass will produce only uniform vertical displacement.

4.4.3.1 Parallel Springing — Describes the suspension of a vehicle in which the *effective static deflections* of the two ends are equal; that is, the *spring center* passes through the center of gravity of the *sprung mass.*

4.5 Damping Devices — As distinct from specific types of damping, damping devices refer to the actual mechanisms used to obtain damping of suspension systems.

4.5.1 SHOCK ABSORBER — A generic term which is commonly applied to hydraulic mechanisms for producing damping of suspension systems.

4.5.2 SNUBBER — A generic term which is commonly applied to mechanisms which employ dry friction to produce damping of suspension systems.

5. Vibrations of Vehicle Suspension Systems

5.1 Sprung Mass Vibrations

5.1.1 RIDE — The *low frequency* (up to 5 Hz) *vibrations* of the sprung mass as a rigid body.

5.1.1.1 Vertical (Bounce) — The translational component of ride *vibrations* of the *sprung mass* in the direction of the vehicle z-axis. (See Fig. 2.)

5.1.1.2 Pitch — The angular component of ride *vibrations* of the *sprung mass* about the vehicle y-axis.

5.1.1.3 Roll — The angular component of ride *vibrations* of the *sprung mass* about the vehicle x-axis.

5.1.2 SHAKE — The intermediate *frequency* (5-25 Hz) *vibrations* of the *sprung mass* as a flexible body.

5.1.2.1 Torsional Shake — A mode of *vibration* involving twisting deformations of *sprung mass* about the vehicle x-axis.

5.1.2.2 Beaming — A mode of *vibration* involving predominantly bending deformations of the *sprung mass* about the vehicle y-axis.

5.1.3 HARSHNESS — The high frequency (25-100 Hz) *vibrations* of the structure and/or components that are perceived tactually and/or audibly.

5.1.4 BOOM — A high intensity *vibration* (25-100 Hz) perceived audibly and characterized as sensation of pressure by the ear.

5.2 Unsprung Mass Vibrations

5.2.1 WHEEL VIBRATION MODES

5.2.1.1 Hop — The vertical oscillatory motion of a wheel between the road surface and the *sprung mass*.

5.2.1.1.1 Parallel hop is the form of wheel hop in which a pair of wheels hop in phase.

5.2.1.1.2 Tramp is the form of wheel hop in which a pair of wheels hop in opposite phase.

5.2.1.2 Brake Hop — An oscillatory hopping motion of a single wheel or of a pair of wheels which occurs when brakes are applied in forward or reverse motion of the vehicle.

5.2.1.3 Power Hop — An oscillatory hopping motion of a single wheel or of a pair of wheels which occurs when *tractive force* is applied in forward or reverse motion of the vehicle.

5.2.2 AXLE VIBRATION MODES

5.2.2.1 Axle Side Shake — Oscillatory motion of an axle which consists of transverse displacement.

5.2.2.2 Axle Fore-and-Aft Shake — Oscillatory motion of an axle which consists purely of longitudinal displacement.

5.2.2.3 Axle Yaw — Oscillatory motion of an axle around the vertical axis through its center of gravity.

5.2.2.4 Axle Windup — Oscillatory motion of an axle about the horizontal transverse axis through its center of gravity.

5.2.3 STEERING SYSTEM VIBRATIONS

5.2.3.1 Wheel Flutter — Forced *oscillation* of steerable wheels about their steering axes.

5.2.3.2 Wheel Wobble — A self-excited *oscillation* of steerable wheels about their steering axes, occurring without appreciable *tramp*.

5.2.3.3 Shimmy — A self-excited *oscillation* of a pair of steerable wheels about their steering axes, accompanied by appreciable *tramp*.

5.2.3.4 Wheelfight — A rotary disturbance of the steering wheel produced by forces acting on the steerable wheels.

6. Suspension Geometry

6.1 Kingpin Geometry

6.1.1   WHEEL PLANE — The central plane of the tire, normal to the *spin axis*.

6.1.2   WHEEL CENTER — The point at which the *spin axis* of the wheel intersects the *wheel plane*.

6.1.3   CENTER OF TIRE CONTACT — The intersection of the *wheel plane* and the vertical projection of the *spin axis* of the wheel onto the road plane. (See Note 1.)

6.1.4   KINGPIN INCLINATION — The angle in front elevation between the steering axis and the vertical.

6.1.5   KINGPIN OFFSET — Kingpin offset at the ground is the horizontal distance in front elevation between the point where the steering axis intersects the ground and the *center of tire contact.*

The kingpin offset at the *wheel center* is the horizontal distance in front elevation from the *wheel center* to the steering axis.

6.2 Wheel Caster

6.2.1   CASTER ANGLE — The angle in side elevation between the steering axis and the vertical. It is considered positive when the steering axis is inclined rearward (in the upward direction) and negative when the steering axis is inclined forward.

6.2.2   RATE OF CASTER CHANGE — The change in *caster angle* per unit vertical displacement of the *wheel center* relative to the *sprung mass.*

6.2.3   CASTER OFFSET — The distance in side elevation between the point where the steering axis intersects the ground, and the *center of tire contact.* The offset is considered positive when the intersection point is forward of the tire contact center and negative when it is rearward.

6.2.4   CENTRIFUGAL CASTER — The unbalance moment about the steering axis produced by a lateral acceleration equal to gravity acting at the combined center of gravity of all the steerable parts. It is considered positive if the combined center of gravity is forward of the steering axis and negative if rearward of the steering axis.

6.3 Wheel Camber

6.3.1   CAMBER ANGLE — The inclination of the *wheel plane* to the vertical. It is considered positive when the wheel leans outward at the top and negative when it leans inward.

6.3.2   RATE OF CAMBER CHANGE — The change of *camber angle* per unit vertical displacement of the *wheel center* relative to the *sprung mass.*

6.3.2.1 Swing Center — That instantaneous center in the transverse vertical plane through any pair of *wheel centers* about which the wheel moves relative to the *sprung mass.*

6.3.2.2. Swing-Arm Radius — The horizontal distance from the *swing center* to the *center of tire contact.*

6.3.3   WHEEL TRACK (WHEEL TREAD) — The lateral distance between the *center of tire contact* of a pair of wheels. For vehicles with dual wheels, it is the distance between the points centrally located between the *centers of tire contact* of the inner and outer wheels. (See SAE J693.)

6.3.4   TRACK CHANGE — The change in *wheel track* resulting from vertical suspension displacements of both wheels in the same direction.

6.3.5   RATE OF TRACK CHANGE — The change in *wheel track* per unit vertical displacement of both *wheel centers* in the same direction relative to the *sprung mass.*

6.4 Wheel Toe

6.4.1   STATIC TOE ANGLE (DEG) — The static toe angle of a wheel, at a specified wheel load or relative position of the *wheel center* with respect to the *sprung mass,* is the angle between a longitudinal axis of the vehicle and the line of intersection of the *wheel plane* and the road surface. The wheel is "toed-in" if the forward portion of the wheel is turned toward a central longitudinal axis of the vehicle, and "toed-out" if turned away.

6.4.2   STATIC TOE (IN [MM]) — Static toe-in or toe-out of a pair of wheels, at a specified wheel load or relative position of the *wheel center* with respect to the *sprung mass,* is the difference in the transverse distances between the *wheel planes,* taken at the extreme rear and front points of the tire treads. When the distance at the rear is greater, the wheels are "toed-in" by this amount; and where smaller, the wheels are "toed-out." (See Note 2.)

6.5   Compression — The relative displacement of *sprung* and *unsprung masses* in the suspension system in which the distance between the masses decreases from that at static condition.

6.5.1   RIDE CLEARANCE — The maximum displacement in *compression* of the *sprung mass* relative to the *wheel center* permitted by the suspension system, from the normal load position.

6.5.2   METAL-TO-METAL POSITION (COMPRESSION) — The point of maximum *compression* travel limited by interference of substantially rigid members.

6.5.3   BUMP STOP — An elastic member which increases the *wheel rate* toward the end of the *compression* travel.

The bump stop may also act to limit the *compression* travel.

6.6 Rebound — The relative displacement of the *sprung* and *unsprung masses* in a suspension system in which the distance between the masses increases from that at static condition.

6.6.1   REBOUND CLEARANCE — The maximum displacement in *rebound* of the *sprung mass* relative to the *wheel center* permitted by the suspension system, from the normal load position.

6.6.2   METAL-TO-METAL POSITION (REBOUND) — The point of maximum *rebound* travel limited by interference of substantially rigid members.

6.6.3   REBOUND STOP — An elastic member which increases the *wheel rate* toward the end of the *rebound* travel.

The rebound stop may also act to limit the *rebound* travel.

6.7 Center of Parallel Wheel Motion — The center of curvature of the path along which each of a pair of *wheel centers* moves in a longitudinal vertical plane relative to the *sprung mass* when both wheels are equally displaced.

6.8 Torque Arm

6.8.1   TORQUE-ARM CENTER IN BRAKING — The instantaneous center in a vertical longitudinal plane through the *wheel center* about which the wheel moves relative to the *sprung mass* when the brake is locked.

6.8.2   TORQUE-ARM CENTER IN DRIVE — The instantaneous center in a vertical longitudinal plane through the *wheel center* about which the wheel moves relative to the *sprung mass* when the drive mechanism is locked at the power source.

6.8.3   TORQUE-ARM RADIUS — The horizontal distance from the *torque-arm center* to the *wheel center.*

7. Tires and Wheels

7.1 General Nomenclature

7.1.1   STANDARD LOADS AND INFLATIONS — Those combinations of loads and inflations up to the maximum load and inflation recommended by the Tire & Rim Association and published in the yearly editions of the Tire & Rim Association Yearbook.

7.1.2   RIM DIAMETER — The diameter at the intersection of the *bead* seat and the flange. (See Tire & Rim Association Yearbook.) Nominal rim diameter (i.e., 14, 15, 16.5, etc.) is commonly used.

7.1.3   RIM WIDTH — The distance between the inside surfaces or the rim flanges. (See Tire & Rim Association Yearbook.)

7.1.4   TIRE SECTION WIDTH — The width of the unloaded new tire mounted on specified rim, inflated to the normal recommended pressure, including the normal sidewalls but not including protective rib, bars, and decorations. (See Tire & Rim Association Yearbook.)

7.1.5   TIRE OVERALL WIDTH — The width of the unloaded new tire, mounted on specified rim, inflated to the normal recommended pressure, including protective rib, bars, and decorations. (See Tire & Rim Association Yearbook.)

7.1.6   TIRE SECTION HEIGHT — Half the difference between the tire *outside diameter* and the nominal *rim diameter.*

7.1.7   OUTSIDE DIAMETER — The maximum diameter of the new unloaded tire inflated to the normal recommended pressure and mounted on a specified rim. (See Airplane Section, Tire & Rim Association Yearbook.)

7.1.8   FLAT TIRE RADIUS — The distance from the *spin axis* to the road surface of a loaded tire on a specified rim at zero inflation.

7.1.9   DEFLECTION (STATIC)—The radial difference between the undeflected tire radius and the static loaded radius, under specified loads and inflation.

7.1.9.1 Percent Deflection — The static deflection expressed as a percentage of the unloaded section height above the top of the rim flange.

7.1.10   TIRE RATE (STATIC) — See 4.3.2.

7.1.11   SIDEWALL — The portion of either side of the tire which connects the *bead* with the *tread.*

7.1.11.1 Sidewall Rib — A raised circumferential rib located on the *sidewall.*

7.1.12   BEAD — The portion of the tire which fits onto the rim of the wheel.

7.1.12.1 Bead Base — The approximately cylindrical portion of the *bead* that forms its inside diameter.

7.1.12.2 Bead Toe — That portion of the *bead* which joins the *bead base* and the inside surface of the tire.

7.1.13   TREAD (TIRE) — The peripheral portion of the tire, the exterior of which is designed to contact the road surface.

7.1.13.1 Tread Contour — The cross-sectional shape of *tread* surface of an inflated unloaded tire neglecting the *tread pattern* depressions.

7.1.13.2 Tread Radius — The radius or combination of radii describing the *tread contour.*

7.1.13.3 Tread Arc Width — The distance measured along the *tread contour* of an unloaded tire between one edge of the *tread* and the other. For tires with rounded tread edges, the point of measurement is that point in space which is at the intersection of the *tread radius* extended until it meets the prolongation of the upper sidewall contour.

7.1.13.4 Tread Chord Width — The distance measured parallel to the *spin axis* of an unloaded tire between one edge of the *tread* and the other. For tires with rounded tread edges, the point of measurement is that point in space which is at the intersection of the *tread radius* extended until it meets the prolongation of the upper sidewall contour.

7.1.13.5 Tread Contact Width — The distance between the extreme edges of road contact at a specified load and pressure measured parallel to the Y' axis at zero *slip angle* and zero *inclination angle.*

7.1.13.6 Tread Contact Length — The perpendicular distance between the tangent to edges of the leading and following points of road contact and parallel to the *wheel plane.*

7.1.13.7 Tread Depth — The distance between the base of a tire *tread* groove and a line tangent to the surface of the two adjacent *tread* ribs or rows.

7.1.13.8 Gross Contact Area — The total area enclosing the pattern of the tire *tread* in contact with a flat surface, including the area of grooves or voids.

7.1.13.9 Net Contact Area — The area enclosing the pattern of the tire *tread* in contact with a flat surface, excluding the area of grooves or other depressions.

7.1.13.10 Tread Pattern — The molded configuration on the face of the *tread*. It is generally composed of ribs, rows, grooves, bars, lugs, and the like.

7.2 Rolling Characteristics

7.2.1 LOADED RADIUS (R_ℓ) — The distance from the *center of tire contact* to the *wheel center* measured in the wheel plane.

7.2.2 STATIC LOADED RADIUS — The *loaded radius* of a stationary tire inflated to normal recommended pressure.

NOTE: In general, static loaded radius is different from the radius of slowly rolling tire. Static radius of a tire rolled into position may be different from that of the tire loaded without being rolled.

7.2.3 SPIN AXIS — The axis of rotation of the wheel. (See Fig. 1.)

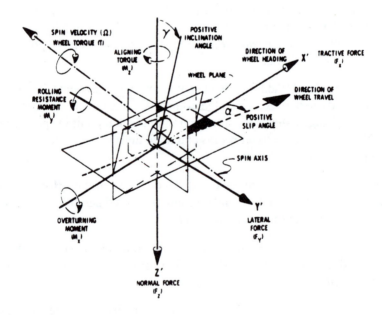

FIG. 1 — TIRE AXIS SYSTEM

7.2.4   SPIN VELOCITY (Ω) — The angular velocity of the wheel on which the tire is mounted, about its *spin axis*. Positive spin velocity is shown in Fig. 1.

7.2.5   FREE-ROLLING TIRE — A loaded rolling tire operated without application of *driving* or *braking torque*.

7.2.6   STRAIGHT FREE-ROLLING TIRE — A *free-rolling tire* moving in a straight line at zero *inclination angle* and zero *slip angle*.

7.2.7   LONGITUDINAL SLIP VELOCITY — The difference between the *spin velocity* of the driven or braked tire and the *spin velocity* of the *straight free-rolling tire*. Both spin velocities are measured at the same linear velocity at the wheel center in the X' direction. A positive value results from *driving torque*.

7.2.8   LONGITUDINAL SLIP (PERCENT SLIP) — The ratio of the *longitudinal slip velocity* to the *spin velocity* of the *straight free-rolling tire* expressed as a percentage.

NOTE: This quantity should not be confused with the slip number that frequently appears in kinematic analysis of tires in which the *spin velocity* appears in the denominator.

7.2.9   EFFECTIVE ROLLING RADIUS (R_e) — The ratio of the linear velocity of the wheel center in the X' direction to the *spin velocity*. (See 7.3.1.)

7.2.10   WHEEL SKID — The occurrence of sliding between the tire and road interface which takes place within the entire *contact area*. Skid can result from braking, driving and/or cornering.

7.3 Tire Forces and Moments

7.3.1   TIRE AXIS SYSTEM (FIG. 1) — The origin of the tire axis system is the *center of tire contact*. The X' axis is the intersection of the wheel plane and the road plane with a positive direction forward. The Z' axis is perpendicular to the road plane with a positive direction downward. The Y' axis is in the road plane, its direction being chosen to make the axis system orthogonal and right-hand.

7.3.2   TIRE ANGLES

7.3.2.1 Slip Angle (α) — The angle between the X' axis and direction of travel of the *center of tire contact*.

7.3.2.2 Inclination Angle (γ) — The angle between the Z' axis and the *wheel plane*.

7.3.3   TIRE FORCES — The external force acting on the tire by the road having the following components:

7.3.3.1 Longitudinal Force (F_x) — The component of the *tire force* vector in the X' direction.

7.3.3.2 Driving Force — The *longitudinal force* resulting from *driving torque* application.

7.3.3.3 Driving Force Coefficient — The ratio of the *driving force* to the *vertical load*.

7.3.3.4 Braking Force — The negative *longitudinal force* resulting from *braking torque* application.

7.3.3.5 Braking Force Coefficient (Braking Coefficient) — The ratio of the *braking force* to the *vertical load*.

7.3.3.6 Rolling Resistance Force — The negative *longitudinal force* resulting from energy losses due to deformations of a rolling tire.

NOTE: This force can be computed from the forces and moments acting on the tire by the road.

$$F_r = (M_y\cos\gamma + M_z\sin\gamma)/R_\ell$$

7.3.3.7 Rolling Resistance Force Coefficient (Coefficient of Rolling Resistance) — The ratio of the *rolling resistance* to the *vertical load*.

7.3.3.8 Lateral Force (F_y) — The component of the *tire force* vector in the Y' direction.

7.3.3.9 Lateral Force Coefficient — The ratio of the *lateral force* to the *vertical load*.

7.3.3.10 Slip Angle Force — The *lateral force* when the *inclination angle* is zero and *plysteer* and *conicity* forces have been subtracted.

7.3.3.11 Camber Force (Camber Thrust) — The *lateral force* when the *slip angle* is zero and the *plysteer* and *conicity* forces have been subtracted.

7.3.3.12 Normal Force (F_z) — The component of the *tire force* vector in the Z' direction.

7.3.3.13 Vertical Load — The normal reaction of the tire on the road which is equal to the negative of the normal force.

7.3.3.14 Central Force — The component of the *tire force* vector in the direction perpendicular to the direction of travel of the *center of tire contact*. Central force is equal to *lateral force* times cosine of *slip angle* minus *longitudinal force* times sine of *slip angle*.

7.3.3.15 Tractive Force — The component of the *tire force* vector in the direction of travel of the *center of tire contact*. Tractive force is equal to *lateral force* times sine of *slip angle* plus *longitudinal force* times cosine of *slip angle*.

7.3.3.16 Drag Force — The negative *tractive force*.

7.3.4 TIRE MOMENTS — The external moments acting on the tire by the road having the following components:

7.3.4.1 Overturning Moment (M_x) — The component of the *tire moment* vector tending to rotate the tire about the X' axis, positive clockwise when looking in the positive direction of the X' axis.

7.3.4.2 Rolling Resistance Moment (M_y) — The component of the *tire moment* vector tending to rotate the tire about the Y' axis, positive clockwise when looking in the positive direction of the Y' axis.

7.3.4.3 Aligning Torque (Aligning Moment) (M_z) — The component of the *tire moment* vector tending to rotate the tire about the Z' axis, positive clockwise when looking in the positive direction of the Z' axis.

7.3.4.4 Wheel Torque (T) — The external torque applied to the tire from the vehicle about the *spin axis*; positive *wheel torque* is shown in Fig. 1.

7.3.4.5 Driving Torque — The positive *wheel torque*.

7.3.4.6 Braking Torque — The negative *wheel torque*.

7.4 Tire Force and Moment Stiffness — (may be evaluated at any set of operating conditions)

7.4.1 CORNERING STIFFNESS — The negative of the rate of change of *lateral force* with respect to change in *slip angle*, usually evaluated at zero *slip angle*.

7.4.2 CAMBER STIFFNESS — The rate of change of *lateral force* with respect to change in *inclination angle*, usually evaluated at zero *inclination angle*.

7.4.3 BRAKING (DRIVING) STIFFNESS — The rate of change of *longitudinal force* with respect to change in *longitudinal slip*, usually evaluated at zero *longitudinal slip*.

7.4.4 ALIGNING STIFFNESS (ALIGNING TORQUE STIFFNESS) — The rate of change of *aligning* torque with respect to change in *slip angle*, usually evaluated at zero *slip angle*.

393

7.5 Normalized Tire Force and Moment Stiffnesses (Coefficients)

7.5.1   CORNERING STIFFNESS COEFFICIENT (CORNERING COEFFICIENT) — The ratio of *cornering stiffness* of a *straight free-rolling tire* to the *vertical load.*

NOTE: Although the term cornering coefficient has been used in a number of technical papers, for consistency with definitions of other terms using the word coefficient, the term cornering stiffness coefficient is preferred.

7.5.2   CAMBER STIFFNESS COEFFICIENT (CAMBER COEFFICIENT) — The ratio of camber stiffness of a *straight free-rolling tire* to the *vertical load.*

7.5.3   BRAKING (DRIVING) STIFFNESS COEFFICIENT — The ratio of braking (driving) stiffness of a *straight free-rolling tire* to the *vertical load.*

7.5.4   ALIGNING STIFFNESS COEFFICIENT (ALIGNING TORQUE COEF-FICIENT) — The ratio of *aligning stiffness* of a *straight free-rolling tire* to the *vertical load.*

7.6 Tire Traction Coefficients

7.6.1   LATERAL TRACTION COEFFICIENT — The maximum value of *lateral force coefficient* which can be reached on a *free-rolling tire* for a given road surface, environment and operating condition.

7.6.2   DRIVING TRACTION COEFFICIENT — The maximum value of *driving force coefficient* which can be reached on a given tire and road surface for a given environment and operating condition.

7.6.3   BRAKING TRACTION COEFFICIENT — The maximum of the *braking force coefficient* which can be reached without locking a wheel on a given tire and road surface for a given environment and operating condition.

7.6.3.1   Sliding Braking Traction Coefficient — The value of the *braking force coefficient* of a tire obtained on a locked wheel on a given tire and road surface for a given environment and operating condition.

7.7 Tire Associated Noise and Vibrations

7.7.1   TREAD NOISE — Airborne sound (up to 5000 Hz) except squeal and slap produced by the interaction between the tire and the road surface.

7.7.1.1   Sizzle — A *tread noise* (up to 4000 Hz) characterized by a soft frying sound, particularly noticeable on a very smooth road surface.

7.7.2   SQUEAL — Narrow band airborne tire noise (150-800 Hz) resulting from either *longitudinal slip* or *slip angle* or both.

7.7.2.1 Cornering Squeal — The *squeal* produced by a *free-rolling tire* resulting from *slip angle*.

7.7.2.2 Braking (Driving) Squeal — The *squeal* resulting from *longitudinal slip*.

7.7.3   THUMP — A periodic vibration and/or audible sound generated by the tire and producing a pounding sensation which is synchronous with wheel rotation.

7.7.4   ROUGHNESS — Vibration (15-100 Hz) perceived tactually and/or audibly, generated by a rolling tire on a smooth road surface and producing the sensation of driving on a coarse or irregular surface.

7.7.5   HARSHNESS — Vibrations (15-100 Hz) perceived tactually and/or audibly, produced by interaction of the tire with road irregularities.

7.7.6   SLAP — Airborne smacking noise produced by a tire traversing road seams such as tar strips and expansion joints.

7.8 Tire and Wheel Non-Uniformity Characteristics

7.8.1   RADIAL RUN-OUT

7.8.1.1 Peak-to-Peak Radial Wheel Run-Out — The difference between the maximum and minimum values of the wheel *bead* seat radius, measured in a plane perpendicular to the *spin axis* (measured separately for each *bead* seat).

7.8.1.2 Peak-to-Peak Unloaded Radial Tire Run-Out — The difference between maximum and minimum undeflected values of the tire radius, measured in plane perpendicular to the *spin axis* on a true running wheel.

7.8.1.3 Peak-to-Peak Loaded Radial Tire Run-Out — The difference between maximum and minimum values of the *loaded radius* on a true running wheel.

7.8.2   LATERAL RUN-OUT

7.8.2.1 Peak-to-Peak Lateral Wheel Run-Out — The difference between maximum and minimum indicator readings, measured parallel to the *spin axis* on the inside vertical portion of a rim flange (measured separately for each flange).

7.8.2.2 Peak-to-Peak Lateral Tire Run-Out — The difference between maximum and minimum indicator readings, measured parallel to the *spin axis* at the point of maximum *tire section,* on a true running wheel (measured separately for each sidewall).

7.8.3   RADIAL FORCE VARIATION — The periodic variation of the *normal force* of a loaded *straight free-rolling tire* which repeats each revolution at a fixed *loaded radius*, given mean *normal force*, constant speed, given inflation pressure and test surface curvature.

7.8.3.1 Peak-to-Peak (Total) Radial Force Variation — The difference between maximum and minimum values of the *normal force* during one revolution of the tire.

7.8.3.2 First Order Radial Force Variation — The peak-to-peak amplitude of the fundamental frequency component of the Fourier series representing *radial force variation.* Its frequency is equal to the rotational frequency of the tire.

7.8.4   LATERAL FORCE VARIATION — The periodic variation of lateral force of a *straight free-rolling tire* which repeats each revolution, at a fixed *loaded radius,* given mean *normal force*, constant speed, given inflation pressure and test surface curvature.

7.8.4.1 Peak-to-Peak (Total) Lateral Force Variation — The difference between the maximum and minimum values of the *lateral force* during one revolution of the tire.

7.8.4.2 First Order Lateral Force Variation — The peak-to-peak amplitude of the fundamental frequency component of the Fourier series representing *lateral force variation.* Its frequency is equal to the rotational frequency of the tire.

7.8.5   LATERAL FORCE OFFSET — The average *lateral force* of a *straight free-rolling tire.*

7.8.5.1 Ply Steer Force — The component of *lateral force offset* which does not change sign (with respect to the *Tire Axis System*) with a change in direction of rotation (positive along positive Y' axis). The force remains positive when it is directed away from the serial number on the right side tire and toward the serial number on the left side tire.

7.8.5.2 Conicity Force — The component of *lateral force offset* which changes sign (with respect to the *Tire Axis System*) with a change in direction of rotation (positive away from the serial number or toward the whitewall). The force is positive when it is directed away from the serial number on the right side tire and negative when it is directed toward the serial number on the left side tire.

8. Kinematics — Force and Moments Notation

8.1 Earth-Fixed Axis System (X,Y,Z) — This system is a right-hand orthogonal axis system fixed on the earth. The trajectory of the vehicle is described with respect to this earth-fixed axis system. The X- and Y-axis are in a horizontal plane and the Z-axis is directed downward.

8.2 Vehicle Axis System (x,y,z) — This system is a right-hand orthogonal axis system fixed in a vehicle such that with the vehicle moving steadily in a straight line on a level road, the x-axis is substantially horizontal, points forward, and is in the longitudinal plane of symmetry. The y-axis points to driver's right and the z-axis points downward. (See Fig. 2.)

8.3 Angular Orientation — The orientation of the *vehicle axis system (x,y,z)* with respect to the *earth-fixed axis system (X,Y,Z)* is given by a sequence of three angular rotations. The following sequence of rotations (see Note 6), starting from a condition in which the two sets of axes are initially aligned, is defined to be the standard:

    (1) A yaw rotation, ψ, about the aligned z- and Z-axis.
    (2) A pitch rotation, Θ, about the vehicle y-axis.
    (3) A roll rotation, ϕ, about the vehicle x-axis.

8.4 Motion Variables

8.4.1   VEHICLE VELOCITY — The vector quantity expressing velocity of a point in the vehicle relative to the *earth-fixed axis system (X,Y,Z)*. The following motion variables are components of this vector resolved with respect to the moving *vehicle axis system (x,y,z)*.

8.4.1.1 Longitudinal Velocity (u) of a point in the vehicle is the component of the vector velocity in the x-direction.

8.4.1.2 Side Velocity (v) of a point in the vehicle is the component of the vector velocity in the y-direction.

8.4.1.3 Normal Velocity (w) of a point in the vehicle is the component of the vector velocity in the z-direction.

8.4.1.4 Forward Velocity of a point in the vehicle is the component of the vector velocity perpendicular to the y-axis and parallel to the road plane.

8.4.1.5 Lateral Velocity of a point in the vehicle is the component of the vector velocity perpendicular to the x-axis and parallel to the road plane.

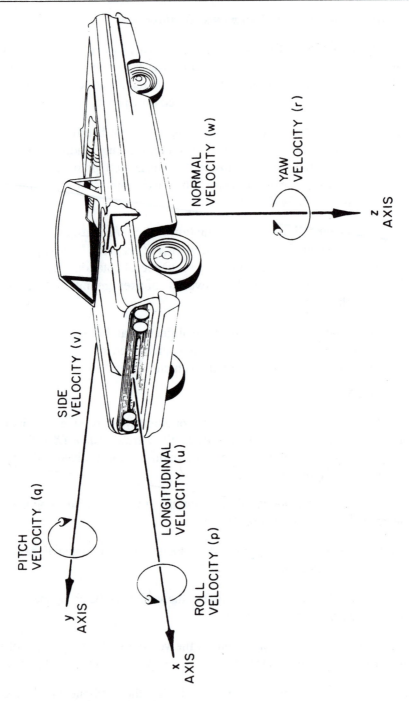

FIG. 2 — DIRECTIONAL CONTROL AXIS SYSTEM

8.4.1.6 Roll Velocity (p) — The angular velocity about the x-axis.

8.4.1.7 Pitch Velocity (q) — The angular velocity about the y-axis.

8.4.1.8 Yaw Velocity (r) — The angular velocity about the z-axis.

8.4.2 VEHICLE ACCELERATION — The vector quantity expressing the acceleration of a point in the vehicle relative to the *earth-fixed axis system (X,Y,Z)*. The following motion variables are components of this vector, resolved with respect to the moving *vehicle axis system*.

8.4.2.1 Longitudinal Acceleration — The component of the vector acceleration of a point in the vehicle in the x-direction.

8.4.2.2 Side Acceleration — The component of the vector acceleration of a point in the vehicle in the y-direction.

8.4.2.3 Normal Acceleration — The component of the vector acceleration of a point in the vehicle in the z-direction.

8.4.2.4 Lateral Acceleration — The component of the vector acceleration of a point in the vehicle perpendicular to the vehicle x-axis and parallel to the road plane. (See Note 7.)

8.4.2.5 Centripetal Acceleration — The component of the vector acceleration of a point in the vehicle perpendicular to the tangent to the path of that point and parallel to the road plane.

8.4.3 HEADING ANGLE (ψ) — The angle between the trace on the X-Y plane of the vehicle x-axis and the X-axis of the *earth-fixed axis system*. (See Fig. 3.)

8.4.4 SIDESLIP ANGLE (ATTITUDE ANGLE) (β) — The angle between the traces on the X-Y plane of the vehicle x-axis and the vehicle velocity vector at some specified point in the vehicle. Sideslip angle is shown in Fig. 3 as a negative angle.

8.4.5 SIDESLIP ANGLE GRADIENT — The rate of change of *sideslip angle* with respect to change in steady-state *lateral acceleration* on a level road at a given *trim* and test conditions.

8.4.6 COURSE ANGLE (υ) — The angle between the trace of the vehicle velocity vector on the X-Y plane and X-axis of the *earth-fixed axis system*. A positive course angle is shown in Fig. 3. Course angle is the sum of *heading angle* and *sideslip angle* ($\upsilon = \psi + \beta$).

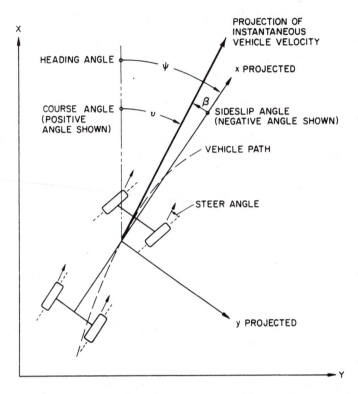

FIG. 3 — HEADING, SIDESLIP, AND COURSE ANGLES

8.4.7   VEHICLE ROLL ANGLE — The angle between the vehicle y-axis and the ground plane.

8.4.8   VEHICLE ROLL GRADIENT — The rate of change in *vehicle roll angle* with respect to change in steady-state *lateral acceleration* on a level road at a given *trim* and test conditions.

8.4.9   VEHICLE PITCH ANGLE — The angle between the vehicle x-axis and the ground plane.

8.5 Forces — The external forces acting on the vehicle can be summed into one force vector having the following components:

8.5.1   LONGITUDINAL FORCE (F_x) — The component of the force vector in the x-direction.

8.5.2   SIDE FORCE (F_y) — The component of the force vector in the y-direction.

8.5.3   NORMAL FORCE (F_z) — The component of the force vector in the z-direction.

8.6 Moments — The external moments acting on the vehicle can be summed into one moment vector having the following components:

8.6.1   ROLLING MOMENT (M_x) — The component of the moment vector tending to rotate the vehicle about the x-axis, positive clockwise when looking in the positive direction of the x-axis.

8.6.2   PITCHING MOMENT (M_y) — The component of the moment vector tending to rotate the vehicle about the y-axis, positive clockwise when looking in the positive direction of the y-axis.

8.6.3   YAWING MOMENT (M_z) — The component of the moment vector tending to rotate the vehicle about the z-axis, positive clockwise when looking in the positive direction of the z-axis.

9. Directional Dynamics

9.1 Control Modes

9.1.1   POSITION CONTROL — That mode of vehicle control wherein inputs or restraints are placed upon the steering system in the form of displacements at some control point in the steering system (front wheels, Pitman arm, steering wheel), independent of the force required.

9.1.2   FIXED CONTROL — That mode of vehicle control wherein the position of some point in the steering system (front wheels, Pitman arm, steering wheel) is held fixed. This is a special case of *position control.*

9.1.3   FORCE CONTROL — That mode of vehicle control wherein inputs or restraints are placed upon the steering system in the form of forces, independent of the displacement required.

9.1.4   FREE CONTROL — That mode of vehicle control wherein no restraints are placed upon the steering system. This is a special case of *force control.*

9.2 Vehicle Response — The vehicle motion resulting from some internal or external input to the vehicle. Response tests can be used to determine the *stability* and control characteristics of a vehicle.

9.2.1   STEERING RESPONSE — The vehicle motion resulting from an input to the steering (control) element. (See Note 8.)

9.2.2   DISTURBANCE RESPONSE — The vehicle motion resulting from unwanted force or displacement inputs applied to the vehicle. Examples of disturbances are wind forces or vertical road displacements.

9.2.3   STEADY-STATE — Steady-state exists when periodic (or constant) vehicle responses to periodic (or constant) control and/or disturbance inputs do not change over an arbitrarily long time. The motion responses in steady-state are referred to as steady-state responses. This definition does not require the vehicle to be operating in a straight line or on a level road surface. It can also be in a turn of constant radius or on a cambered road surface.

9.2.4   TRANSIENT STATE — Transient state exists when the motion responses, the external forces relative to the vehicle, or the control positions are changing with time. (See Note 9.)

9.2.5   TRIM — The *steady-state* (that is, equilibrium) condition of the vehicle with constant input which is used as the reference point for analysis of dynamic vehicle *stability* and control characteristics.

9.2.6   STEADY-STATE RESPONSE GAIN — The ratio of change in the *steady-state* response of any motion variable with respect to change in input at a given *trim*.

9.2.7   STEERING SENSITIVITY (CONTROL GAIN) — The change in *steady-state lateral acceleration* on a level road with respect to change in *steering wheel angle* at a given *trim* and test conditions.

9.3 Stability — (See Note 10.)

9.3.1   ASYMPTOTIC STABILITY — Asymptotic stability exists at a prescribed *trim* if, for any small temporary change in disturbance or *control input,* the vehicle will approach the motion defined by the *trim.*

9.3.2   NEUTRAL STABILITY — Neutral stability exists at a prescribed *trim* if, for any small temporary change in disturbance or *control input*, the resulting motion of the vehicle remains close to, but does not return to, the motion defined by the *trim.*

9.3.3   DIVERGENT INSTABILITY — Divergent instability exists at a prescribed *trim* if any small temporary disturbance or *control input* causes an ever-increasing *vehicle response* without *oscillation.* (See Note 11.)

9.3.4 OSCILLATORY INSTABILITY — Oscillatory instability exists if a small temporary disturbance or *control input* causes an oscillatory *vehicle response* of ever-increasing amplitude about the initial *trim*. (See Note 12.)

9.4 Suspension Steer and Roll Properties — (Fig. 4) (See Note 13.)

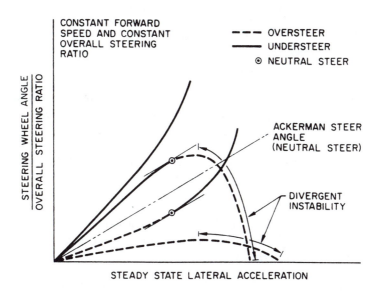

FIG. 4 — STEER PROPERTIES (SEE NOTE 17)

9.4.1 STEER ANGLE (δ) — The angle between the projection of a longitudinal axis of the vehicle and the line of intersection of the *wheel plane* and the road surface. Positive angle is shown in Fig. 3.

9.4.2 ACKERMAN STEER ANGLE (δ_a) — The angle whose tangent is the wheelbase divided by the radius of turn.

9.4.3 ACKERMAN STEER ANGLE GRADIENT — The rate of change of *Ackerman steer angle* with respect to change in *steady-state lateral acceleration* on a level road at a given *trim* and test conditions. (See Note 14.)

9.4.4 STEERING WHEEL ANGLE — Angular displacement of the steering wheel measured from the straight-ahead position (position corresponding to zero average *steer angle* of a pair of steered wheels).

403

9.4.5   STEERING WHEEL ANGLE GRADIENT — The rate of change in the *steering wheel angle* with respect to change in *steady-state lateral acceleration* on a level road at a given *trim* and test conditions.

9.4.6   OVERALL STEERING RATIO — The rate of change of *steering wheel angle* at a given steering wheel *trim* position, with respect to change in average *steer angle* of a pair of steered wheels, assuming an infinitely stiff steering system with no roll of the vehicle (see Note 15).

9.4.7   UNDERSTEER/OVERSTEER GRADIENT — The quantity obtained by subtracting the *Ackerman steer angle gradient* from the ratio of the *steering wheel angle gradient* to the *overall steering ratio*.

9.4.8   NEUTRAL STEER — A vehicle is neutral steer at a given *trim* if the ratio of the *steering wheel angle gradient* to the *overall steering ratio* equals the *Ackerman steer angle gradient*.

9.4.9   UNDERSTEER — A vehicle is understeer at a given *trim* if the ratio of the *steering wheel angle gradient* to the *overall steering ratio* is greater than the *Ackerman steer angle gradient*.

9.4.10   OVERSTEER — A vehicle is oversteer at a given *trim* if the ratio of the *steering wheel angle gradient* to the *overall steering ratio* is less than the *Ackerman steer angle gradient*.

9.4.11   STEERING WHEEL TORQUE — The torque applied to the steering wheel about its axis of rotation.

9.4.12   STEERING WHEEL TORQUE GRADIENT — The rate of change in the *steering wheel torque* with respect to change in *steady-state lateral acceleration* on a level road at a given *trim* and test conditions.

9.4.13   CHARACTERISTIC SPEED — That forward speed for an understeer vehicle at which the steering sensitivity at zero *lateral acceleration* trim is one-half the steering sensitivity of a *neutral steer* vehicle.

9.4.14   CRITICAL SPEED — That forward speed for an *oversteer* vehicle at which the steering sensitivity at zero *lateral acceleration trim* is infinite.

9.4.15   NEUTRAL STEER LINE — The set of points in the x-z plane at which external lateral forces applied to the *sprung mass* produce no *steady-state yaw velocity*.

9.4.16   STATIC MARGIN — The horizontal distance from the center of gravity to the *neutral steer line* divided by the wheelbase. It is positive if the center of gravity is forward of the *neutral steer line*.

9.4.17 SUSPENSION ROLL — The rotation of the vehicle *sprung mass* about the x-axis with respect to a transverse axis joining a pair of *wheel centers.*

9.4.18 SUSPENSION ROLL ANGLE — The angular displacement produced by *suspension roll.*

9.4.19 SUSPENSION ROLL GRADIENT — The rate of change in the *suspension roll angle* with respect to change in *steady-state lateral acceleration* on a level road at a given *trim* and test conditions.

9.4.20 ROLL STEER — The change in *steer angle* of front or rear wheels due to *suspension roll.*

9.4.20.1 Roll Understeer — *Roll steer* which increases vehicle *understeer* or decreases vehicle *oversteer.*

9.4.20.2 Roll Oversteer — *Roll steer* which decreases vehicle *understeer* or increases vehicle *oversteer.*

9.4.21 ROLL STEER COEFFICIENT — The rate of change in *roll steer* with respect to change in *suspension roll angle* at a given *trim.*

9.4.22 COMPLIANCE STEER — The change in *steer angle* of front or rear wheels resulting from compliance in suspension and steering linkages and produced by forces and/or moments applied at the tire-road contact.

9.4.22.1 Compliance Understeer — *Compliance steer* which increases vehicle *understeer* or decreases vehicle *oversteer.*

9.4.22.2 Compliance Oversteer — *Compliance steer* which decreases vehicle *understeer* or increases vehicle *oversteer.*

9.4.23 COMPLIANCE STEER COEFFICIENT — The rate of change in *compliance steer* with respect to change in forces or moments applied at the tire-road contact.

9.4.24 ROLL CAMBER — The camber displacements of a wheel resulting from *suspension roll.*

9.4.25 ROLL CAMBER COEFFICIENT — The rate of change in wheel *inclination angle* with respect to change in *suspension roll angle.*

9.4.26 COMPLIANCE CAMBER — The camber motion of a wheel resulting from compliance in suspension linkages and produced by forces and/or moments applied at the tire-road contact.

9.4.27   COMPLIANCE CAMBER COEFFICIENT — The rate of change in wheel *inclination angle* with respect to change in forces or moments applied at the tire-road contact.

9.4.28   ROLL CENTER — The point in the transverse vertical plane through any pair of *wheel centers* at which lateral forces may be applied to the *sprung mass* without producing *suspension roll*. (See Note 16.)

9.4.29   ROLL AXIS — The line joining the front and rear *roll centers.*

9.4.30   SUSPENSION ROLL STIFFNESS — The rate of change in the restoring couple exerted by the suspension of a pair of wheels on the *sprung mass* of the vehicle with respect to change in *suspension roll angle.*

9.4.31   VEHICLE ROLL STIFFNESS — Sum of the separate *suspension roll stiffnesses.*

9.4.32   ROLL STIFFNESS DISTRIBUTION — The distribution of the *vehicle roll stiffness* between front and rear suspension expressed as percentage of the *vehicle roll stiffness.*

9.5 Tire Load Transfer

9.5.1 · TIRE LATERAL LOAD TRANSFER — The *vertical load* transfer from one of the front tires (or rear tires) to the other that is due to acceleration, rotational, or inertial effects in the lateral direction.

9.5.2   TIRE LATERAL LOAD TRANSFER DISTRIBUTION — The distribution of the total *tire lateral load transfer* between front and rear tires expressed as the percentage of the total.

9.5.3   TIRE LONGITUDINAL LOAD TRANSFER — The *vertical load* transferred from a front tire to the corresponding rear tire that is due to acceleration, rotational, or inertial effects in the longitudinal direction.

9.5.4   OVERTURNING COUPLE — The overturning moment on the vehicle with respect to a central, longitudinal axis in the road plane due to lateral acceleration and roll acceleration.

9.5.5   OVERTURNING COUPLE DISTRIBUTION — The distribution of the total *overturning couple* between the front and rear suspensions expressed as the percentage of the total.

10. Aerodynamic Nomenclature

10.1 Aerodynamic Motion Variables

10.1.1   AMBIENT WIND VELOCITY (v_a) — The horizontal component of the air mass velocity relative to the *earth-fixed axis system* in the vicinity of the vehicle.

10.1.2   AMBIENT WIND ANGLE (v_a) — The angle between the X axis of the *earth-fixed axis system* and the *ambient wind velocity* vector.   A positive ambient wind angle is shown in Fig. 5.

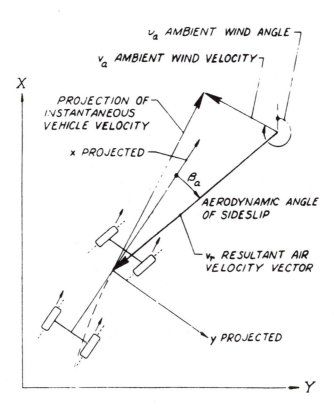

FIG. 5 — WIND VECTORS

10.1.3   RESULTANT AIR VELOCITY VECTOR (v_r) — The vector difference of the *ambient wind velocity* vector and the projection of the velocity vector of the vehicle on the X-Y plane.

407

10.1.4   AERODYNAMIC SIDESLIP ANGLE (β_a) — The angle between the traces on the vehicle x-y plane of the vehicle x-axis and the resultant air velocity vector at some specified point in the vehicle.

10.1.5   AERODYNAMIC ANGLE OF ATTACK (α_a) — The angle between the vehicle x-axis and the trace of the resultant air velocity vector on a vertical plane containing the vehicle x-axis.

10.2 Aerodynamic Force and Moment Coefficients

10.2.1   REFERENCE DIMENSIONS

10.2.1.1 Vehicle Area (A) — The projected frontal area including tires and underbody parts.

10.2.1.2   Vehicle Wheelbase (ℓ) — The characteristic length upon which aerodynamic moment coefficients are based.

10.2.2   STANDARD AIR PROPERTIES

10.2.2.1 The density of standard dry air shall be taken as 2378 x 10^{-6} slugs/ft at 59°F and 29.92 in Hg.

10.2.2.2 The viscosity of standard dry air shall be taken as 373 x 10^{-9} slugs/ft-sec.

10.2.3   FORCE COEFFICIENTS

10.2.3.1 The Longitudinal Force Coefficient (C_x) is based on the aerodynamic force acting on the vehicle in the x-direction (as established by 8.2) and is defined as:

$$C_x = F_x/(qA)$$

where q is dynamic pressure at any given relative air velocity as given by the formula

$$q = (pv^2)/2$$

10.2.3.2 Side Force Coefficient (C_y) is based on the aerodynamic force acting on the vehicle in the y-direction (as established by 8.2) and is defined as:

$$C_y = F_y/(qA)$$

where q is the dynamic pressure at any given relative air velocity as given by the formula

$$q = (pv^2)/2$$

10.2.3.3 The Normal Force Coefficient (C_z) is based on the aerodynamic force acting in the z-direction (as established by 8.2) and is defined as:

$$C_z = F_z/(qA)$$

10.2.4 MOMENT COEFFICIENTS

10.2.4.1 The Rolling Moment Coefficient (C_{Mx}) is based on the rolling moment deriving from the distribution of aerodynamic forces acting on the vehicle and is defined as:

$$C_{Mx} = M_x/(qA\ell)$$

10.2.4.2 The Pitching Moment Coefficient (C_{My}) is based on the pitching moment deriving from the distribution of aerodynamic forces acting on the vehicle and is defined as:

$$C_{My} = M_y/(qA\ell)$$

10.2.4.3 The Yawing Moment Coefficient (C_{Mz}) is based on the yawing moment deriving from the distribution of aerodynamic forces acting on the vehicle and is defined as:

$$C_{Mz} = M_z/(qA\ell)$$

NOTES

1. The *center of tire contact* may not be the geometric center of the tire contact area due to distortion of the tire produced by applied forces.

2. The *static toe (inches)* is equal to the sum of the toe angles (degrees) of the left and right wheels multiplied by the ratio of tire diameter (inches) to 57.3.

   If the toe angles on the left and right wheels are the same and the outside diameter of tire is 28.65 in (727.7 mm), the *static toe (inches)* is equal to *static toe angle (degrees)*.

3. It is important to recognize that to make axis transformations and resolve these forces with respect to the direction of vehicle motion, it is essential to measure all six force and moment components defined in 7.3.3.1-7.3.3.6 and 7.3.4.1-7.3.4.6.

4. This *rolling resistance force* definition has been generalized so that it applies to wheels which are driven or braked. The *wheel torque* can be expressed in terms of the *longitudinal force, rolling resistance force,* and *loaded radius* by the equation

$$T = (F_x + F_r)R_\ell$$

For a *free-rolling wheel,* the *rolling resistance force* is therefore the negative of the *longitudinal force.*

5. For small *slip* and *inclination angles,* the *lateral force* developed by the tire can be approximated by

$$F_y = -C_\alpha \alpha + C_\gamma \gamma$$

6. Angular rotations are positive clockwise when looking in the positive direction of the axis about which rotation occurs.

7. In *steady-state* condition, *lateral acceleration* is equal to the product of *centripetal acceleration* times the cosine of the vehicle's *sideslip angle.* Since in most test conditions the *sideslip angle* is small, for practical purposes, the *lateral acceleration* can be considered equal to *centripetal acceleration.*

8. Although the steering wheel is the primary directional control element it should be recognized that *longitudinal forces* at the wheels resulting from driver inputs to brakes or throttle can modify directional response.

9. Transient responses are described by the terminology normally employed for other dynamic systems. Some terminology is described in the "Control Engineers' Handbook,"[1] but a more complete terminology is contained in ANS C85.1-1963.[2]

10. Passenger vehicles exhibit varying characteristics depending upon test conditions and *trim.* Test conditions refer to vehicle conditions such as wheel loads, front wheel alignment, tire inflation pressure, and also atmospheric and road conditions which affect vehicle parameters. For example, temperature may change shock absorber damping characteris-

---

[1] John G. Truxal (Ed.), <u>Control Engineers' Handbook</u>, New York: McGraw Hill.

[2] "Terminology for Automatic Control," ANS C85.1-1963, published by American Society of Mechanical Engineers.

tics and a slippery road surface may change tire cornering properties. *Trim* has been previously defined as the vehicle operating condition within a given environment, and may be specified in part by *steer angle, forward velocity,* and *lateral acceleration.* Since all these factors change the vehicle behavior, the vehicle *stability* must be examined separately for each environment and *trim.*

For a given set of vehicle parameters and particular test conditions, the vehicle may be examined for each theoretically attainable *trim.* The conditions which most affect *stability* are the *steady-state* values of *forward velocity* and *lateral acceleration.* In practice, it is possible for a vehicle to be stable under one set of operating conditions and unstable in another.

11. Divergent instability may be illustrated by operation above the *critical speed* of an *oversteer vehicle.* Any input to the steering wheel will place the vehicle in a turn of ever-decreasing radius unless the driver makes compensating motions of the wheel to maintain general equilibrium. This condition represents *divergent instability.* A linear mathematical analog of a vehicle is *divergently unstable* when its characteristic equation has any positive real roots.

12. Oscillatory instability may be illustrated by the *free control response* following a pulse input of displacement or force to the steering wheel. Some vehicles will turn first in one direction, and then the other, and so on, until the *amplitude* of the motion increases to the extent that the vehicle "spins out." In this event, the vehicle does not attempt to change its general direction of motion, but does not achieve a *steady-state* condition and has an oscillatory motion. A linear mathematical analog of a vehicle is *oscillatorily unstable* when its characteristic equation has any complex roots with positive real parts.

13. It is possible for a vehicle to be *understeer* for small inputs and *oversteer* for large inputs (or the opposite), as shown in Fig. 4, since it is a nonlinear system and does not have the same characteristics at all *trims.* Consequently, it is necessary to specify the range of inputs and velocities when making a determination of the vehicle's *steer characteristics.*

There is a set of equivalent definitions in terms of *yaw velocity* or curvature (reciprocal of radius of curvature), which can be used interchangeably with these definitions. These definitions only apply to two-axle vehicles, since the *Ackerman steer angle* only applies to two-axle vehicles.

14. *Ackerman Steer Angle Gradient* is equal to the wheelbase divided by the square of the vehicle speed (rad/ft/s^2).

15. For nonlinear steering systems, this ratio should be presented as a function of *steering wheel angle* in order to be compatible with the definition of *understeer/oversteer gradient.*

16. The *roll center* defined in 9.4.28 constitutes an idealized concept and does not necessarily represent a true instantaneous center of rotation of the sprung mass.

17. Illustration applies only for constant *overall steering ratio.* For other steering systems, refer directly to definitions for interpretation of data.

Appendix B
SAE J6a
Ride and Vibration Data Manual

Report of Riding Comfort Research Committee approved July 1946 and last revised by the Vehicle Dynamics Committee October 1965.

Table of Contents

Section 3. Energy Absorption and Impact

Section 4. Vibration Limits for Passenger Comfort

Section 5. Bibliography

Foreword

This is the third edition of Ride and Vibration Data to be issued by the SAE Vehicle Dynamics Committee, formerly the Riding Comfort Research Committee.

The first edition, prepared in 1945, was confined to basic relationships involved in vehicle suspension and impact energy absorption. However, the treatment of vibration did not go beyond the characteristics of undamped simple harmonic motion.

The second edition, issued in 1950, consisted essentially of the original material with the addition of a section on human vibration tolerance.

In this new edition, the editorial subcommittee has attempted to include a graphical presentation of damped vibrating system characteristics, aimed at applications to vehicle ride and vibration problems. A detailed description of the scope of this subject matter is given in the Introduction to Section 2.

SAE Vehicle Dynamics Committee:
William Le Fevre, Chairman

Editorial Subcommittee:
R. N. Janeway, Chairman, Janeway Engineering Co.
William Le Fevre, Le Fevre Co.
W. C. Oswald, Bostrom Research Labs.
John Versace, Ford Motor Co.

1. Basic Relationships

1.1 Acceleration versus Static Deflection

Fig. 1 shows the acceleration of a mass to which a force is applied through a spring. It expresses the relation $A = x/\delta$, based on Newton's second law:

$$a = \frac{xk}{m}$$

$$A = \frac{a}{g} = \frac{xk}{mg} = \frac{x}{\delta}$$

where:

> A = Acceleration, g units
> a = Acceleration, in./sec^2
> g = Acceleration of gravity, 386 in./sec^2
> k = Spring rate, lb/in.
> $m = w/g$ = mass, lb-sec^2/in.
> w = Weight of mass, lb
> $\delta = w/k$ = effective static deflection, in.
> x = Amplitude = deflection of spring from static equilibrium, in.

Referring to Fig. 1A, note that the dynamic deflection (amplitude) of the spring is the relative displacement of the ends of the spring and can be caused by the displacement of the mass or of the support or of both. Also note that the effective static deflection is not necessarily equal to the total loaded deflection. The relations expressed in this chart are not restricted to cyclic vibrations. They apply equally to the acceleration produced by a single sudden displacement. The figure refers to instantaneous values of acceleration and deflection, not to maxima only, for zero damping force.

The scales of Fig. 1B can be changed in conformity with the relation $A = x/\delta$. Either x and δ can be multiplied by the same number, leaving A as given; or A and x can be multiplied by the same number, leaving δ as given.

> EXAMPLE: A vehicle is sprung with 8 in. effective static deflection. The wheel is suddenly pushed up 2 in. What is the resulting acceleration?

> ANSWER: At the intersection of the 8 in. vertical with the 2 in. amplitude curve read 0.25 g acceleration.

416

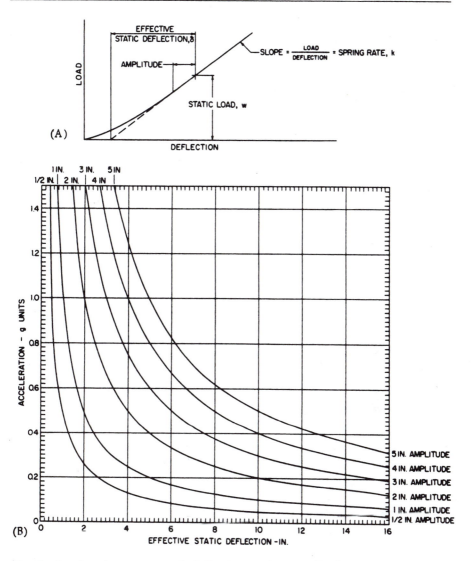

Fig. 1 - Acceleration versus static deflection at given amplitudes

1.2 Undamped Natural Frequency of Sprung Mass*

The fundamental physical quantity which determines the natural vibration frequency of an undamped single degree of freedom system is the acceleration of the suspended mass produced by unit displacement from its static position.

*See Refs. 1, 2, and 3.

417

This is developed to give the basic frequency equation in Fig. 2A. Note that the spring rate must be measured in the direction of the motion and at the static loaded position. The simplified system diagram in Fig. 2A assumes a constant spring rate. However, the effective static deflection is not necessarily equal to the total spring deflection at the static load, as illustrated in Fig. 1A.

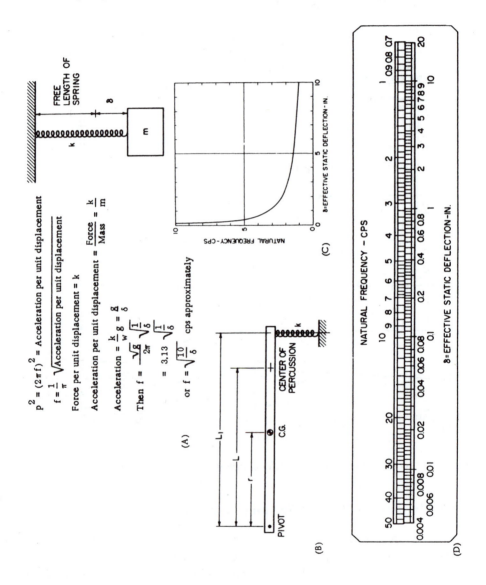

Fig. 2 - Undamped natural frequency versus static deflection

418

Fig. 2C shows the relation between static deflection and natural frequency in ordinary (Cartesian) coordinates; it give a visual impression of the characteristic.

Fig. 2D shows the same relationship by means of fixed logarithmic scales. It can be used to obtain quantitative values with satisfactory accuracy up to 50 cycles per second.

Fig. 2B illustrates another important type of system with single degree of freedom; namely, angular motion about a fixed pivot. In this case the acceleration per unit displacement is:

$$\ddot{\alpha} = \frac{\text{Restoring force moment per radian}}{\text{Mass moment of inertia}}$$

The deflection at the spring per radian $= L_1 =$ Distance from pivot to spring center.

The restoring force at the spring per radian $= kL_1$.

The restoring force moment per radian $= kL_1{}^2$.

The moment of inertia of the mass, m, about the pivot $= m\,(i^2 + r^2)$,

where:

$\quad\quad$ i = Radius of gyration about the c.g.
$\quad\quad$ r = Distance of c.g. from pivot

In this case it is useful to apply the concept of center of percussion which is located at the distance, L, from the pivot such that $L = r + i^2/r$.

Therefore, $r^2 + i^2 = rL$, and the moment of inertia $= mrL$.

Then, the frequency,

$$f = \frac{1}{2\pi}\sqrt{\ddot{\alpha}} \;=\; \frac{1}{2\pi}\sqrt{\frac{kL_1{}^2}{mrL}} \;=\; \frac{\sqrt{g}}{2\pi}\sqrt{\frac{kL_1{}^2}{wrL}}$$

Since the load on the spring, $R = \dfrac{wr}{L_1}$

the static deflection, $\delta = \dfrac{wr}{L_1 k}$

and, $f = \dfrac{\sqrt{g}}{2\pi}\sqrt{\dfrac{L_1}{\delta L}}$

Therefore, the frequency is again dependent on the static spring deflection but modified by the relation of the spring center to the center of percussion. It is important to note, however, that when $L_1 = L$, i.e., spring is located at the center of percussion, the frequency is determined by the static deflection alone, as in the case of rectilinear motion. Figs. 2C and 2D apply also to this special case of angular motion.

1.3 Relations in Simple Harmonic Motion

In simple harmonic motion, the distance, x, a vibrating mass is displaced from the static position at a given time, t, can be approximated closely by the equation:

$$x = x_0 \sin(2\pi f)t = x_0 \sin\omega t$$

where:

x = Instantaneous displacement from static position, in.
x_0 = Maximum displacement from static position (amplitude), in.
t = Time from zero displacement, sec
f = Frequency of oscillation, cps
ω = $2\pi f$, radians per sec

From this equation, the following relationships for peak values can also be derived:

Maximum velocity: $v_m = 2\pi f x_0 = \omega x_0$ in./sec

Maximum acceleration: $a_m = 4\pi^2 f^2 x_0 = \omega^2 x_0$ in./sec^2

Maximum jerk (rate of change of acceleration): $j_m = 8\pi^3 f^3 x_0 = \omega^3 x_0$ in./sec^3

The interrelationships of frequency, velocity, amplitude, and acceleration are shown in Fig. 3.

Any deviation from true simple harmonic motion shows up more and more as higher derivatives of the motion are taken (or, as we go from displacement to velocity, to acceleration, to jerk).

EXAMPLE: A mass has an amplitude of 1 in. at a frequency of 1 cycle/sec. What is the maximum velocity and the maximum acceleration?

ANSWER: Follow the vertical line at frequency 1 to the intersection with the diagonal line for 1 in. amplitude. Reading horizontally to the left hand scale gives approximately 6.2 in./sec maximum velocity. Reading diagonally on the maximum acceleration scale gives approximately 0.1 g.

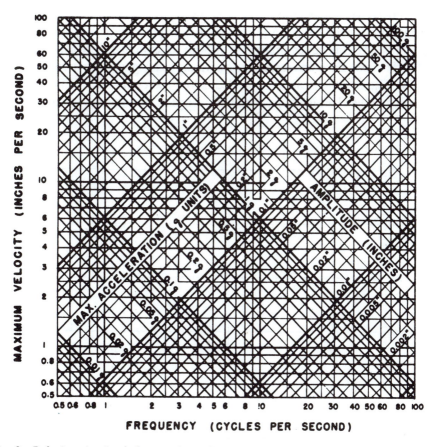

Fig. 3 - Relations in simple harmonic motion

1.4 Resonant Speed on Uniformly Spaced Road Disturbances*

Periodic force impulses, such as are produced by uniformly spaced road joints or washboard surfaces, can excite resonant vibration in a vehicle system if any natural frequency of the system is equal to, or an even multiple of, the impulse frequency. In the latter case, however, the induced vibration is usually of minor intensity because of the greater time available for damping between impulses.

It should be noted that the occurrence of this type of resonance is not related to the time duration of the individual impulse, but is determined by the uniform time period between impulses. Thus, impulse frequency, cpm = mph x 88/spacing (ft).

*See Ref. 5, p. 2-36.

For convenient reference, Fig. 4 has been prepared to relate spacing distance between disturbances and vehicle speed to impulse frequency. The lines are plotted with frequency as ordinate, from 50 to 700 cpm, and speed as abscissa, from 10 to 80 mph. Each line represents a periodic spacing, varying in steps from 3 to 60 ft. The ordinate scale is expanded below 100 cpm to enable more accurate reading of the chart in the passenger car range of sprung mass frequency.

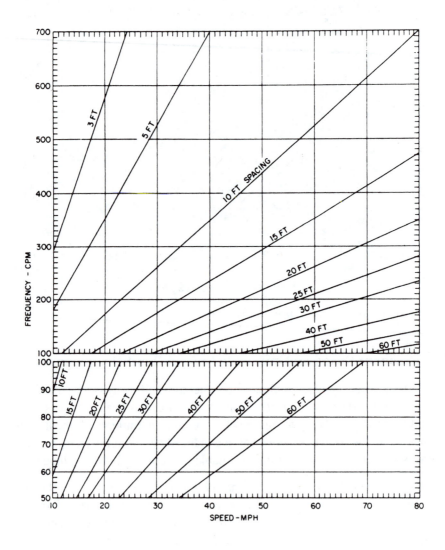

Fig. 4 - Resonant speed on uniformly spaced road disturbances

The following examples of the use of this chart have been selected to bring out some of the situations in which disturbing resonances are commonly experienced (all values are based on resonance with the fundamental of the road disturbance):

(a)　A truck has a natural pitching frequency of 300 cpm of the entire mass on the tires alone, due to friction in the suspension. At what speed will resonance be excited by road joints uniformly spaced at 15 ft?

　　　At intersection of the 15 ft line with the 300 cpm horizontal line, read 50 mph.

(b)　A passenger car has a front end unsprung mass natural frequency of 11.7 cps or 700 cpm. At what speed will unsprung mass resonance occur on a road with 10 ft joint spacing?

　　　At intersection of the 10 ft line with the horizontal line through 700 cpm, read 79.8 mph.

(c)　The natural frequency in vertical bounce of a driver on the seat cushion of a truck is 240 cpm. At what speed is he likely to experience resonant bouncing on the cushion on a road with 20 ft joint spacing?

　　　At intersection of the 240 cpm horizontal with the 20 ft line, read 55 mph.

It will be noted that none of these cases of resonance could occur in the operating speed range with uniform joint spacing of 25 ft or more. Unevenly spaced joints would largely eliminate the possibility of resonance.

2. Vibration Systems

This section presents the basic quantitative relationships in generalized vibration systems with a single degree of freedom. Both free and forced steady-state vibrations are described, including damping effects. Undamped conditions are automatically covered by the same equations, corresponding to zero value of the damping constant.

With the exception of the coulomb damping condition, only linear systems with viscous damping are considered. Such systems must meet the following conditions (refer to SAE J670, Vehicle Dynamics Terminology).

(a)　The mass or masses have only one mode of vibratory motion, along a prescribed path. This is seldom strictly true, but the relations shown are sufficiently accurate if the constraining supports or guides are much stiffer than the elastic restraint along the path of motion.

(b) The elastic or spring restoring forces produced by displacement of the mass must be proportional to the displacement. Even though the spring rate may not be constant over the entire range of load, sufficient accuracy can usually be obtained by taking the average rate over the actual range of displacement.

(c) Since viscous damping is assumed, the damping force, by definition, is proportional to the relative velocity between the points of attachment of the damper.

In each case of forced vibration, a sinusoidal driving force is assumed. All the forces acting upon the system, then, can be represented graphically by rotating vectors. This type of diagram is used throughout to derive the basic relationships. The important advantages of this procedure are:

(a) Differential equations are entirely eliminated.
(b) The reader can readily visualize the mechanics of the vibration.

2.1 List of Symbols

2.1.1 Vibrating System Parameters

m = Mass, lb-sec^2/in.
k = Spring constant, lb/in.
F = Coulomb friction, lb
c = Viscous damping constant, lb-sec^2/in.
c_c = Critical damping constant, lb-sec^2/in.
  = $2\sqrt{km}$
b = Damping ratio
  = c/c_c

2.1.2 Vibratory Motion

t = Time, sec
f = Forcing frequency, cycles/sec
f_n = Undamped natural frequency, cps
f_d = Damped natural frequency, cps
ω = Vector angular velocity at forcing frequency = $2\pi f$, radians/sec

ω_n = Vector angular velocity at undamped natural frequency = $2\pi f_n$, radians/sec

ω_d = Vector angular velocity at damped natural frequency = $2\pi f_d$, radians/sec

a_0 = Amplitude of sinusoidal excitation of spring support, in.

X_0 = Initial mass amplitude in free vibration, in.

X_0' = Mass amplitude after one-half cycle in free vibration, in.

$X_1, X_2,..., X_n$ = Mass amplitude after successive cycles in free vibration, in.

ΔX = Mass amplitude decrement per cycle in coulomb damping, in.

x_0 = Mass amplitude in steady-state forced vibration, in.

x_{st} = Static deflection of mass corresponding to peak force steadily applied, in.

x'_{st} = Static deflection of mass corresponding to peak force at undamped natural frequency, in.

y_0 = Amplitude of relative motion between mass and vibrating spring support, in.

P_m = Peak value of sinusoidal exciting force, lb

P_0 = Peak value of sinusoidal exciting force, at undamped natural frequency, lb

P = Instantaneous variable sinusoidal force, lb

P_t = Peak value of force transmitted to the spring support, lb

ϕ = Phase angle

2.2 Free Vibration with Coulomb Damping*

Coulomb damping is produced when a constant friction force, F, opposes the vibratory motion. Referring to the force-displacement diagram (Fig. 5B), the decrement in amplitude can readily be derived from the energy relationships:

$$\left[\text{The work done in accelerating the mass}\right]^5_1 = k(X_0)^2/2 - FX_0$$

$$\left[\text{The work done in decelerating the mass}\right]^2_5 = k(X_0')^2/2 + FX_0'$$

Since the vibrating mass starts from rest and again comes to rest after each half cycle:

$$k(X_0)^2/2 - FX_0 = kX_0'^2/2 + FX_0'$$

$$(k/2)[X_0^2 - X_0'^2] = F(X_0 + X_0')$$

$$X_0 - X_0' = 2F/k \text{ per 1/2 cycle}$$

*See Ref. 4.

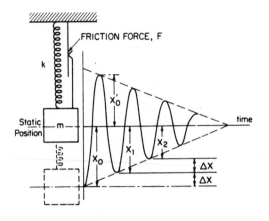

Conditions: initial deflection (upward) = 0.5 in.

Friction, lb = 0.10 x spring rate (k)

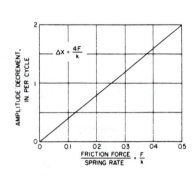

(A) Amplitude decrement

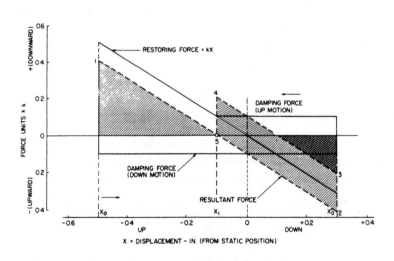

(B) Force - displacement diagram

Fig. 5 - Free vibration with coulomb damping

Thus, in free vibration, coulomb damping produces a constant decrement of amplitude per cycle, $\Delta X = 4F/k$, which depends only on the ratio of friction force to spring rate. Amplitude decrement in inches as a function of the ratio of damping friction force to spring rate is shown in Fig. 5A.

426

Referring again to the force-displacement diagram (Fig. 5B) the damping force is assumed to be 1/10 the spring force at unit-deflection or 1/10 the spring rate, k. The friction force opposes the spring force to produce the resultant force line 1-5-2 which, of course, determines the motion of the vibrating mass. Since the vibrating mass starts from rest, and comes to rest again after each half cycle, the positive work of acceleration must be equal to the negative work of deceleration represented by the shaded areas above and below the baseline. The same condition must be met on the return motion so that at the end of one cycle, the displacement is 1/10 in. (point 4) with a reduction in amplitude of 4/10 in. equal to four times the ratio of friction to spring rate in accordance with the above equation. The mass has, in one cycle, come to rest at a displacement of 1/10 in. because the spring restoring force and the friction force are equal at this position, and velocity is zero.

The force-displacement diagram also shows that the mass will come to rest at zero elongation of the spring, only if the initial displacement is an even multiple of 4F/k. The natural frequency of a free vibration determined by the equation in Fig. 3 is unchanged by friction damping.

2.3 Free Vibration with Viscous Damping*

Viscous damping is produced when the force opposing the motion is proportional to the velocity of the vibrating mass relative to the spring support. This type of damping in free vibration forms an amplitude decrement in which the amplitude between any two consecutive cycles is reduced by a constant ratio, according to the equation:

$$\frac{X_1}{X_0} = \frac{X_2}{X_1} = \frac{X_{(n+1)}}{X_n} = e^{-2\pi b/\sqrt{1-b^2}}$$

where:

$b = c/c_c$ = Damping factor which is the ratio of damping constant to the critical damping value, $c_c = 2\sqrt{km}$

$\dfrac{X_{(n+1)}}{X_n}$ = Amplitude ratio between any two consecutive cycles

---

*See Ref. 4.

A typical displacement trace is illustrated in the graph in Fig. 6A. The amplitude decrement ratio as a function of percent of critical damping is shown plotted in the curve 6A.

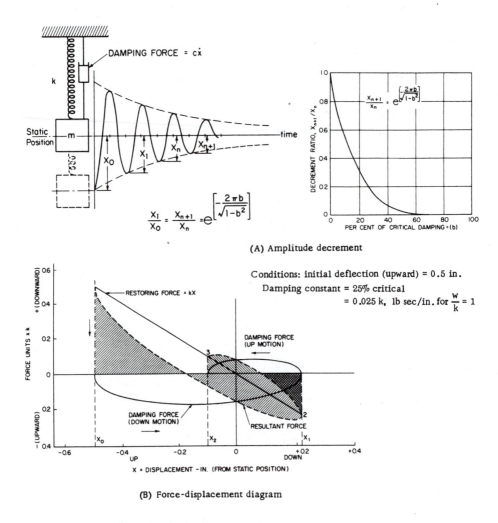

(A) Amplitude decrement

Conditions: initial deflection (upward) = 0.5 in.
Damping constant = 25% critical
= 0.025 k, lb sec/in. for $\frac{w}{k} = 1$

(B) Force-displacement diagram

Fig. 6 - Free vibration with viscous damping

Fig. 6B shows a force-displacement diagram for free vibration with viscous damping at 25% of critical. The damping force, in this case, is variable in contrast to a constant force for friction damping. The resultant force curve meets the condition that the work of acceleration equals the work of deceleration represented by the shaded areas in the diagram. Damping force is a

428

maximum at maximum velocity of the vibrating mass and is zero at points of maximum displacement where velocity is zero. Therefore, the maximum resultant deceleration force acting on the mass is out of phase with the maximum displacement.

After one complete cycle, the mass has momentarily come to rest at point 3, but oscillation will continue with a constant ratio in reduction of amplitude until displacement is small enough for extraneous friction to stop the motion.

Viscous damping reduces the natural frequency (f_d) below that of undamped free vibration (f_n) according to the equation:

$$f_d = \frac{1}{2\pi} \sqrt{\frac{k(1 - b^2)}{m}} \quad \text{or} \quad \frac{f_d}{f_n} = \sqrt{1 - b^2}$$

The effect on frequency is slight in the ordinary damping range. For example, if $b = 0.25$ (25% of critical damping) the frequency is reduced only 3%. However, when $b = 1$ (critical damping) the frequency becomes zero, indicating the absence of vibratory motion. Instead, the mass, when displaced, will gradually creep back to its neutral position without a change in direction of the spring force.

2.4 Forced Vibration with Viscous Damping*

2.4.1 Force Applied to Suspended Mass

2.4.1.1 Peak Exciting Force Constant — This case is represented by the system diagram (Fig. 7A) in which the suspended mass, m, is driven directly by a sinusoidal force, and is also acted upon by a viscous damper interposed between the mass and the spring support.

If the frequency of the driving force is f, and its maximum value is P_m, the generalized force can be written $P = P_m \sin(\omega t + \phi)$, where $\omega = 2\pi f$ and $\phi =$ phase angle by which maximum force leads the maximum displacement. The variable driving force can be graphically generated by a rotating vector of magnitude P_m, having a constant angular velocity $= 2\pi f$, as in the Fig. 7B. Then, the vector component along the coordinate of vibratory motion at any instant equals the relative driving force. For steady-state vibration, the spring restoring force, inertia force, and damping force can also be represented by

*See Ref. 1, p. 48.

similar vectors having the proper relative magnitudes and phase relations. Note that the damping force, being proportional to, and in phase with, the mass velocity, is represented by a vector that lags the displacement by 90 deg.

For simplicity in deriving the equations, the vector diagram is drawn for the instant when the mass displacement is at its maximum positive value, $(\omega t = \pi/2)$. Then for force equilibrium, the sum of vertical vector components and sum of horizontal vector components must each equal zero. Therefore,

Sum of vertical components

$$= kx_0 - m\omega^2 x_0 - P_m \cos\phi = 0 \tag{1}$$

Sum of horizontal components

$$= c\omega x_0 - P_m \sin\phi = 0 \tag{2}$$

In vertical force equation and Fig. 7B, note that inertia force is always in phase with displacement, while spring restoring force is opposed to displacement.

Combining Eqs. 1 and 2:

$$x_0 = \frac{P_m/k}{\sqrt{\left(1 - \frac{\omega^2}{\omega_n^2}\right)^2 + \left(2\frac{c}{c_c}\frac{\omega}{\omega_n}\right)^2}} \tag{3}$$

Since P_m/k is the deflection x_{st}, corresponding to the maximum force if steadily applied, the magnification factor is:

$$\frac{x_0}{x_{st}} = \frac{1}{\sqrt{\left(1 - \frac{\omega^2}{\omega_n^2}\right)^2 + \left(2\frac{c}{c_c}\frac{\omega}{\omega_n}\right)^2}} \tag{4}$$

We may consider the undamped condition a special case in which $c = 0$. Then:

$$\frac{x_0}{x_{st}} = \frac{1}{1 - (\omega/\omega_n)^2} \tag{5}$$

Eq. 4 is the generalized magnification factor and is independent of the

430

magnitude of the applied force. Values of this factor are plotted in Fig. 7C against frequency ratio, for various constant damping intensities, expressed as the ratio, b, to the critical value.

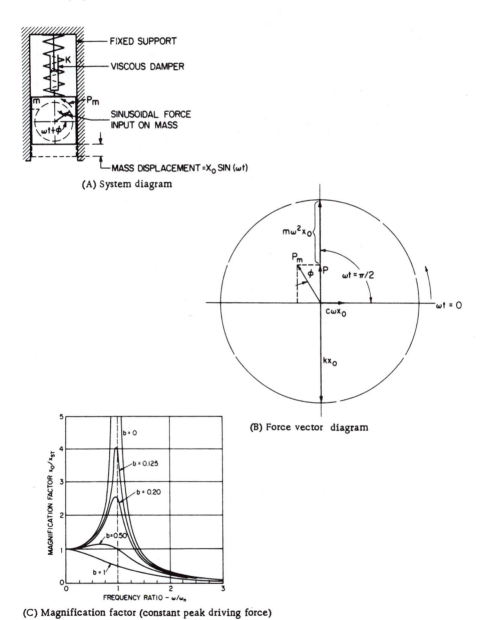

(A) System diagram

(B) Force vector diagram

(C) Magnification factor (constant peak driving force)

Fig. 7 - Forced vibration with viscous damping (driving force applied to suspended mass)

431

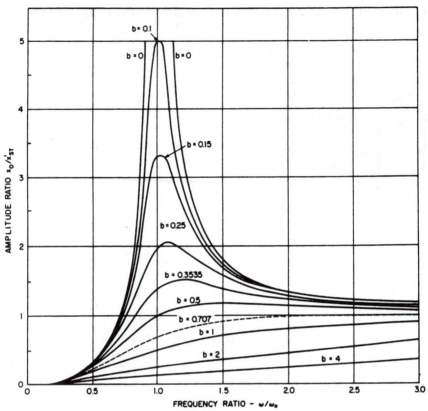

(D) Amplitude ratio when peak driving force is proportional to $(\omega/\omega_n)^2$

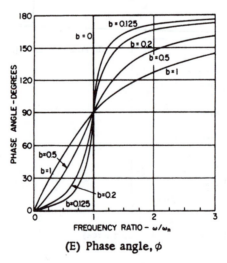

(E) Phase angle, ϕ

Fig. 7 - Continued

2.4.1.2 <u>Exciting Force Increasing as Square of Frequency</u> — Note that the magnification factor curves also define the amplitude ratio for the specific case where the maximum force remains constant as the frequency varies. However, another important type of forced vibration encountered in practice is that in which the maximum exciting force increases as the square of frequency. The most common situation of this kind occurs in machines having unbalanced rotating and/or reciprocating masses.

In this case, the maximum force may be written $P_m = P_0(\omega/\omega_n)^2$, where P_0 = maximum force at resonance, $(\omega/\omega_n = 1)$. Substituting for P_m in Eq. 3, we get for amplitude ratio:

$$\frac{x_0}{x'_{st}} = \frac{(\omega/\omega_n)^2}{\sqrt{\left(1 - \frac{\omega^2}{\omega_n^2}\right)^2 + \left(2b\frac{\omega}{\omega_n}\right)^2}}$$

where:

$$x'_{st} = P_0/k.$$

Fig. 7D shows how the amplitude ratio varies with frequency for different damping intensities, in this case. It will be seen that, regardless of damping, the amplitude ratio is zero at zero frequency and approaches unity at infinite frequency.

Since ω_n is the undamped natural frequency by definition, resonance occurs at $\omega/\omega_n = 1$ only for very low damping values. It will be observed in Figs. 7C and 7D that the resonant frequency (corresponding to peak amplitude) departs progressively further from the undamped resonant frequency as damping intensity increases. Table 1 summarizes the equations which define the amplitude ratio at nominal resonance $(\omega/\omega_n = 1)$, the maximum amplitude, and the frequency of maximum amplitude as functions of damping ratio, for the two cases considered previously.

Referring again to Eqs. 1 and 2, the phase angle, ϕ, is readily derived by evaluating $\sin \phi$ and $\cos \phi$, and dividing $(\sin \phi)$ by $(\cos \phi)$. (x_0) drops out and we get:

$$\tan\phi = \frac{2b\,\omega/\omega_n}{1 - (\omega^2/\omega_n^2)}$$

433

Table 1

Force Condition	Amplitude Ratio at $\omega/\omega_h = 1$	Amplitude Ratio, Max	$\dfrac{\omega}{\omega_h}$ at Max Amplitude	Range of Damping Ratio
P_m = const.	$1/2b$	$\dfrac{1}{2b\sqrt{1-b^2}}$	$\sqrt{1-2b^2}$	$b < .707$
		1	0	$b \geq .707$
$P_m = P_0\left(\dfrac{\omega}{\omega_n}\right)^2$	$1/2b$	$\dfrac{1}{2b\sqrt{1-b^2}}$	$\dfrac{1}{\sqrt{1-2b^2}}$	$b < .707$
		1	∞	$b \geq .707$

Note that the phase angle is independent of the magnitude of the driving force, so that the curves of Fig. 7E apply equally to both cases considered. At $\omega/\omega_n = 1$, ϕ becomes 90 deg regardless of the degree of damping.

For values of ω/ω_n greater than 1, tan ϕ becomes negative, corresponding to phase angles between 90 and 180 deg.

2.4.1.3 Equivalent Impedance — The impedance concept, as applied to mechanical vibrating systems, expresses the ratio of applied force to the induced motion at the point of force application. Thus, displacement impedance is the force required per unit of peak displacement; velocity impedance is the force required per unit of peak velocity.

The term "mobility" is used to designate the reciprocal of impedance. Thus, displacement mobility is the peak displacement per unit of driving force.

These quantities are obviously variable with frequency ratio, but are independent of the magnitude of the driving force.

For the case of vibration excited by a force applied directly to a suspended mass, the impedance and mobility functions become:

Displacement Impedance, $Z_d = P_m/x_0$

Velocity Impedance, $Z_v = P_m/x_0\omega$

Substituting the value of x_0 as a function of frequency ratio (Eq. 3):

$$x_0 = \frac{P_m}{k \sqrt{\left[1-\left(\frac{\omega}{\omega_n}\right)^2\right]^2 + \left(2b\frac{\omega}{\omega_n}\right)^2}}$$

$$Z_d = k \sqrt{\left[1-\left(\frac{\omega}{\omega_n}\right)^2\right]^2 + \left(2b\frac{\omega}{\omega_n}\right)^2}$$

$$Z_v = \frac{k}{\omega} \sqrt{\left[1-\left(\frac{\omega}{\omega_n}\right)^2\right]^2 + \left(2b\frac{\omega}{\omega_n}\right)^2}$$

Likewise, Displacement Mobility,

$$M_d = \frac{x_0}{P_m} = \frac{1}{k \sqrt{\left[1-\left(\frac{\omega}{\omega_n}\right)^2\right]^2 + \left(2b\frac{\omega}{\omega_n}\right)^2}}$$

Velocity Mobility,

$$M_v = \frac{\omega x_0}{P_m} = \frac{\omega}{k \sqrt{\left[1-\left(\frac{\omega}{\omega_n}\right)^2\right]^2 + \left(2b\frac{\omega}{\omega_n}\right)^2}}$$

Note that the displacement mobility varies with frequency in exactly the same way as the peak displacement for a constant harmonic force intensity.

2.4.2 Excitation Applied to Spring Support

2.4.2.1 Absolute Amplitude Ratio - When the system shown in Fig. 8A is excited by sinusoidal vibration of the spring support at a constant amplitude,

a_0, instead of by a force applied directly to the suspended mass, the absolute amplitude of the mass is obtained by deriving the equivalent force acting on the mass. It is important to note, in this case, that the spring force component due to the support motion acts on the mass in the same direction as the support displacement, as opposed to the spring force due to mass displacement.

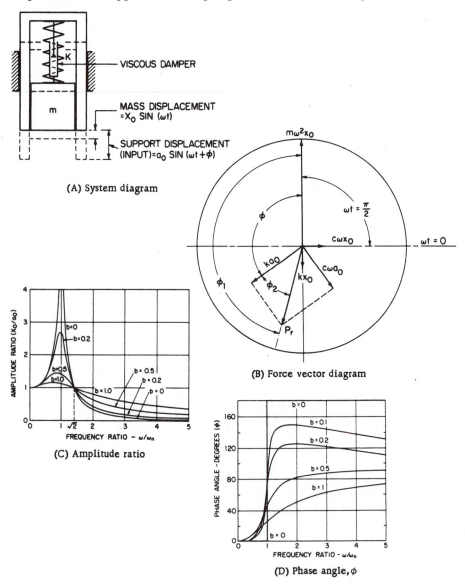

(A) System diagram

(B) Force vector diagram

(C) Amplitude ratio

(D) Phase angle, ϕ

Fig. 8 - Forced vibration of spring support at constant amplitude (absolute amplitude ratio)

436

Referring to the vector diagram Fig. 8B (for the condition $\omega/\omega_n = 2$, $b = 0.2$), it will be seen that the rotating force vector, P_m, of Fig. 7B has been replaced by two rotating force vectors at 90 deg to each other, ka_0 and $c\omega a_0$, corresponding respectively to the spring force and to the damping force components of the exciting amplitude. The resultant of these forces is:

$$P_r = \sqrt{(ka_0)^2 + (c\omega a_0)^2} = a_0 k \sqrt{1 + \left(2b\frac{\omega}{\omega_n}\right)^2}$$

and the absolute amplitude of the mass, x_0, is determined by the same magnification factor as in Eq. 4, paragraph 2.4.1.1.

$$x_0 = P_r/k \quad \text{(M.F.)}$$

Substituting in this expression the equation for P_r, the magnification factor yields:

$$\frac{x_0}{a_0} = \sqrt{\frac{1 + \left(2b\frac{\omega}{\omega_n}\right)^2}{\left(1 - \frac{\omega^2}{\omega_n^2}\right)^2 + \left(2b\frac{\omega}{\omega_n}\right)^2}}$$

Referring to Fig. 8C, the values of this amplitude ratio are plotted for various damping intensities over a range of frequency ratios. It is significant to note that damping is beneficial in reducing amplitude only for frequency ratios lower than $\sqrt{2}$. At higher ratios the amplitude increases progressively with the damping intensity. However, with zero damping the amplitude ratio reduces to the same equation as that for constant force excitation applied directly to the mass (Eq. 4). Thus, the curves for $b = 0$ are identical in Figs. 7C and 8C.

The general expression for the phase angle between the mass and support motions is derived as follows (refer to Fig. 8B):

ϕ_1 = Phase angle between P_r and x_0

ϕ_2 = Phase angle between P_r and a_0

ϕ = Phase angle between a_0 and $x_0 = \phi_1 - \phi_2$

$$\tan \phi_1 = \frac{2b \left(\frac{\omega}{\omega_n}\right)}{1 - \left(\frac{\omega}{\omega_n}\right)^2} \qquad \text{(See paragraph 2.4.1.2)}$$

$$\tan \phi_2 = \frac{ca_0\omega}{a_0 k} = \frac{c\omega}{k} = 2b \left(\frac{\omega}{\omega_n}\right) \qquad \text{(See Fig. 8B)}$$

$$\tan \phi = \tan (\phi_1 - \phi_2) = \frac{\tan \phi_1 - \tan \phi_2}{1 + \tan \phi_1 \tan \phi_2}$$

Substituting,

$$\tan \phi = \frac{2b \left(\frac{\omega}{\omega_n}\right)^3}{1 - \left(\frac{\omega}{\omega_n}\right)^2 + \left(2b \frac{\omega}{\omega_n}\right)^2}$$

This equation is shown plotted in Fig. 8D for various frequency ratios and for different damping intensities.

The frequency ratio at which maximum amplitude occurs is given by the expression:

$$\left(\frac{\omega}{\omega_n}\right)_m = \frac{1}{2b} \sqrt{\sqrt{1 + 8b^2} - 1}$$

For finite values of b, $(\omega/\omega_n)_m$ is less than 1, but approaches unity as b approaches zero.

The maximum amplitude ratio for a given damping intensity, b, may be determined directly from the relation:

$$\left(\frac{x_0}{a_0}\right)_m = \sqrt{\frac{\sqrt{1+8b^2}}{\left\{1-\frac{1}{4b^2}\left[\sqrt{1+8b^2}-1\right]\right\}^2 + \sqrt{1+8b^2}-1}}$$

For small values of b (less than 0.15) this equation can be reduced to the approximate form:

$$\left(\frac{x_0}{a_0}\right)_m = \sqrt{\frac{\sqrt{1+8b^2}}{\sqrt{1+8b^2}-1}}$$

2.4.2.2 Relative Amplitude Ratio — Fig. 9 illustrates the method of deriving the equations for the relative motion between vibrating mass and support in the same system as in Fig. 8, where the excitation is applied to the support. Knowledge of the relative motion is needed in such important practical problems as designing vibration instruments and providing for clearance between components of a vibrating system.

Referring to the vector diagram in Fig. 8B, it can be seen that the separate spring and damping force vectors associated with the respective motions of mass and support can be combined into resultants as shown in Fig. 9B, in terms of the relative motion, y, between the mass and support. The resultant spring and damping forces are necessarily in 90 deg phase relationship, while the inertia force on the mass remains in phase with the absolute mass displacement. In Fig. 9C, the complete force vector diagram is seen to reduce to the simple form in which the only forces acting on the mass are the inertia force due to its absolute motion, the spring restoring force due to the relative displacement between mass and support, and the damping force due to the relative velocity between the mass and support. It is clear that for equilibrium the resultant between the spring and damping forces must be equal and opposite to the inertia force. The equations are readily derived as follows:

Denoting the phase angle between the relative motion and absolute motion as ϕ_3 and summing up the vertical and horizontal forces when the relative displacement is a maximum (y_0), (when $\omega t = \pi/2$):

439

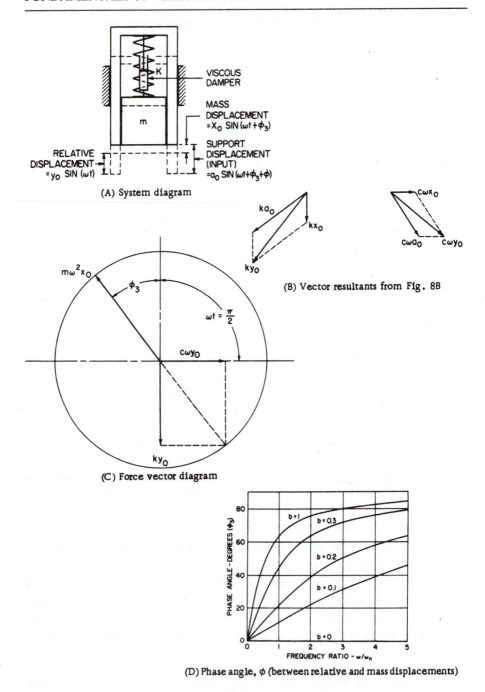

(A) System diagram

(B) Vector resultants from Fig. 8B

(C) Force vector diagram

(D) Phase angle, ϕ (between relative and mass displacements)

Fig. 9 - Forced vibration of spring support at constant amplitude (relative motion between mass and support)

440

Sum of the vertical forces

$$-m\omega^2 x_0 \cos \phi_3 + k y_0 = 0$$

therefore:

$$\cos \phi_3 = \frac{k y_0}{m\omega^2 x_0}$$

Sum of the horizontal forces

$$-m\omega^2 x_0 \sin \phi_3 + c\omega y_0 = 0$$

therefore:

$$\sin \phi_3 = \left[\frac{c\omega y_0}{m\omega^2 x_0} \right]$$

Since $\sin^2 \phi_3 + \cos^2 \phi_3 = 1$:

$$\left[\frac{c\omega y_0}{m\omega^2 x_0} \right]^2 + \left[\frac{k y_0}{m\omega^2 x_0} \right]^2 = 1$$

$$y_0^2 = \frac{(x_0 m\omega^2)^2}{k^2 [1 + (c\omega/k)^2]}$$

$$y_0 = \frac{x_0 \omega^2/\omega_n^2}{\sqrt{1 + (c\omega/k)^2}}$$

From paragraph 2.4.2.1, the absolute amplitude ratio is:

$$\frac{x_0}{a_0} = \sqrt{\frac{1 + \left(2b \dfrac{\omega}{\omega_n} \right)^2}{\left(1 - \dfrac{\omega^2}{\omega_n^2} \right)^2 + \left(2b \dfrac{\omega}{\omega_n} \right)^2}}$$

so:

$$\frac{y_0}{a_0} = \frac{\omega^2/\omega_n^2}{\sqrt{(1 - \omega^2/\omega_n^2)^2 + (2b\omega/\omega_n)^2}}$$

Note that the expression for relative amplitude ratio is identical with that in paragraph 2.4.1.2, when the mass is excited directly by a force varying as the square of the frequency ratio. Thus, the curves of Fig. 7D apply to this case.

$$\tan \phi_3 = \sin \phi_3 / \cos \phi_3 = c\omega/k = 2b \, \omega/\omega_n$$

This phase angle relationship is shown in Fig. 9D for various damping intensities and frequency ratios.

2.5 Forced Vibration with Coulomb Damping

As pointed out earlier in this section, a system with coulomb (constant friction) damping is nonlinear and the solution of the differential equations becomes complicated. Den Hartog (Ref. 1) has arrived at an exact solution, which gives amplification factors and phase angles as shown in Fig. 10A and B, for specified ratios of friction to maximum driving force (F/P_0), over a range of frequency ratios (ω/ω_n).

Den Hartog points out that the dotted line in Fig. 10A defines the limits below which the vibratory motion is not continuous, but involves a "stop" or dwell every half-cycle. His analysis also emphasizes an important limitation of coulomb damping in that it will not restrict the amplitude of resonant vibration unless the friction force is greater than $\pi P_0/4$. This can be seen from the fact that the work done by a harmonic driving force at resonance is equal to $\pi P_0 x/4$, whereas the work done by the constant friction force, F, is equal to Fx. Consequently, if F is less than $\pi P_0/4$, the amplitude will increase on successive cycles to infinity. Coulomb damping is not applicable to a system which is subjected to steady-state vibration at resonance unless the maximum value of the driving force has a known limitation.

An approximate solution for this case is also given in the cited reference, but its use is not recommended because of the very limited range of conditions for which it is valid.

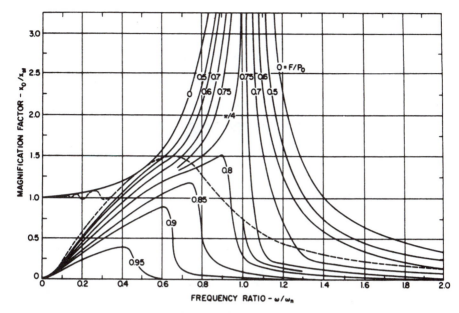

(A) Magnification factor

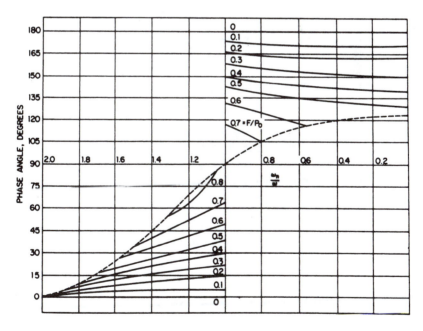

(B) Phase angle

Fig. 10 - Forced vibration with coulomb damping

An interesting approximate solution for the use of combined viscous and friction damping is given by Timoshenko (Ref. 3, p. 96). This solution is applicable when the friction damping is small relative to the viscous component.

2.6 Force Transmission Through Suspension

2.6.1 Direct Excitation of Mass with Viscous Damping

2.6.1.1 Constant Peak Driving Force — From the standpoint of vibration isolation, when a directly excited mass is flexibly mounted, we are interested in the magnitude of the force transmission to the support. Thus, referring to the force vector diagram Fig. 7B for the forced vibration system with constant maximum driving force and viscous damping, it will be seen that the spring force, kx_0, and damping force $c\omega x_0$, both react on the support. Since the two force components are 90 deg out of phase, their vector resultant as shown in Fig. 11A is

$$P_t = \sqrt{(x_0 k)^2 + (c\omega x_0)^2} = x_0 k \sqrt{1 + (c\omega/k)^2}$$

$$= x_0 k \sqrt{1 + (2b\omega/\omega_n)^2}$$

Since (from Eq. 3, paragraph 2.4.1.1):

$$x_0 = \frac{P_m/k}{\sqrt{\left(1 - \frac{\omega^2}{\omega_n^2}\right)^2 + \left(2b\frac{\omega}{\omega_n}\right)^2}}$$

$$P_t = \frac{P_m \sqrt{1 + (2b\omega/\omega_n)^2}}{\sqrt{\left(1 - \frac{\omega^2}{\omega_n^2}\right)^2 + \left(2b\frac{\omega}{\omega_n}\right)^2}}$$

This resultant is obviously the maximum force transmitted to the support when $\omega t = \tan^{-1} k/c\omega = \tan^{-1} 1/2b \, (\omega/\omega_n)$. This is the position shown in Fig. 11.

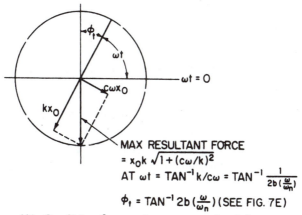

MAX RESULTANT FORCE

$= x_0 k \sqrt{1 + (c\omega/k)^2}$

AT $\omega t = \text{TAN}^{-1} k/c\omega = \text{TAN}^{-1} \dfrac{1}{2b(\frac{\omega}{\omega_n})}$

$\phi_t = \text{TAN}^{-1} 2b(\dfrac{\omega}{\omega_n})$ (SEE FIG. 7E)

(A) Condition for maximum transmitted force

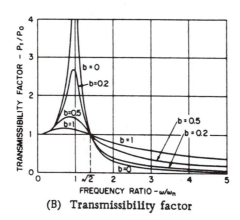

(B) Transmissibility factor

Fig. 11 - Force transmission through suspension
CASE 1: Constant maximum driving force on mass, with viscous damping

Thus, the transmissibility, defined as the ratio of maximum transmitted force to driving force is,

$$\frac{P_t}{P_m} = \sqrt{\frac{1 + (2b\omega/\omega_n)^2}{\left(1 - \dfrac{\omega^2}{\omega_n^2}\right)^2 + \left(2b \dfrac{\omega}{\omega_n}\right)^2}}$$

This equation is shown plotted in Fig. 11B and will be seen to be identical with that for absolute amplitude ratio, when a constant vibration amplitude is applied to the support, Fig. 8C.

It will be noted that damping reduces the transmissibility for frequency ratios less than $\sqrt{2}$ but increases it for higher ratios. Thus, the undamped system has smaller transmissibility for frequency ratios larger than $\sqrt{2}$.

The force vector diagram, Fig. 11A, shows that the transmitted force reaches its maximum value at $\omega t = \tan^{-1} 1/2b(\omega/\omega_n)$, and at 180 deg intervals thereafter. Since the mass displacement is a maximum at $\omega t = \pi/2$ or $3\pi/2$, the phase angle, ϕ_t, by which the transmitted force leads the mass displacement is:

$$\phi_t = \frac{\pi}{2} - \tan^{-1} \frac{1}{2b\left(\dfrac{\omega}{\omega_n}\right)} = \tan^{-1} 2b\left(\frac{\omega}{\omega_n}\right)$$

This is identical with the phase relationship previously derived between relative displacement and mass displacement under constant amplitude excitation of the support. Therefore, the curves of Fig. 9D also define the value of ϕ_t, as a function of frequency ratio and damping intensity.

2.6.1.2 <u>Driving Force on Mass Proportional to Square of Speed with Viscous Damping</u>* — Instead of a constant driving force, let the driving force on the mass increase with the square of the frequency ratio so that $P = P_0 (\omega/\omega_n)^2$, where P_0 is the maximum driving force at resonance ($\omega/\omega_n = 1$).

Although the transmissibility (ratio of transmitted force to applied force) remains the same as in Case 1 (Fig. 11), the absolute value of transmitted force obviously varies in the same ratio as the applied force, or as $(\omega/\omega_n)^2$. The transmitted force thus becomes

$$P_t = P_0 (\omega/\omega_n)^2 \times \text{transmissibility}$$

$$\frac{P_t}{P_0} = \left(\frac{\omega}{\omega_n}\right)^2 \sqrt{\frac{1 + (2b\omega/\omega_n)^2}{\left(1 - \dfrac{\omega^2}{\omega_n^2}\right)^2 + \left(2b\dfrac{\omega}{\omega_n}\right)^2}}$$

*See Ref. 2.

Fig. 12 shows the ratio of force transmission, P_t/P_0, plotted against frequency ratio, ω/ω_n, for various damping intensities. Comparison with Fig. 7D shows that the amplitude gives no indication of the detrimental effect of even a small amount of viscous damping at frequency ratios greater than $\sqrt{2}$, when the applied force is a function of the square of frequency. It is evident that at high values of ω/ω_n, the force transmitted to the support is primarily through the damper.

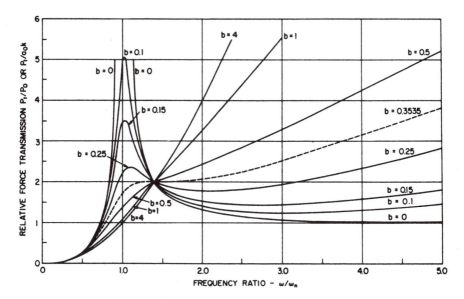

Fig. 12 - Transmission through suspension
CASE 2: Driving force on mass proportional to square of speed (T.F. = P_t/P_0) or spring support excited at constant amplitude (T.F. = P_t/a_0k)

In the case of vibration excited by oscillation of the spring support at constant amplitude, the transmitted force vector is likewise equal to the resultant of the relative spring and damping force vectors. In Fig. 9C this resultant, P_t, is shown to be equal to the inertia force, $m\omega^2 x_0$. By substituting for x_0, its value as derived in paragraph 2.4.2.1, we get:

$$P_t = m\omega^2 a_0 \sqrt{\frac{1 + (2b\omega/\omega_n)^2}{\left(1 - \dfrac{\omega^2}{\omega_n^2}\right)^2 + \left(2b\dfrac{\omega}{\omega_n}\right)^2}}$$

447

The maximum force transmission, relative to $(a_0 k)$ as nominal input force, becomes:

$$\frac{P_t}{a_0 k} = \frac{\omega^2}{\omega_n^2} \sqrt{\frac{1 + (2b\omega/\omega_n)^2}{\left(1 - \frac{\omega^2}{\omega_n^2}\right)^2 + \left(2b\frac{\omega}{\omega_n}\right)^2}}$$

It is evident that the force reaction formula so derived is identical with the formula given above for directly applied excitation to the suspended mass variable as the square of the frequency. Therefore, the curves of Fig. 12 apply equally when the spring support is excited at constant amplitude. The corresponding maximum acceleration of the sprung mass is:

$$A_m = \frac{P_t}{m} = a_0 \omega^2 \sqrt{\frac{1 + (2b\omega/\omega_n)^2}{\left(1 - \frac{\omega^2}{\omega_n^2}\right)^2 + \left(2b\frac{\omega}{\omega_n}\right)^2}}$$

This relationship is fundamental to the ride problem, since it covers the condition of a vehicle traversing a series of periodic sine wave irregularities of constant amplitude at variable speed.

2.6.1.3 Equivalent Impedance — When excitation is applied to the spring support, the input force must be an equal reaction to the force transmitted to the mass. The impedance relationships, thus, become:

(Displacement)

$$Z_d = \frac{P_t}{a_0} = m\omega^2 \sqrt{\frac{1 + (2b\omega/\omega_n)^2}{\left(1 - \frac{\omega^2}{\omega_n^2}\right)^2 + \left(2b\frac{\omega}{\omega_n}\right)^2}}$$

(Velocity)

$$Z_v = \frac{P_t}{a_0\omega} = m\omega \sqrt{\frac{1 + (2b\omega/\omega_n)^2}{\left(1 - \frac{\omega^2}{\omega_n^2}\right)^2 + \left(2b\frac{\omega}{\omega_n}\right)^2}}$$

The corresponding mobility equations, as reciprocals, are:

$$M_d = \frac{1}{m\omega^2} \sqrt{\frac{\left(1 - \frac{\omega^2}{\omega_n^2}\right)^2 + \left(2b\frac{\omega}{\omega_n}\right)^2}{1 + (2b\omega/\omega_n)^2}}$$

$$M_v = \frac{1}{m\omega} \sqrt{\frac{\left(1 - \frac{\omega^2}{\omega_n^2}\right)^2 + \left(2b\frac{\omega}{\omega_n}\right)^2}{1 + (2b\omega/\omega_n)^2}}$$

It is important to note that the above equations are entirely different from the corresponding expressions in paragraph 2.4.1.3 for the case of driving force applied directly to the mass. This comparison brings out the wide variability of the impedance characteristic of a vibrating system, depending on where the driving force is applied.

2.6.2 Comparison of Viscous and Coulomb Damping*

To compare the effects of viscous and coulomb damping, assume the following alternative conditions (driving force on mass assumed proportional to square of frequency, $P_m = P_0(\omega/\omega_n)^2$ in both cases):

(a) The friction force of coulomb damping is equal to the driving force at resonance ($\omega/\omega_n = 1$), $F = P_0$

*Ref. 4

(b) Viscous damping is 25% of critical

These conditions are chosen to give about the same transmissibility in the resonant frequency range.

In Fig. 13A, the undamped and viscous damping curves have been transposed from Fig. 7D. The coulomb damping curve is obtained by interpolation from the curves of Chart 10A, as follows:

Up to $\omega/\omega_n = 1$, for $F/P_0 = 1$, amplitude $= 0$.

Taking $F = P_0$:

$$\text{For } \frac{\omega}{\omega_n} > 1, \ P_m = P_0 \left(\frac{\omega}{\omega_n} \right)^2 \text{ and } \frac{F}{P_m} = \frac{P_0}{P_0 \left(\frac{\omega}{\omega_n} \right)^2} = \left(\frac{\omega_n}{\omega} \right)^2$$

From Fig. 10A, value of x_0/x_{st} is obtained for a given value of ω/ω_n at intersection with curve corresponding to $F/P_0 = F/P_m = (\omega_n/\omega)^2$.

$$\text{Since } x'_{st} = P_0/k, \ x_{st} = P_m/k = (\omega/\omega_n)^2 \ P_0/k = (\omega/\omega_n)^2 \ x'_{st}$$
$$\text{Then } x_0/x'_{st} = (x_0/x_{st}) \ (\omega/\omega_n)^2$$

At frequency ratios above 2.5, there is no difference in displacement amplitude for either viscous or coulomb damping or for no damping. This comparison gives no indication of the force transmission to the support through the suspension, especially for viscous damping.

As with viscous damping, the same curves apply to vibration excited by sinusoidal motion of the support at constant amplitude a_0. In this case, however, the ordinate scale gives amplitude ratio in terms of (y_0/a_0), where y_0 is the relative displacement amplitude between mass and support (see paragraph 2.4.2.2).

Fig. 13B shows the comparative force transmission. For direct excitation of the mass, the ordinate scale defines the force transmitted to the support in terms of the ratio P_t/P_c; for excitation of the support the ordinate scale gives the ratio $P_t/a_0 k$, as in Fig. 12. It can be seen that the undamped suspension, although capable of infinite amplitude at resonance, gives the lowest force transmission at high frequency ratios.

With viscous damping at 25% of critical, the force transmission at high frequencies increases continuously above a frequency ratio of about 2.0.

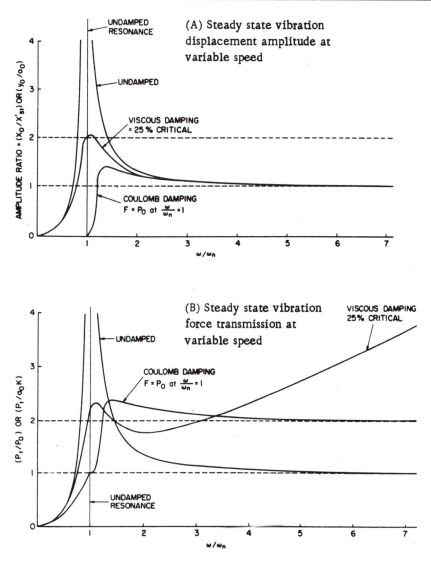

Fig. 13 - Comparison of viscous and coulomb damping on displacement amplitude and force transmission

With the specified coulomb damping, the maximum force transmission at resonance is nearly the same as for viscous damping but thereafter reduces continuously with increase in frequency ratio, approaching $2P_0$ as a limit. Thus at high frequency ratios, coulomb damping adds to the force transmission only by the amount of the damping friction force.

451

In general, it can be said that the optimum damping characteristic for any given suspension is a compromise between a minimum penalty of force transmission and the required amplitude control.

2.6.3 Vibrating System with Two Degrees of Freedom

The previous section is confined to consideration of systems having a single degree of freedom. A great many practical vibration problems, however, involve two degrees of freedom. In keeping with the definition in the introduction to Section 2, this means that the system either contains two masses, each capable of a single type of independent motion, or a single mass capable of two independent types of motion.

A primary example of the latter configuration, which is basic to ride problems, is a vehicle mass having two spaced spring supports. Considering only degrees of freedom in the longitudinal plane, the mass has two independent modes of motion, namely, vertical displacement of the center of gravity and angular displacement about the center of gravity.

In order to present in a useful form the effects of the principal parameters, the standard equations for frequencies and centers of oscillation (Ref. 3, p. 201) have been reduced to a few dimensionless terms.

Thus, if r = ratio of static spring deflection at the supports = δ_2/δ_1 where $\delta_2 \geq \delta_1$

$$d = i^2/AB = \text{dynamic index}$$

where:

$\quad\quad i\ =$ Radius of gyration about c.g.
$\quad\quad A =$ Distance of c.g. from spring support at δ_1
$\quad\quad B =$ Distance of c.g. from spring support at δ_2
$\quad\quad a\ =$ A/(A+B)

Then, if

$\quad\quad f_1 =$ Bounce frequency (about center outside of spring supports), cycles per second
$\quad\quad f_2 =$ Pitch frequency (about center between spring supports), cycles per second
$\quad\quad$ m/B = Relative distance from c.g. to pitch center ((-) toward δ_2)
$\quad\quad$ n/A $\ =$ Relative distance from c.g. to bounce center ((+) toward δ_1)

NOTE: For values of $d > 1$ bounce and pitch frequencies and centers are reversed.

Then

$$f_1{}^2 = 4.9/\delta_2 \; [C - \sqrt{D^2 + E}\,]$$
$$f_2{}^2 = 4.9/\delta_2 \; [C + \sqrt{D^2 + E}\,]$$

where:

$$C = a \left(\frac{1}{d} - 1\right)(r - 1) + \frac{1}{d} + r$$

$$D = a \left(\frac{1}{d} + 1\right)(r - 1) + \frac{1}{d} - r$$

$$E = \frac{4}{d} a (1 - a)(r - 1)^2$$

Likewise,

$$\frac{m}{B} = \frac{2a(1 - r)}{D + \sqrt{D^2 + E}} \qquad\qquad \frac{n}{A} = \frac{2(1 - r)(1 - a)}{D - \sqrt{D^2 + E}}$$

Note that $-(m/B \bullet n/A) = d = i^2/AB$, a useful relationship for checking purposes.

The curves of Fig. 14 show in relative terms the characteristic changes in frequencies and centers of oscillation with variation in the ratio δ_2/δ_1, for four values of dynamic index, namely, $d = 1.2, 1, 0.8$, and 0.6. The chart values have been calculated for a center of gravity position equidistant from the spring centers $(a = 0.5)$. Nevertheless, these curves are applicable to the usual passenger car range of c.g. position $(a = 0.5 \text{ to } 0.6)$, with negligible error as to frequencies, and without exceeding $\pm 5\%$ error in centers of oscillation. However, for accurate determination in commercial vehicles, where the c.g. usually falls outside this range, the values should be calculated from the general equations.

It will be seen in Fig. 14 that only the pitch frequency is sensitive to the suspension deflection ratio and to the dynamic index. However, both frequencies increase as the dynamic index diminishes and are a minimum when the static spring deflections are equal. When $d = 1$, each of the two modes consists of an independent vibration at one spring about the other spring as a center of oscillation.

453

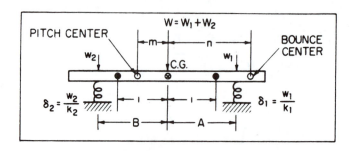

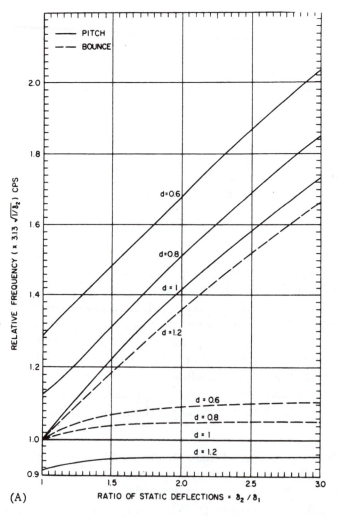

(A)

RATIO OF STATIC DEFLECTIONS = δ_2 / δ_1

Fig. 14 - Vibrating system with two degrees of freedom. Natural frequencies and centers of oscillation versus ratio of static deflections, δ_2/δ_1, and dynamic index, d

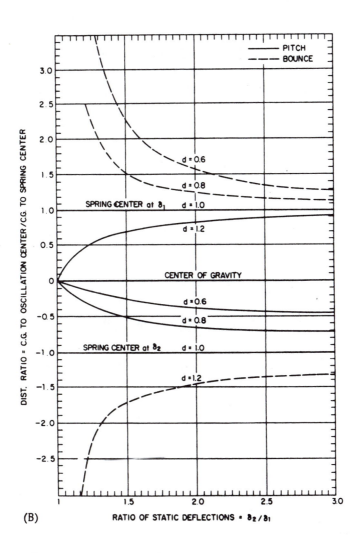

(B)

Fig. 14 - Continued

It is important to note that when d = 1, and the springs have the same static deflection ($\delta_2/\delta_1 = 1$), there are two other possible modes of vibration, since the two principal vibrations have the same frequency. One is a uniform bounce of the mass, when both springs act in the same phase, the other a pitch about the c.g. when both springs act in opposite phase. The curves of pitch center show that this condition exists whenever $\delta_2/\delta_1 = 1$, regardless of the dynamic index. This situation is definitely undesirable because it produces the maximum angular amplitude for a given vertical displacement at the springs. Consequently, this condition should be avoided in practice.

As long as there is a differential of static deflection between the two springs, a dynamic index of 1 gives the best vehicle ride because angular amplitudes are minimized and there is no interaction between springs.

2.7 References

1. J.P. Den Hartog, Mechanical Vibrations, McGraw-Hill, New York, N.Y., (4th Ed.).

2. N.O. Myklestad, Fundamentals of Vibration Analysis, McGraw-Hill, New York, N.Y.

3. S. Timoshenko, Vibration Problems in Engineering, D. Van Nostrand Co., Princeton, N.J., (3rd Ed.).

4. R. N. Janeway, "Problems in Shock and Vibration Control," *Industrial Mathematics*, Vol. 5 (1955).

3. Energy Absorption and Impact

Figs. 15-17 are designed to facilitate the solution of problems involving the absorption of energy in bringing a moving mass to rest. Each chart provides an essential step in relating the resistance characteristic of the absorption gear to the displacement and resultant deceleration of the mass for any given initial velocity. These charts are particularly applicable to extreme operating conditions of vehicle suspensions and to normal operating conditions of airplane landing gears.

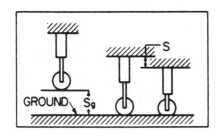

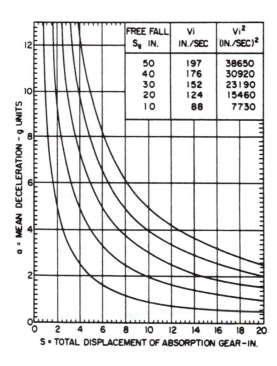

FREE FALL S_g IN.	Vi IN./SEC	Vi² (IN./SEC)²
50	197	38650
40	176	30920
30	152	23190
20	124	15460
10	88	7730

Fig. 15 - Mean deceleration versus displacement for constant impact velocity

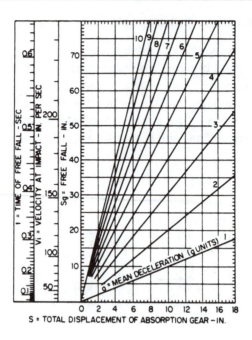

Fig. 16 - Freefall conditions

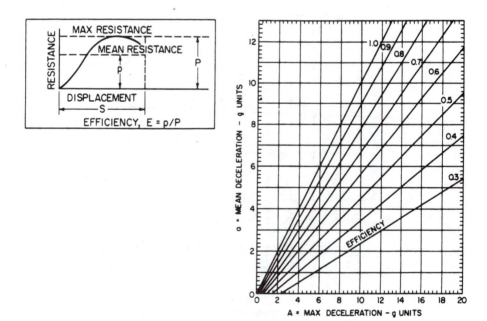

Fig. 17 - Freefall conditions, mean versus maximum deceleration at given efficiencies

458

Absorption systems usually combine the elastic resistance of springs with friction of damping forces, so that the resultant variation in resistance with displacement becomes complex. In order to make the charts readily applicable to any resistance characteristic, the resistance is reduced to a simple efficiency value. As shown diagrammatically in Fig. 17, the efficiency is taken as the ratio of mean to maximum resistance over the total displacement of the absorption gear.

The relations between velocity at impact, total displacement and mean deceleration, shown in Figs. 15 and 16, apply equally whether gravity acceleration is present or not. But, in relating the resistance of the absorption gear to the resultant deceleration, the force of gravity, if present, must be reckoned with as an additional component. Cases with and without gravity must, therefore, be treated differently as shown on the following pages. Where gravity is not involved the relations between mean and maximum deceleration and efficiency of the gear are simple and direct.

Note that, while Fig. 17 applies only to freefall conditions, it extends all the way to zero impact, i.e., a mass falling freely from rest in contact with the absorption gear. Since both the initial and final velocities are zero, the mean deceleration is also zero. Therefore, the maximum deceleration for any efficiency value is defined by the intersection of the constant efficiency line with the abscissa. Thus, Fig. 17 can be applied to a sprung mass, initially at rest but displaced from its static position.

3.1 List of Symbols

S_g = Distance of freefall to point of contact, in.

V_i = Velocity at impact, in./sec

t = Time of free fall, sec

S = Total displacement of absorption gear in bringing mass velocity to zero, in.

a = Mean deceleration of moving mass, g units

A = Maximum deceleration of moving mass, g units

p = Mean resistance of absorption gear over total displacement, lb

P = Maximum resistance of absorption gear, lb

E = Efficiency of absorption gear, p/P

W = Total moving weight, lb

g = 386 in./sec^2

3.2 Mathematical Relations

3.2.1 Impact from Free Fall

$$V_i^2 = 2g\,S_g \qquad\qquad V_i = 27.8\,\sqrt{S_g}$$

$$t = \sqrt{2S_g/g} = 0.072\,\sqrt{S_g}$$

$$a = S_g/S \ \text{ or } \ a = V_i^2/(2gS)$$

$$a = (p/W) - 1 \qquad\qquad A = (P/W) - 1$$

$$E = p/P = (a+1)/(A+1)$$

$$a = E(A+1) - 1 \qquad A = [(a+1)/E] - 1$$

Example:

Given:　　　Freefall, $S_g = 40$ in.
　　　　　　Maximum displacement of gear, $S = 10$ in.
　　　　　　Maximum deceleration $= 7g$

To Find:　　Required efficiency of gear and maximum resistance

From Figs. 15 or 16 for $S_g = 40$ in. and $S = 10$ in. (read) $a = 4g$.

Required efficiency: From Fig. 17 for $a = 4g$ and $A = 7g$ (read) $E = 62.5\%$.

Maximum resistance:　　　$A = (P/W) - 1 = 7$
　　　　　　　　　　　　$P = 8W$

3.2.2 Impact Without Gravity Acceleration

$$a = V_i^2/(2gS)$$
$$a = p/W \qquad A = P/W$$
$$E = p/P = a/A$$

Example:

Given:　　　Velocity at impact, $V_i = 176$ in./sec
　　　　　　Maximum Displacement of gear, $S = 10$ in.
　　　　　　Maximum Deceleration $= 7g$

To Find:　　Required efficiency of gear and maximum resistance

From Fig. 15, for $V_i = 176$ and $S = 10$ in. (read) $a = 4g$.

Required efficiency:　　　$E = p/P = a/A = 4/7 = 57.2\%$

Maximum resistance:　　　$P = AW = 7W$

4. Vibration Limits for Passenger Comfort

A study of the subject indicates that there is no absolute standard of human comfort or discomfort expressed in physical terms such as amplitudes or acceleration at a given frequency. However, there is enough agreement among the test data from various investigators so that a zone may be outlined, above which vibration is certainly intolerable and below which it is immaterial. A limit of acceptable vibration can only be drawn by combining a judgment factor with analysis of the laboratory test data.

Three independent analyses appear in this section, each based on data from many investigators, and each with the author's own comments. The Burton-Douglas chart (Fig. 19) presents the conclusions of aircraft engineers, the Janeway recommendations (Fig. 20) are directed at automobile and railroad practice, and the Goldman presentation (Fig. 21) represents a broad biological viewpoint.

A comparison of these three presentations shows that above a frequency of 4 cps, the disagreement is minor, so that all analyses can be considered identical. Below this frequency, the presentations differ, partly because the data used are not entirely the same, but also because each author takes a different point of view based upon his particular field of interest. Note that all three are superimposed in Fig. 18 to show how they are related to each other in the 1-10 cps frequency range. It is significant that Janeway's recommended limits agree very closely with Goldman's mean discomfort threshold, while both fall well inside the very disagreeable threshold of the Burton-Douglas chart. For practical use, that chart should be selected which has been derived with the particular field of application in mind.

It will be noted that these criteria apply specifically to vertical vibration. For lateral vibrations, recently published experimental evidence from railway tests indicates that tolerable amplitudes for passenger comfort are 30% lower in the 1-2 cps frequency range.

The latest publications* on the subject strongly confirm the validity of constant peak jerk intensity as the comfort criterion for vertical vibration in the 1-6 cps frequency range. (See Fig. 20.)

---

*See Bibliography, Refs. 1, 4, 10, and 11.

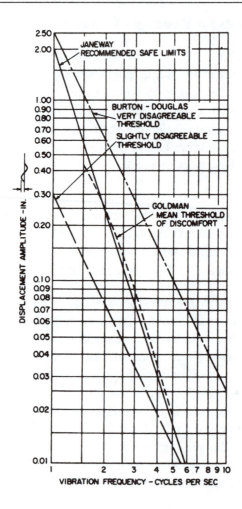

Fig. 18 - Discomfort thresholds — comparison at low frequencies

4.1 Response to Vertical Vibrations

E. F. Burton, Douglas Aircraft Co.

Passengers and crew in aircraft are subject to complex oscillatory disturbances, the effects of which must be evaluated in planning for comfort. The data on which such evaluations can be based are not extensive and are on sinusoidal oscillations, chiefly in a vertical plane, for a seated or standing person.

The accompanying chart (Fig. 19) is a simplified graph based on a careful comparison and evaluation of ten original studies ranging from motion

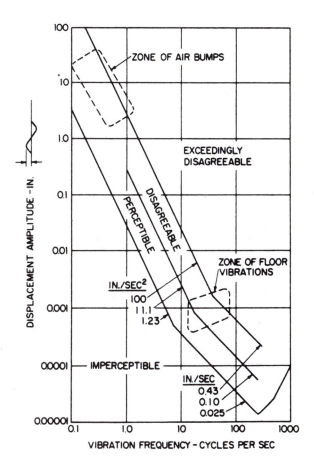

Fig. 19 - Response to vertical vibrations

sickness and ordinary vibration table studies down to fingertip sensitivity experiments. The assumption that the comfort classification is continuous over a wide range of frequencies and amplitudes, even though the physiological responses may differ within these ranges, permits the engineer to employ a single set of standards in judging the vibration experience in aircraft.

The graph applies to simple harmonic motions, but, if each major component of the complex steady-state vibration in an airplane is separately evaluated as if it were the only motion, a useful first approximation of the experience will be obtained. In the absence of other data, the graph may also serve as an approximate guide to the evaluation of fore and aft and lateral oscillations.

463

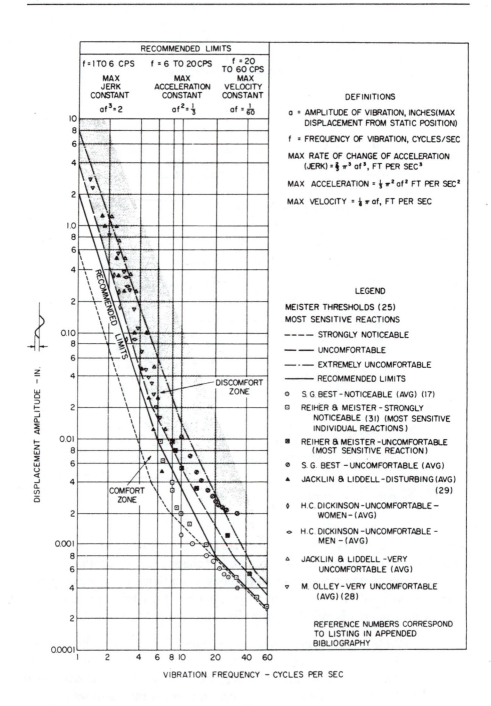

Fig. 20 - Human reaction to vertical vibration

The family of three curves shown in the graph is taken from Lippert*. Janeway's description in terms of maximum acceleration and maximum velocity is applied to each curve.

4.2 Vertical Vibration Limits for Passenger Comfort**

R. N. Janeway, Janeway Engineering Co.

In dealing with human reactions to vibration, limits for comfort cannot be fixed objectively, but must be derived by interpreting the available experimental data. The problem is complicated by the variations in individual sensitivity, and the diversity of test method and sensation level adopted by different investigators.

All the pertinent data available are correlated in Fig. 20 in terms of vibration amplitude versus frequency. The three broken lines define the sensation thresholds, respectively, of "strongly noticeable," "uncomfortable" and "very uncomfortable" for the most sensitive subjects at frequencies from 1 to 60 cycles per second, according to F. J. Meister***. Adjacent threshold lines encompass the results on at least 90% of all individuals tested and, therefore, also define the probable range of variation. The average results obtained by the other investigators confirm these boundaries remarkably well. The probable zones of "comfort" and "discomfort" are indicated on both sides of Meister's "uncomfortable" threshold.

The heavy line shows recommended limits which include a judgment factor and should be well within the comfort range even for the most sensitive person. It is important to note that the experimental data are all based on short exposure time of five minutes or less. Vibration, as actually encountered in vehicles, tends to be more prolonged the higher the frequency. For this reason, the recommended limits allow a greater margin of safety as the frequency increases.

As indicated at the top of Fig. 20, the recommended criterion consists of three simple relationships, each covering a portion of the frequency range, as shown in Table 2.

---

*See Bibliography, Refs. 14 and 15.
**Based on Bibliography, Ref.13.
***See Bibliography, Ref. 25.

TABLE 2

Frequency Range, cps	Amplitude versus Frequency, in
1-6	$af^3 = 2$ (constant peak jerk)
6-20	$af^2 = 1/3$ (constant peak acceleration)
20-60	$af = 1/60$ (constant peak velocity)

Example: At $f = 2$, the recommended amplitude limit is: $a = 1/4$ in.

The following should be noted in studying this figure:

1. All values are based on vertical sinusoidal vibration of a single frequency. Where two or more components of different frequencies are present, there is no established basis on which to evaluate the resultant effect. It is probable, however, that the component, which taken alone represents the highest sensation level, will govern the sensation as a whole.

2. A relatively low noise level is assumed so that the vibration alone determines the sensation. No data are available to indicate how to evaluate the resultant sensation where vibration and noise are of comparable intensities in terms of human reaction.

3. All data used were obtained with subjects standing (Refs. 25 and 31) or sitting on a hard seat. In any case where a cushion is used, it must be evaluated independently. Only the resultant vibration transmitted to the passenger should be related to the chart values.

4.3 Subjective Responses of the Human Body to Vibratory Motion*

David E. Goldman, Naval Medical Research Inst.

Fig. 21 shows mean amplitudes of simple harmonic vibration at frequencies up to 60 cycles per sec, at which subjects: I) perceive vibration, II) find it unpleasant, or III) refuse to tolerate it.

Each point on the chart represents an average of 4-9 different values based on measurements reported by various investigators. The data are for short duration exposure (a few minutes) of the body standing or sitting on a vibrating support subjected to vertical oscillation.

*Based on Bibliography, Ref. 12.

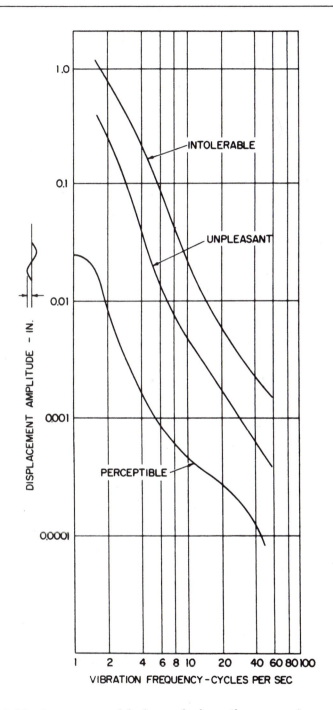

Fig. 21 - Subjective response of the human body to vibratory motion

5. Bibliography

1. Richard Fine, "Correlation of Vertical Acceleration and Human Comfort in a Passenger Car," SAE Paper No. 630314 (733C), Sept. 1963.

2. C.B. Notess, "A Triangle-Flexible Airplanes, Gusts, Crew," Cornell Aero. Lab. Memo No. 343, May 1963.

3. D.E. Goldman, "The Effects of Shock and Vibration on Man," ASA Report, June 1963, Shock and Vibration Handbook, Chap. 44, McGraw-Hill, 1961.

4. H.C.A. von Eldik Thieme, "Passenger Riding Comfort Criteria and Methods of Analyzing Ride and Vibration Data," SAE Paper No. 610173 (295A), January 1961.

5. G.H. Ziegenruecker and E. G. Magid, "Short Time Human Tolerance to Sinusoidal Vibrations," WADC, TR 59-391, July 1959.

6. A.O. Radke, Proceedings of the ASME, New York, December 1957.

7. A.K. Simons, "Mechanical Response of Human Body to Wheeled Vehicle Vibration, 1 - 6 cps," Paper presented at 40th Annual Conference of Human Engineers, N.Y.U., Sept. 1956.

8. G.L. Getline, Shock and Vibration Bulletin, No. 22. Supplement DOD, Washington, D.C., 1955.

9. J.E. Steele, "Motion Sickness," Shock and Vibration Bulletin, No. 22. Supplement DOD, Washington, D.C., 1955.

10. M. Haack, "Tractor Seat Suspension for Easy Riding," SAE Transactions, Vol. 63 (1955).

11. M. Haack, "Stressing of Human Being by Vibration of Tractor and Farm Machinery," Grundlagen der Landtechnik (Germany), Vol. 4, 1953.

12. David E. Goldman, "A Review of Subjective Responses to Vibratory Motion of Human Body," Report No. 1, Naval Medical Research Institute, March 16, 1948.

13. R.N. Janeway, "Vehicle Vibration Limits to Fit the Passenger," Paper presented before SAE, March 5, 1948.

14. S. Lippert, "Comprehensive Graph for the Collection of Noise and Vibration Data," Jrl. of Aviation Medicine, August 1948.

15. S. Lippert, "Human Response to Vertical Vibration," Paper presented before SAE, October 1946.

16. E.E. Dart, "Effects of High Speed Vibrating Tools on Operators Engaged in the Aircraft Industry," *Occupational Medicine*, Vol. I, (1946) p. 515.

17. S.G. Best, "Propeller Balancing Problems," *SAE Journal*, November 1945.

18. S.J. Alexander, *et al.*, "Effects of Various Accelerations upon Sickness Rates," *Journal of Psychology*, July 1945.

19. F. Postlethwaite, "Human Susceptibility to Vibration," *Engineering* (British), Vol. 157, (1944) p. 61.

20. F.A. Geldard, "The Perception of Mechanical Vibration," Part III, *Jrl. of Gen. Psychology*, Vol. 22, (1940) p. 281-289.

21. G. Von Bekesy, "The Sensitivity of Standing and Sitting Human Beings to Sinusoidal Shaking," *Akustische Zeitschrift,* Vol. 4, (1939) p. 360.

22. G. Von Bekesy, "Vibration Sensation," *Akustische Zeitschrift*, Vol. 4, (1939) p. 316.

23. R. Coermann, "Investigation of the Effect of Vibrations on the Human Organism," *Jahrbuch der Deutschen Luftfahrtforschung*, Vol. III, (1938), p. 142.

24. H.M. Jacklin, "Human Reactions to Vibration," SAE Trans., Vol. 39, (1936) p. 401.

25. F.J. Meister, "Sensitivity of Human Beings to Vibration," *Forschung* (V.D.I. Berlin), May-June, 1935.

26. H. Constant, "Aircraft Vibration," British Air Ministry Report No. 1637, October 1934.

27. L.D. Goodfellow, "Experiments on the Senses of Touch and Vibration," *A.S.A. Journal,* July 1934.

28. Maurice Olley, "Independent Wheel Suspension — Its Whys and Wherefores," *SAE Journal*, March 1934.

29. H.M. Jacklin and G.J. Liddell, "Riding Comfort Analysis," Engineering Bulletin, Research Series - No. 44, Purdue Univ., May 1933.

30. S.J. Zand, "Vibration of Instrument Boards and Airplane Structures," *SAE Journal*, November 1932.

31. H. Reiher and F.J. Meister, "Sensitiveness of the Human Body to Vibrations," *Forschung* (V.D.I. Berlin), November 1931.

32. F.A. Moss, "Measurement of Riding Comfort (General)," *SAE Journal*, April 1932, May 1931, July 1930, April 1930, and September 1929.

33. G.B. Upton, "Report on Lecture by Prof. Lemaire (France) at Cornell Univ," *Sibley Journal of Engineering*, Vol. 34, (1925) p. 354.

Index

472